Applied Ecology and Environmental Management

Applied Ecology and Environmental Management

EDWARD I. NEWMAN
School of Biological Sciences, University of Bristol, England

SECOND EDITION

Editorial Offices:
Osney Mead, Oxford OX2 0EL
25 John Street, London WC1N 2BS
23 Ainslie Place, Edinburgh EH3 6AJ
350 Main Street, Malden
MA 02148-5018, USA
54 University Street, Carlton
Victoria 3053, Australia
10, rue Casimir Delavigne
75006 Paris, France

Other Editorial Offices:
Blackwell Wissenschafts-Verlag GmbH
Kurfürstendamm 57
10707 Berlin, Germany

Blackwell Science KK
MG Kodenmacho Building
7–10 Kodenmacho Nihombashi
Chuo-ku, Tokyo 104, Japan

First published 1993
Reprinted 1994, 1995, 1996
Second Edition 2000

Set by Jayvee, Trivandrum, India
Printed and bound in Great Britain
at the Alden Press Ltd, Oxford
and Northampton

DISTRIBUTORS

Marston Book Services Ltd
PO Box 269
Abingdon, Oxon OX14 4YN
(*Orders*: Tel: 01235 465500
Fax: 01235 465555)

USA
Blackwell Science, Inc.
Commerce Place
350 Main Street
Malden, MA 02148-5018
(*Orders*: Tel: 800 759 6102
781 388 8250
Fax: 781 388 8255)

Canada
Login Brothers Book Company
324 Saulteaux Crescent
Winnipeg, Manitoba R3J 3T2
(*Orders*: Tel: 204 837 2987)

Australia
Blackwell Science Pty Ltd
54 University Street
Carlton, Victoria 3053
(*Orders*: Tel: 3 9347 0300
Fax: 3 9347 5001)

A catalogue record for this title
is available from the British Library

ISBN 0-632-04265-6

Library of Congress
Cataloging-in-Publication Data

Newman, E. I.
Applied ecology and environmental management / Edward I. Newman.—
2nd ed.
p. cm.
Includes bibliographical references (p.).
ISBN 0-632-04265-6
1. Environmental sciences.
2. Ecology. I. Title.
GE105 .N48 2000
333.95—dc21 00-029782

For further information on
Blackwell Science, visit our website:
www.blackwell-science.com

Contents

Preface

The world faces very serious environmental problems. This book is about what science—and especially biological science—can do to help. The book deals with a wide range of topics which are usually covered in separate books by different experts, who may well have been trained in different university departments, including biological science, environmental science, forestry, agriculture, range science, fisheries and wildlife, marine science and others. Here the topics are covered in a single book written by one person. I have written such a book because I believe the world needs people who have studied a wide range of environmental problems, who understand how they relate to each other and how they are based on underlying principles of ecological science.

The human population of the world will continue to increase for at least some decades in the new millennium. This is one of the reasons why there are bound to be pressures on resources. This book assumes that in the future there will be increased demand for energy, water, food, timber, and also for new chemicals for many uses. The ecological challenge is to meet these needs in a sustainable way, yet at the same time reducing as far as possible harmful effects on wild species, communities, landscapes and the quality of the environment on which they depend.

Sometimes science can suggest solutions to ecological problems: for example, ways of controlling diseases or minimizing the effects of pollution. Sometimes it can answer practical questions, such as how many fish we can take from an ocean this year without reducing the catch in future years. Sometimes it can help with resolution of conflicts, for example over alternative uses of land. This book is concerned with each of these aspects of applied ecology.

This second edition is much more than an update of the first edition, it is a major rewrite. In the seven intervening years there has been tremendous research activity in many of the relevant subject areas. There have also been important events, such as the successful re-establishment of wolves in Yellowstone National Park and the collapse of the Newfoundland cod stock. These events and research discoveries have not only increased our knowledge and understanding, but have suggested new priorities and led to changes in attitudes. Hence the need for a major rewrite.

This book is not dedicated to my parents, my wife, my children or to any of the other people who have given me personal support during my life. It is dedicated to everyone who is concerned about the future of our world.

Edward I. Newman

Acknowledgements

First I want to thank my dear wife Edna for her continued love and support.

Susan Sternberg at Blackwell Science played an important part in guiding me through the stages of writing the first edition of this book. Ian Sherman encouraged me to write a second edition, and was at all stages available with helpful advice and suggestions.

Many other people have helped by their comments on parts of the first edition or on earlier versions of this second edition, by making suggestions or supplying information. In particular, I want to thank Ian Cowx, John Grace, Ted Gullison, Steve Hopkin, Mike Hutchings, Andrew Illius, Michel Kaiser, Jane Memmott, William Newman, Adrian Newton, Julian Partridge, Clare Robinson, Colin Walker and Richard Wall.

I thank the following copyright holders, and also many authors, who kindly gave permission for material to be used in this book.

American Association for the Advancement of Science: Fig. 9.5(b).
American School of Classical Studies at Athens: Fig. 4.2.
American Society of Agronomy: Fig. 4.5(a).
American Society of Limnology and Oceanography: Fig. 5.2.
British Trust for Ornithology: Fig. 10.11(a).
Cambridge University Press: Figs 2.1, 2.3, 4.1.
CSIRO Publishing: Figs 3.2, 4.4, 6.4.
Elsevier Science: Figs 3.3, 4.3, 6.11, 9.3, 10.11(b), 10.12, 11.2.
Dr M. Hulme: Fig. 3.4.
Intergovernmental Panel on Climate Change: Figs 2.3, 2.4, 2.7(a,c).
Kluwer Academic Publishers: Figs 3.10, 4.9, 6.9, 7.2.
Dr T.E. Lovejoy: Fig. 10.3.
National Research Council of Canada: Figs 5.9, 5.10, 5.11.
Nature (Macmillan Magazines): Fig. 2.7(b), 8.5, 10.13(c).
New Phytologist Trust: Figs 3.9, 4.6, 9.2, 9.9.
Prof. D.M. Newbery: Fig. 7.4.
The Royal Society: Fig. 7.7.
Society for Range Management: Fig. 8.7.
Springer-Verlag: Figs 2.6, 2.8, 2.9, 6.8, 9.5(a), 10.2.
Dr S.C. Tapper: Fig. 8.9.
University of Chicago Press: 10.13(a).
University of Illinois: Fig. 4.5(b).
University of Washington: Fig. 7.5.
John Wiley & Sons: Fig. 9.8.

Chapter 1: Introduction

This chapter explains what this book is about and how it is organized

The size of our world is fixed, but the number of people in it is increasing (see Table 1.1 and Fig. 1.1). This conflict is a basic driving force underlying many of the problems discussed in this book.

Increasing human population . . .

Figure 1.1 shows how the human population has increased during past decades, in the whole world and in two continents: Africa, which had the fastest percentage increase, and Europe, which had the slowest. The graph does not predict populations in the future. There are many alternative predictions for population change during the 21st century, which differ widely (see Chapter 2), but there can be no serious doubt that the total number of people will rise substantially higher than the figure of about 6 billion at the start of the new millennium. Increasing human population puts further pressure on basic resources, including land and soil, oceans, fresh water and energy sources. It will become more difficult to provide adequate amounts of food and timber, creating pressures for more intensive management of soil and pests, and for changes in land use from the present allocation (Table 1.1). This will result in more risks to wild species and to the areas where they live. More people almost certainly means more production of polluting chemicals. These are the principal topics of this book. So one message, right at the start, is that some things are bound to change.

. . . puts pressures on resources

Table 1.1. Area of water, land and principal land uses on the Earth. Land use data for 1992–94, from World Resources 1998/9.

	Area		
	million km^2	% of whole world	% of land area
Whole world	510		
Oceans	376	74	
Fresh water	3	<1	
Land	131	26	
Crops	14.7		11
Permanent grazing land	34.1		26
Forest and woodland	41.8		32
Other land*	39.9		31

* Includes ice, tundra, desert, towns.

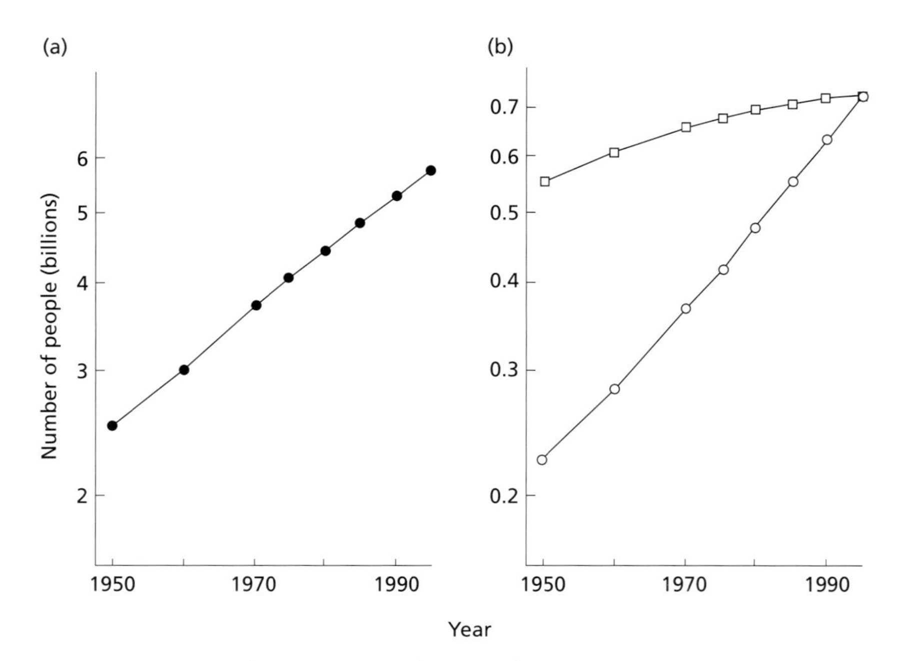

Fig. 1.1 Human population 1950–95. The vertical axes are on log scales. (a) Whole world. (b) □ Europe, ○ Africa. Data from UN Statistics Yearbook 1995.

We should like, if possible, to make things better, for all the people of the world and for the other species in it too. Some topics in this book do aim for that. Can we grow crops in arid areas by using salt water for irrigation? Can we find better ways of breaking down polluting chemicals, using microbes? Can we restore wild species and communities to areas where they formerly occurred? However, much of this book is about how to prevent things getting worse. Can we halt global warming? Can we maintain the fish stocks in the oceans? Can we maintain the productivity of grazing lands and forests long term, but without harming the wildlife? Sometimes we must accept that harmful changes will occur, and the most useful thing ecologists can do may be to give advice on how to minimize the harm. If some of the forests of Amazonia or the US northwest have to be lost, how can we best preserve the species in the remnants?

Difficult choices

Because of these pressures on resources we shall have to make choices, often difficult ones. Should we continue with current or increasing rates of fossil fuel use, in spite of the effect this will have on climate? Should we use a new pesticide to prevent crop loss, even though there is a danger that it may harm other (non-target) species? Should we destroy a community of native species to make more room for food production? Should we extract timber from tropical rainforest, if this will put wild species at risk? These and other choices are discussed in this book. Such choices involve value judgements: how serious is it if a particular species becomes extinct, or if a particular piece of landscape is changed? How do we decide between the needs of people now and in future generations? between food for people and the survival of wild species? This book tries to avoid value judgements. Its aim, instead, is to show how science can help when such decisions have to be made. One of the advantages of a book that covers so many of the major environmental problems of the world is that we can look at these difficult conflicts and choices in a balanced way. Up to now topics such as agriculture, fisheries management, timber production, pollution and conservation have each been dealt with in separate books, and naturally the authors each think their own subject is very important.

Economics and politics applied to environmental problems

Two principal ways of making choices and resolving conflicts are politics and economics. A third is war. All three have been applied to conflicts over natural resources. This book considers economics briefly in a few chapters: for example, how long timescales influence the economics of forestry, and hence decisions about forest management (Chapter 7); and whether we can put a monetary value on wild species (Chapter 10). Politics and regulations also feature occasionally: alternative types of rule for controlling ocean fish catches are explained and discussed in Chapter 5; the US Endangered Species Act is mentioned in Chapter 10. These appearances of economics and regulations are intended as examples, to show how they can interact with science in decisions about management of biological resources. They are deliberately kept few and short. Chapter 2 (on climate change) could, for example, have said much about the negotiations between countries about future carbon dioxide emissions, what they agreed, and how far they have kept their promises; but it does not. The application of politics and economics to management of the environment is very important. This book aims to provide scientific information that will be helpful to politicians and economists, but it does not aim to tell them how to operate politics and economics.

Sustainable systems

One underlying assumption in this book is that we must be prepared to think long term. The word *sustainable* occurs many times in the text. A sustainable system is one that can continue indefinitely, or at least for a long time. A system of growing wheat on a farm is not sustainable unless it can continue to produce as high a yield as it does now. If the farming system results in soil being lost by erosion or the soil structure becoming less favourable for root growth, or an increase in insects harmful to the wheat plant, so there is a long-term decline in yields, then the system is not sustainable. One definition of sustainable grain production requires only that yield be maintained long term. Alternatively, we may also take into account what inputs are needed. If the farming system requires inputs that come from non-renewable sources, for example phosphate fertilizer or fossil fuel, then it can be regarded as not sustainable. A third possible definition requires that the system should not do harm outside its boundaries, for example not put so much nitrate into well waters that they become harmful to people, not use insecticides that kill insects or birds in nearby woods. This book does not confine itself to any one of these definitions of sustainable: we should bear all of them in mind.

Applied science, but based on fundamentals

This book is about *applied* science. The structure of each chapter is designed around a set of environmental problems. So, this book is not pure science with applications tacked on at the end: the applied problems are at the heart of it. Nevertheless, *fundamental* science is crucial to tackling these problems. Why this must be so can be illustrated by Table 10.1 (p. 284), which shows how many thousand species are known in some major groups of animal and plant. We wish to preserve as many as possible of these species, but we do not have the time or resources to do research on every one of them. If we adopt the attitude that we can do nothing about preserving any species until we have performed detailed research on it, almost certainly some species will become extinct before we get round to investigating them. So, their conservation and management must be based substantially on fundamental scientific understanding. That is why Chapter 10, on conservation, considers questions such as: 'How can we decide which species should have higher priority in conservation?'; 'Why can particular species not survive in habitat patches smaller than a certain size?'; 'Can we alter conditions to promote high biodiversity? How?' Or consider biological control of pests and diseases. Some books deal with this case by case, describing in turn each pest species and its successful biological control. Here, Chapter 8 instead considers basic questions such as: 'Can we decide which species are likely to be effective biological control agents, before elaborate testing?'; 'Will a species that initially provides good control evolve to become less effective?'; 'Is biological control safe? How can we be sure it will not harm other, non-target species?'. In these and other chapters, the questions are answered with the aid of examples—particular ecosystems, particular species, particular pollutant chemicals—examples chosen to illuminate the fundamental question, to provide scientific evidence, but never aiming to be a complete list of all those that have been studied.

The fundamental science used in this book covers the whole range of scales in biology, from landscapes and ecosystems, through communities and populations, animal behaviour, physiology and biochemistry, down to single genes; and from the physics of rain formation to the chemistry of pollutant breakdown in soil. Applied ecologists need to be mentally agile. This book has been written primarily for undergraduates studying biological science. It should also be useful to students studying other subjects, such as environmental science, and to many other people who want to find out about the scientific background to current ecological problems, provided they accept the book's strong biological emphasis. For example, in Chapter 2 the section on global climate change passes rapidly over the difficulties of predicting how increases in greenhouse gases will affect future climate, and pays much more attention to how plants and animals will respond to increases in temperature and atmospheric carbon dioxide. Chapter 9 (Pollution) says little about how pollutant chemicals are produced and dispersed, but much about their effects on living things and how to minimize them.

How this book is organized

Following this short introductory chapter there are 10 main chapters. Chapters 2–4 are about basic resources: energy (from the sun and from fossil fuels), water, soil. Then there are three chapters about exploitation and management of biological resources—fish from the oceans, grazing lands, forests; followed by two chapters about things we do not want—weeds, pests, diseases, chemical pollutants—and how to reduce their harmful effects; and finally two chapters on wild species—how to conserve them where they still exist and how to restore them where they have been lost. So, there is a logical progression through the chapters.

Links between chapters

There is also much interaction between chapters: as indicated earlier, this is a key advantage of dealing with so many environmental problems in one book. For example, pest control by chemicals (Chapter 8) produces potential pollutants (Chapter 9). Rainfall (Chapter 3) may be affected by global climate change (Chapter 2), also by overgrazing (Chapter 6) and changes to forests (Chapter 7). The forest chapter considers the effects of different methods of forest management on wildlife as well as on timber production, but there is also further relevant information in Chapter 10 (Conservation), for example on how fragmentation of remaining forests affects wild species. There are also links between chapters at a more fundamental level: there are, for example, fundamental similarities between the population control of fish and pasture foliage (compare Figs 5.4 and 6.3), and between the population biology of disease-causing organisms and of wild animals living in habitat fragments (Chapters 8 and 10). So if you understand one it will help you to understand the other. Thus every chapter contains cross-references to other chapters. If you want to read just one chapter on its own you should be able to understand it well enough, but I hope it will encourage you to read others.

How to find out what is in a chapter

There are no lists of chapter contents, nor does each chapter have a summary in the normal sense. If you want to find out what is in a chapter you can begin by looking at the *Questions* list at the start, which introduces the main problems to be considered. Then follows a list headed *Background science*, but in the text the background science does not come after the problems, nor before them: it is interwoven with them in the chapter. Within the text, headings are sparse: instead, there are many *side headings*, which I hope will guide you through the text without breaking its flow. At the end of each chapter there are *Conclusions*. These are only a selection of the conclusions from the chapter, and they are gross simplifications of what was said earlier in the text. So, if you read the Conclusions and nothing else you will miss a lot.

What I expect you to know already

What do you need to know already in order to understand this book? I have assumed some prior knowledge of biology, such as would occur in an introductory course at university. You also need some knowledge of basic physics and chemistry, such as any biology or environmental science student at university should have. What about mathematics? Ecology is a quantitative subject. Every chapter of this book contains graphs and numbers which are essential to the subject matter. But the mathematics in the book is sparse and simple. A textbook on ocean fisheries by Hilborn & Walters (1992) says near the start: 'Quite frankly, if you are not comfortable writing computer programs and playing with numbers, you should not be interested in fisheries management'. Their book contains more than 300 equations. I have written a chapter on fisheries management for this book which contains three equations, and you certainly do not need to write any computer programs to understand it. Mathematical models are important in ecology: they feature here in many of the chapters, but they are usually presented by words and graphs rather than by equations. The densest mass of equations is in Box 8.3 (p. 218); if you can cope with that, the maths elsewhere in the book should be no problem for you. You also need to know a little about statistics, enough to understand what a correlation coefficient shows and what is meant by 'this difference is statistically significant ($P < 0.001$)'.

There is a glossary near the end of the book, which gives the meanings of technical and specialist words, and of abbreviations. You are expected to know the meanings of more basic scientific terms: if you do not, one of the dictionaries listed below may help you, but they cannot replace the requirement for a groundwork of scientific knowledge. In the text I call a species by its English name, if it has one that is widely used and precise enough. If not, the Latin name is used; this applies to some plants, most invertebrate animals and most microbial species. If the Latin name is used the glossary may give you an English name, or else tell you what major group the species belongs to. If the English name on its own has been used in the text the glossary will give the Latin name.

I have enjoyed writing this book. I hope you will enjoy reading it.

Further reading and reference

Ecology textbooks:
Begon, Harper & Townsend (1996)
Brewer (1994)
Krebs (1994)
Stiling (1996)

Dictionaries:
Allaby (1998)
Lincoln, Boxshall & Clark (1998)
Waites (1998)

Chapter 2: Energy, Carbon Balance and Global Climate Change

Questions

- How many people per hectare can various food production systems support?
- Could low-input systems, on their own, feed the present world population?
- The concentration of carbon dioxide in the world's atmosphere is increasing. What is causing that?
- Could this increase in CO_2 be significantly slowed by using more biomass fuel instead of fossil fuels? or by growing more forest? or by getting the oceans to absorb more?
- What effect will future increases in CO_2 and other gases have on world climate?
- How will future increases in CO_2 and temperature affect (a) crops? (b) wild plants and animals?

Background science

- Energy from the sun reaching the Earth, and what happens to it.
- Primary production of oceans, natural vegetation on land, crops. How crop productivity has been increased.
- The carbon cycle of the Earth: processes, amounts, rates.
- The greenhouse effect. The principal greenhouse gases and their sources.
- How rapidly temperatures changed in the past.
- How fast plants and animals spread in the past, in response to climate change. How fast they can migrate today.

All life depends on energy. Nearly all of that energy comes ultimately from the sun: chlorophyll-containing plants and microorganisms capture solar energy by photosynthesis, and almost all of the remaining living things obtain energy from them, along food chains. This chapter considers how much solar energy is captured by crops and pastureland and is made available to people in their food, and hence how many people different farming systems can support. Many people also use energy obtained by burning fossil fuels—coal, oil and gas—which has increased the concentration of carbon dioxide in the world's atmosphere. Much of this chapter is about the carbon balance of the world, the effects of

Box 2.1 Radiation from the sun and what happens to it.

Radiation emitted by the sun (*solar radiation*) mostly has wavelengths within the range *0.2–3 μm*. This is called *short-wave radiation*.

Fate of the solar radiation reaching the top of the Earth's atmosphere:

reflected by clouds;
absorbed by gases, especially ozone, carbon dioxide and water vapour, which then reradiate it; or
reaches the Earth's surface.

Fate of short-wave radiation hitting plants:

reflected;
passes through to reach soil; or
absorbed by plant. Fate of absorbed energy:
radiated, as *long-wave radiation* (wavelength > 3 μm);
used in transpiration;
used in photosynthesis (primary production); or
warms plants and surrounding soil and air.
Of the short-wave solar radiation reaching the Earth's surface, about half is *photosynthetically active radiation*, i.e. within the wavelength range 0.4–0.7 μm which can be absorbed by photosynthetic pigments.

Further information: Nobel (1991a); Houghton *et al.* (1996); Robinson & Henderson-Sellers (1999).

increases in CO_2 and other gases, and how living things are likely to be affected in the future.

Solar radiation and primary production

Energy balance of vegetation

Box 2.1 summarizes what happens to the energy in solar radiation that reaches the Earth. Most of the energy in the radiation absorbed by plants is (1) lost as long-wave radiation, (2) used to convert liquid water to vapour, or (3) ends up warming the nearby air. The same is true of radiation absorbed by soil. Plants affect the relative proportion of the incoming energy going into these three 'sinks', which can in turn affect air temperature and rainfall. Chapter 3 (Water) explains how this happens, and considers whether people can alter vegetation sufficiently to have a significant effect on climate.

Primary production

A small but important proportion of the short-wave radiation hitting plants is used in photosynthesis. On the ecological scale this is measured as *net primary production* (or net primary productivity, meaning *rate* of production). *Primary* means production by photosynthetic organisms, as opposed to secondary production by non-photosynthetic (heterotrophic) organisms. *Net* means excluding organic matter used by the green plants for respiration; so the net production is new organic matter that is potentially available to heterotrophs. The net primary production over a year is rarely all still present as extra standing biomass at the end of the year: plants or parts of them are eaten by herbivorous animals, attacked by parasites, or die and are degraded by decomposer organisms. In a true climax ecosystem we should expect that on average the biomass present now is the same as that a year ago: the reproduction and growth of some individuals is on average equalled by the death and decomposition of others.

Table 2.1 shows net primary productivities for some major natural vegetation types. Measuring the productivity of natural vegetation on land poses problems, for example how to measure the amount of primary production eaten by herbivores, and how to measure root growth. Much attention was paid to measuring the productivity of terrestrial vegetation during the 1960s and early 1970s , but not so much since. That is why textbooks, including this one, still quote the summary figures drawn together by Whittaker (1975). These were, inevitably, based on the sites where measurements had been made, which were not evenly distributed across the world and may not be representative. There has been continued research on the primary productivity of the oceans, so more recent data are available. Methods are being developed for estimating primary productivity across large areas of land and ocean by measurements from satellites (see Box 3.2, p. 59; also Chapter 5, Fig. 5.2).

In spite of the uncertainties attached to the figures in Table 2.1, they give us a clear indication of the order of magnitude for primary productivity. It may seem surprising that the figures are so similar for very different ecosystems. It is worth noting the very large variation within the oceans. Much of the area of the world's oceans has productivity less than 3 tonnes ha^{-1} $year^{-1}$ (Behrenfeld & Falkowski 1997); ocean regions with productivities much above that are quite localized, and this has important implications for the management of ocean fish production, as Chapter 5 will explain.

Energy content of plant material

Net primary production is often expressed in terms of the dry weight of the plant biomass produced, as in Table 2.1. However, if we take account of the energy content of the plant material, production can be expressed in energy terms. The energy content of most plant materials, when dry, differs little: it is usually within the range 17–21 kJ/g (FAO 1979; Lawson,

Table 2.1 Range of net primary productivities found among some major terrestrial vegetation types, and in the oceans

Environment	t ha^{-1} $year^{-1}$	Source of data
Tropical rainforest	10–35	1
Savanna	2–20	1
Temperate grassland	2–15	1
Boreal forest (= northern conifer forest)	4–20	1
Oceans	0.2–10	2

1. Whittaker (1975); 2. Barnes & Hughes (1999); Behrenfeld & Falkowski (1997).

Table 2.2 Basic energy data for the world. Values are accurate to only one significant figure, except for fossil fuels

	Total energy per year (Joules × 10^{20})	Source of data
Incoming short-wave radiation reaching surfaces of oceans or land cover	30 000	1
Net primary production	30–50	2
Human food consumption	0.2	3
Human energy use		
fossil fuels	3.1	4
fuelwood	0.2	4
others*	0.4	4
total	4	4

* Includes nuclear and hydroelectricity.
Sources of data. (1) Harte (1985). (2) Values within this range given by Whittaker (1975), Vitousek *et al.* (1986). (3) 5–6 billion people × mean food energy supply per person 1980–92 (FAO Production Yearbook 1994). (4) Data for 1995, from UN Energy Statistics Yearbook 1995, World Resources 1998/9.

Callaghan & Scott 1984), though a few storage tissues such as oil-rich seeds give higher values. The net primary production of the whole Earth, land plus sea, is probably within the range 30–50 × 10^{20} J $year^{-1}$. This is about 0.1% of the incoming short-wave radiation (Table 2.2). The energy content of the food consumed by the world's human population is only about 0.5% of the world's net primary production. Wood for fuel comprises about another 0.5% of the net primary production. But even taking into account all plant and animal materials used today, their energy content is far less than that of the fossil fuels we use.

Food production per hectare

Table 2.3 shows the energy content of the food produced, per hectare per year, by various contrasting systems. The figures in column (b) range over more than four orders of magnitude. Obtaining fish from the oceans is clearly a very inefficient way of converting solar energy to food energy. However, fish and meat are usually eaten for their protein content rather than primarily as energy sources. Chapter 5 considers in detail the fish stocks of the world's oceans and whether we can exploit them in a sustainable way.

Among the land-based food production systems listed in Table 2.3, the lowest energy capture is by Turkana pastoralists in northern Kenya (line 2). They keep a mixture of animals, migrating with them in relation to the seasonal rainfall. They are almost entirely dependent on their animals for food, milk forming a major component of their diet. Further information on their system of exploiting this unfavourable environment is given in Chapter 6.

Lines 3–5 of Table 2.3 show data from three farming systems which produced crops without inputs such as inorganic fertilizers or synthetic

Table 2.3 Energy content of food produced per hectare by various systems, and number of people that could be supported by that food

Production system	Energy in food ($GJ\ ha^{-1}\ year^{-1}$) (a)	(b)	People supported (per ha) (c)
Low-input systems			
1. Fish from oceans, 1986–95		0.004	
2. Migratory pastoralists, Kenya, 1981–82.		0.025	0.005
3. Shifting cultivation, Papua New Guinea, 1962–63	19	1.4	0.3
4. European open-field system, England, 1320–40	12	5	1
5. Southern India, 1955	18	8	2
Modern high-input systems			
6. Beef cattle, lowland England		5	1
7. Wheat, Canada		31	6
8. Wheat, UK		106	21

Notes on columns: (a) Calculated by (energy in food from arable crops)/(land area under arable crops that year). (b) Calculated by (total food energy produced)/(total land area of farm or village). (c) Assumes: energy production as in column (b); population limited by food energy supply; mean food energy use per person 14 $MJ\ day^{-1}$ (typical for developed countries; FAO Production Yearbook 1994).

Notes on rows: **1**. (Total annual fish catch)/(total area of ocean). See Chapter 5. **2**. Most of the food came from herded cattle, sheep, goats, donkeys and camels, plus a little from growing sorghum and from wild plants and animals. From Coughenour *et al.* (1985); see Chapter 6. **3**. About one-tenth of the area usable by the village was cultivated at any one time, the remainder was regenerating forest fallow. Meat was obtained by feeding some of the crop produce to pigs, plus a small amount of hunting in the forest. From Bayliss-Smith (1982). **4**. One farm in Oxfordshire. Arable mostly cereals; three-field rotation, one field uncultivated each year. Also some pasture and haymeadow, giving some animal produce. Production data from farm records (Newman & Harvey 1997), energy per g from Altman & Dittmer (1968). **5**. Irrigated rice + unirrigated millet. No fertilizers or other inputs apart from irrigation. Cattle grazed on rough pasture, provided milk. From Bayliss-Smith (1982). **6**. Fertilized pasture, producing herbage equal to 50 kg dry matter $ha^{-1}\ day^{-1}$ (see Fig. 6.3) for 6 months of the year; plus an equal area to provide winter feed. Cattle growth per feed intake based on Snaydon (1987, Chapter 9). **7,8**. Mean production for 1995–97; data from FAO Production Yearbook 1997. Energy per g from Altman & Dittmer (1968).

pesticides. Column (a) shows the energy content of the plant food (mostly cereal grain) per hectare of the arable fields on which it was grown. On that basis their production is lower than modern high-input wheat farming (lines 7 and 8), but compared with countries such as Canada by a factor of only 2 or 3. However, that is not the most useful comparison: the low-input systems of lines 3–5 could only continue by having some land each year that was not producing crops. *Shifting cultivation* involves abandoning the cropland after a few years to allow forest to regenerate, and clearing another patch of forest to cultivate. The European open-field system also involved a rotation, though the fallow was usually only for 1 year. Grazing land was also an essential component of the system, and this was also the case in the traditional system of southern India. The grazing animals provided some food but also their manure, which was crucial in maintaining crop production. In all three of these systems the extra land was essential for maintaining the fertility of the soil and for control of weeds, pests and diseases. For further information on this, see Chapters 4 and 8. Column (b) shows the energy in all the food (including animal produce) per total area needed to keep the system operating. This is the true food energy capture per hectare of these systems, and it greatly increases the gap between them and modern wheat farming.

Lines 6–8 show energy capture by modern animal and arable farming systems using modern crop and animal varieties, inorganic fertilizers and synthetic pesticides, thereby not requiring land to be left fallow. Meat production is about an order of magnitude lower than that of cereals in its food energy per hectare. This is commonly the case, and results from the extra trophic level in the system. Modern beef production is, as might be expected, vastly higher in food production than that of migratory pastoralists in a semiarid climate; and modern wheat produces far more than the three low-input farming systems.

How many people per hectare?

Column (c) shows how many people could be supported per hectare, for their energy requirements, by each system. These figures may be compared with the number of people that the world needs to feed. At the start of the new millennium there are about 6 billion people in the world (Fig. 1.1). Various projections of future human population have been made (Fischer & Heilig 1997): it is extremely likely that the population will exceed 7 billion during the 21st century, and it could well reach 11 billion or more. However, if we just consider the present population of 6 billion, the world's total arable area of about 1.5 billion hectares (Table 1.1) requires four people to be supported by each hectare. It is clear from column (c) of Table 2.3 that none of the traditional systems could support the world's population on that arable area: only modern crop production systems can produce the required yield. The world also has 3.4 billion hectares of grazing land which, if evenly shared, means about two people to each hectare. Much of that land has low productivity, e.g. because of low rainfall, but even the high-input cattle system of Table 2.3 line 6 cannot support two people per hectare. So, meat production could not on its own feed the world's future population, though it can make a contribution by supplementing food from arable crops. This book does not dismiss low-input farming systems as worthless: they feature substantially in several later chapters and there is much we can learn from them. But Table 2.3 makes clear that systems that were adequate in the past can no longer support the total world population of the present or the future.

Energy captured in food can be compared with the productivity of natural ecosystems. Taking 10 tonnes ha^{-1} $year^{-1}$ as an example (a productivity figure within the ranges given in Table 2.1): as the energy content of most plant materials is not far from 20 kJ/g, productivity of 10 tonnes ha^{-1} $year^{-1}$ is equivalent to about 200 GJ ha^{-1} $year^{-1}$. All the productivity values in Table 2.3 are below that, even modern wheat. However, Table 2.3 refers only to the energy in edible parts, Table 2.1 to the whole plant. Nevertheless, it is a fact that on farms in developed countries using modern methods, productivities are lower than in some natural ecosystems. Primary productivity is limited by the efficiency of photosynthesis, which is able to convert only a small proportion of the solar energy falling on the plant into chemical energy. Total incoming short-wave radiation in temperate regions is mostly within the range $3–7 \times 10^4$ GJ ha^{-1} $year^{-1}$ (Sims *et al.* 1978; Monteith & Unsworth 1990), so a productivity of 100 GJ ha^{-1} $year^{-1}$ by wheat represents an efficiency of energy conversion of about 0.2%.

How crop yields have been increased

Since the middle of the 20th century there has been much research activity devoted to photosynthesis, which has transformed our understanding of how it operates—the mechanisms of capture of light and CO_2, the biochemical reactions and their control. One might hope that this knowledge would allow us to increase the efficiency of photosynthesis in crop species, but so far it has not. Plant breeding has increased the yields of crop plants, but by changes other than the efficiency of the photosynthetic process (Lawlor 1995; Evans 1997). Breeding has produced varieties where a larger proportion of the total plant weight goes into the edible parts, where the foliage expands more rapidly at the start of the season and stays green longer at the end. Alternatively, in some tropical crops the growing season has been shortened, allowing two, three or even four crops to be grown per year. Modern varieties can benefit from larger amounts of fertilizer: older varieties of cereals tend to 'lodge' if heavily fertilized, i.e. they are easily blown over, whereas modern, short-strawed varieties lodge less readily. Ample supplies of nitrogen lead to a higher rate of photosynthesis per unit weight of leaf, mainly because there is more chlorophyll and more of key enzymes. Apart from the breeding of new varieties, increased crop yields since the mid-19th century have been mainly due to increased use of irrigation and inorganic fertilizers, and to improved control of weeds, pests and diseases (see Chapters 3, 4 and 8).

Lawlor (1995) discussed why selection for high-yielding varieties has not led to higher efficiency of photosynthesis, and whether this is something we may achieve in the future. Genetic engineering techniques provide potential new methods of manipulating steps in the photosynthetic process. One possibility is to improve the efficiency of *Rubisco*, the enzyme of the initial CO_2 capture step in plants with C3 photosynthesis. It is not 100% specific for CO_2: it also reacts with O_2, and the resulting *photorespiration* is a wasteful process which reduces C capture. It may be possible to improve the specificity of Rubisco for CO_2. More rapid removal of products of photosynthesis, from the cells where they are formed to other parts of the plant, could also speed up the process.

Fossil fuels and the carbon balance of the world

The lifestyle of the world's richer countries is much dependent on fossil fuels. Table 2.2 shows that our worldwide use of energy for heating, cooking, transport, operating factories and so on, is about 20 times that of the food we eat. Most of it comes from fossil fuels. The world's resources of fossil fuels are finite, but predicting how long they will last is notoriously difficult. If the present rate of use of coal and oil is compared with known reserves that are likely to be extractable, this suggests that coal will last 1–2 centuries and oil about half a century (UN Energy Statistics Yearbook 1995). However, the world's total coal is estimated to be at least 10 times as much as the 'known recoverable'. The size of known stocks of oil tends to depend on how much money and effort the oil companies spend on exploration, so there are likely to be reserves not yet discovered. In any case, it may never be possible to use all these reserves, because of the effect the released CO_2 would have on the world's climate. This chapter considers that topic in detail, first the changing carbon balance of the world and the increase in atmospheric CO_2, then the predicted effects of increases in CO_2 and other gases on climate. That section makes substantial use of a fat book called *Climate Change 1995*, written by numerous experts belonging to the Intergovernmental Panel on Climate Change (Houghton *et al.* 1996). A slimmer book by Houghton (1997) summarizes many of the key facts. The final main section of this chapter will then draw on many sources of information to consider how living things (crops and wild species) may respond to these changes in CO_2 and climate.

During the 19th and early 20th centuries it was obvious that burning coal released soot and other pollutants, which affected the atmosphere of cities. It was known that CO_2 was released as well, but there was no obvious reason to worry about it. The world's atmosphere is so large, surely any extra CO_2 would be so much diluted it could not possibly have any effect? This assumption has proved to be incorrect. To measure whether the CO_2 concentration in the atmosphere is changing requires very accurate equipment, carefully used. Reliable continuous measurements started in 1958, on Mount Mauna Loa in Hawaii and subsequently at other sites. We also now know CO_2 concentrations before 1958, back over more than 200 000 years, by measurements on small bubbles of air extracted from ice cores several kilometres deep from Greenland and Antarctica (Moore *et al.* 1996, Fig. 3.21). In these cores there are annual layers visible, caused by the different falls of snow in winter and summer, so the bubbles can be dated accurately. Figure 2.1 shows how the CO_2 concentration has changed since 1750. In 1750–1800, in the early years of the industrial revolution, the concentration was about 280 $\mu l\ l^{-1}$ and rising slowly. During the 1990s it was rising at about 1.5 $\mu l\ l^{-1}$ per year, and by 2000 it has passed 360 $\mu l\ l^{-1}$.

Measuring CO_2 in the atmosphere

Because this increase will affect living things (as will be explained

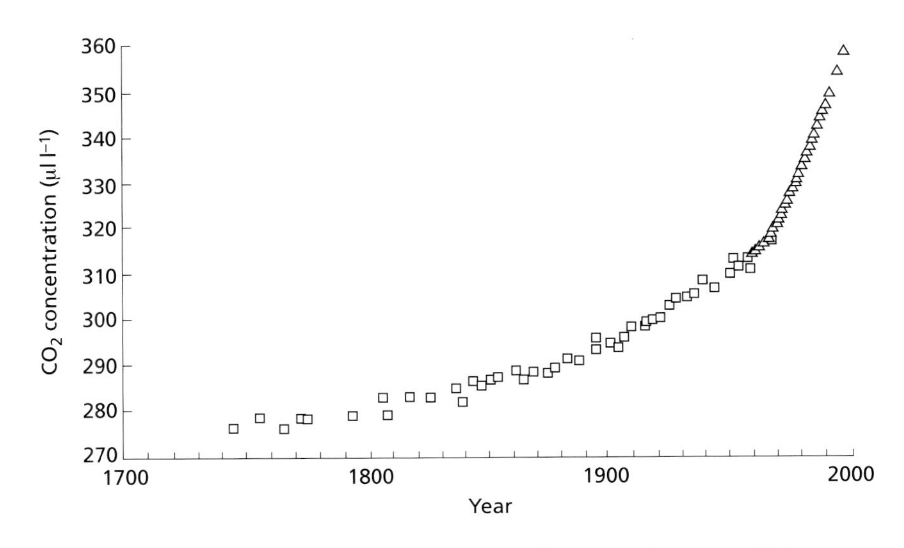

Fig. 2.1 Concentration of carbon dioxide in the Earth's atmosphere since 1750. □ Bubbles in Antarctic ice cores; Δ air at Mauna Loa, Hawaii. From Houghton (1997).

later), it is important to know what is causing it. This will give us a basis for predicting how fast CO_2 concentrations will rise in the future, and how various possible actions by people might affect that. Figure 2.2 gives estimates of the amount of carbon in major global pools, and the rates of transfer between them. To estimate such figures for the whole world is difficult, and they are expected to be accurate only to one significant figure. Nevertheless, it is informative to compare the size of each pool with the amount in the atmosphere. The amount of carbon in the world's fossil fuel reserves is probably more than 10 times as great as the amount in the atmosphere's CO_2: therefore, if we keep on burning it we have the potential to increase atmospheric CO_2 greatly. Terrestrial plants, organic matter on land and in the oceans each have a C pool of the same order of magnitude as the atmosphere, so a change in any one of those three could influence how much is in the atmosphere. Compared with these, the amount of C dissolved in the oceans as inorganics (mainly HCO_3^-) is enormous, so even a small percentage change in that could have a large effect on the atmospheric CO_2 pool. The world's rocks contain enormous amounts of C, in the $CaCO_3$ of limestone and as organic matter in sedimentary rocks. The recycling of that C, by natural weathering, operates on a much longer timescale than concerns us here, though a small amount of CO_2 is released from limestone during the manufacture of cement (Table 2.4). Another small release from the deep Earth is by volcanic eruptions.

Carbon storage: pools

C transfers between pools

Figure 2.2 also shows rates of transfer between pools, and Table 2.4 shows more precise figures for the 1980s for transfers to and from the atmosphere. The rate of increase of CO_2 in the atmosphere (3.3 Gt year^{-1}) was less than the input from fossil fuels plus cement (5.5 Gt year^{-1}).

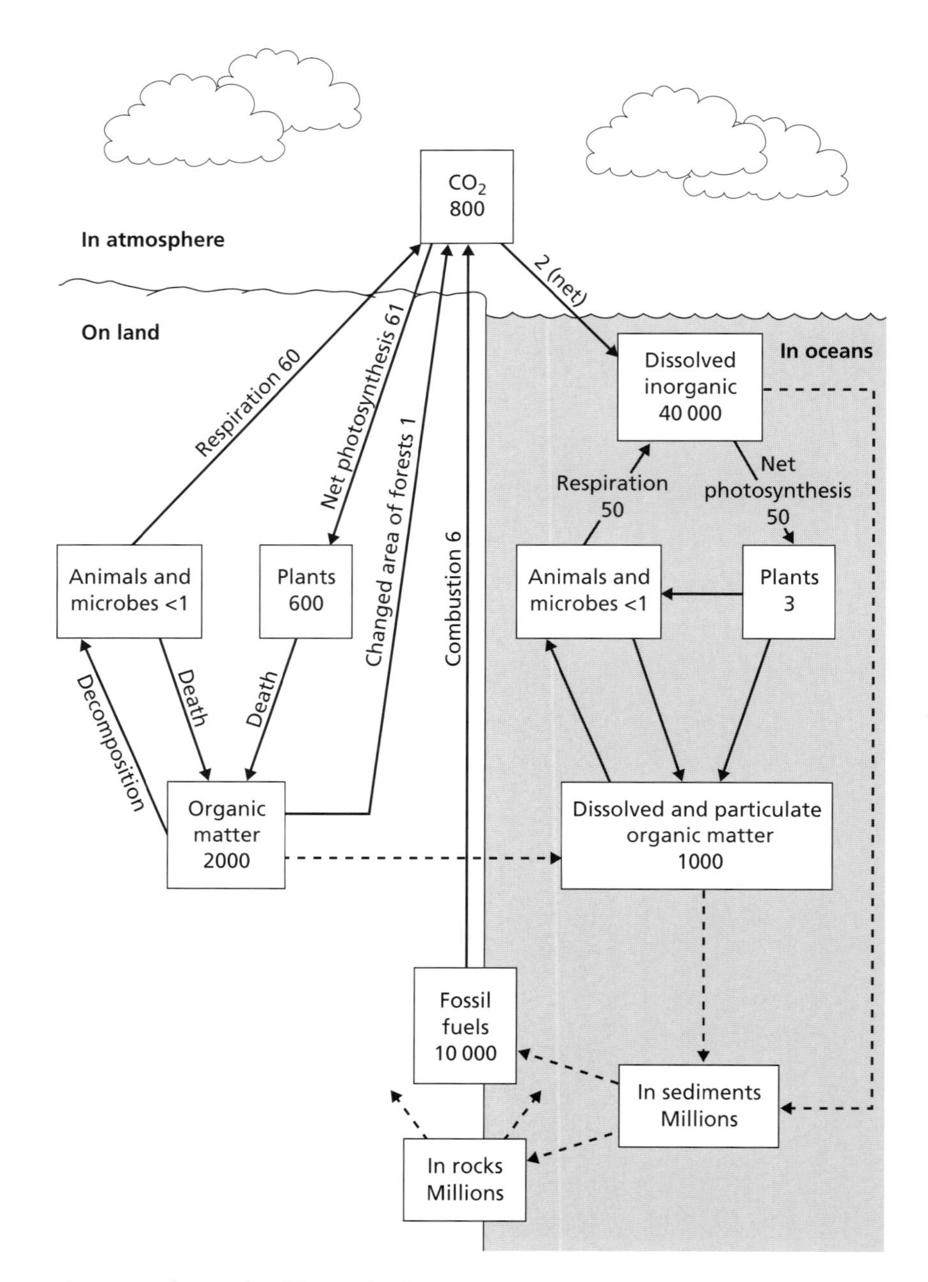

Fig. 2.2 Carbon cycle of the Earth, showing amounts of C in pools and rates of transfer between pools. Dashed lines: rates uncertain. Units: pools, Gt; rates, Gt year^{-1}. (1 Gt = 10^9 tonnes = 10^{15} g.) Most rates are for the 1980s, but fossil fuel combustion rate and amount of C in the atmosphere are for the 1990s. Based on Houghton *et al.* (1996, Chapters 2 and 10); Houghton (1997); Berner (1998).

Both these figures are fairly accurately known. The difference is approximately accounted for by CO_2 transferred into the oceans: as the atmospheric concentration increases, some of the CO_2 dissolves in the

Table 2.4 Rates of transfer of CO_2 to and from the world's atmosphere during the 1980s, expressed as Gt C $year^{-1}$

Inputs to atmosphere	
Burning fossil fuels	5.3
Released during manufacture of cement	0.2
Tropical forest converted to other land use	1.6
Total	**7.1**
Removed from atmosphere	
Into oceans	2.0
Temperate zone forest regrowth after felling	0.5
Increased biomass of existing vegetation	1.3
Total	**3.8**
Increased concentration in atmosphere	**3.3**

Based on Houghton *et al.* (1996), Houghton (1997); see also Dixon *et al.* (1994), Phillips *et al.* (1998).

oceans and adds to the HCO_3^- pool there. The rate of this transfer is known with fair confidence, thanks to a ^{14}C pulse-labelling experiment in the 1950s and early 1960s. Tests of nuclear bombs during that period increased the concentration of radioactive $^{14}CO_2$ in the atmosphere, and following the subsequent fate of that pulse allows us to estimate the rate at which CO_2 is entering the oceans (Houghton *et al.* 1996). Changes in concentrations of the natural stable isotope ^{13}C have also provided independent estimates which agree (Quay *et al.* 1992).

Are living things a source or sink for C?

We need also to consider how living things make a net contribution to changing the CO_2 concentration. Figure 2.2 shows the photosynthetic capture of C each year exactly equalled in the oceans by C loss through respiration, and on land almost equalled. If you have been trained as a physiologist this may surprise you: plants take up CO_2 when they photosynthesize. However, we must think of the whole ecosystem, not just the plants. We should expect that in an ecosystem at steady state, C uptake and loss will balance. In a forest, trees and other plants are growing and so are storing C in new tissue; but animals are eating parts of them; other parts (and sometimes whole trees) are dying and being decomposed. So, heterotrophs are returning C to the atmosphere. Wheat plants on a farm absorb CO_2 while they are growing; but when they are harvested the stubble and roots are left to rot; the grain is made into bread, which is eaten and respired by people. So, again, the C gets back into the atmosphere. Living things can only act as net sources or sinks for C if their mass changes significantly. This will have to be mass of plants or of dead organic matter: the total biomass of animals and microbes is too small to have any significant effect.

If forest is cut down and replaced by vegetation with a smaller biomass per hectare, there is a release of CO_2 by burning or decomposition of the forest plants. There may also be net release of C from soil over some years, if the amount of organic matter declines (see Chapter 4 for information on soil organic matter turnover). In recent decades there has been loss of forest in the tropics, as the land is converted to other uses. On the other hand, in temperate regions there has been a net increase of forested land, as forests regrow after previous felling (see Chapter 7). The amounts of C involved in these changes are difficult to estimate; Table 2.4 gives figures near the centre of likely ranges. Tropical deforestation is a substantial contributor to total CO_2 production by human activity. It is only partly offset by net uptake by regrowth forests in the northern temperate zone.

To balance, Table 2.4 must have a further sink for 1.3 Gt year^{-1}, not accounted for by changed area of forest. One possibility is that in ecosystems which we have assumed to be in steady state the vegetation is in fact increasing in biomass. Some evidence does support this. Phillips *et al.* (1998) analysed data from 120 long-term plots in forests in the humid tropics of South and Central America. The standing biomass has evidently increased, and if these plots are representative of the whole of humid tropical America this would provide a C sink of 0.6 Gt year^{-1}. However, data from Africa, Asia and Australia (from fewer plots) showed no consistent biomass increase. There are several possible reasons why standing plant biomass could be increasing at the moment:

1 A response to increasing atmospheric CO_2 (see later);

2 A response to increased N deposition, as gases, aerosols and dissolved in rain (see Chapter 4);

3 Regrowth after past disturbance, e.g. abandoned shifting cultivation in the tropics.

There may be a major C sink in the vegetation of North America (Fan *et al.* 1998), but so far it has not been identified. Another possibility is that organic matter is increasing in soil and as peat, or is being washed into the oceans and joining the deep sediment (Woodward *et al.* 1998).

Thus there are various sources and sinks, known or possible, that are large enough to have a significant effect on the rate of C increase in the atmosphere. One message is that the way we manage forests in future could be important.

The greenhouse effect and climate change

In spite of its increase since 1800, carbon dioxide is still a rare gas—less than 0.04% of all the gas in each litre of air. Could it possibly have any effect on the world's climate? The answer is yes.

As explained in Box 2.1, radiation from the sun is short-wave (wavelength less than 3 μm), whereas radiation from plants and any other object at a temperature that occurs on Earth is long-wave (>3 μm). Short-wave radiation mostly passes through the glass of a greenhouse. Inside, much of it is absorbed by the plants, benches, floor and other objects, which reradiate some of it as long-wave. The glass is less transparent to

Box 2.2. The principal greenhouse gases.

	Main sources of origin
Water vapour	Evaporation from water surfaces. Transpiration by plants.
Carbon dioxide	See Fig. 2.2
Methane (CH_4)	Produced by microorganisms in natural wetlands, rice paddy fields, guts of ruminant mammals (including sheep and cattle). Fossil natural gas, leaking from gas wells, oil wells and coal mines.
Nitrous oxide (N_2O)	Produced by microorganisms in soil (denitrifiers). N fertilizers. Burning fossil fuels and plant materials.
Ozone (O_3)	Photochemical reactions between other gases.
Halocarbons	No natural sources. Manufactured for use in refrigerators, as aerosol propellants, and for other purposes.

Further information: Houghton *et al.* (1996); Moore *et al.* (1996); Houghton (1997).

long-wave than to short-wave, so it absorbs some of the outgoing long-wave and reradiates some of it back inwards. This *greenhouse effect* keeps the greenhouse warmer than the outside air during daylight hours. There are gases in the atmosphere whose molecules act in a similar way to the glass of a greenhouse, letting much short-wave radiation pass through but absorbing more outgoing long-wave and radiating it back again. These are known as *greenhouse gases* (see Box 2.2). The principal natural greenhouse gases are water vapour, carbon dioxide, methane, nitrous oxide and ozone. If all these were removed from the atmosphere the temperature near the ground would quickly become about 21°C colder than it is at present (Houghton 1997). So, the greenhouse effect is undoubtedly a Good Thing for human beings and for life on Earth. What we are concerned about here is a potential *change* in the greenhouse effect: if the concentration of greenhouse gases increases we should expect the world to get warmer. In addition to the known increase in CO_2 (Fig. 2.1), methane and nitrous oxide are increasing. Ozone is decreasing in some parts of the upper atmosphere but increasing in the lower atmosphere. In addition to the natural greenhouse gases there are synthetic gases, manufactured by people and then released, which can have a significant greenhouse effect. Of these, CFCs (chlorofluorocarbons, e.g. $CFCl_3$) were found to be destroying ozone in the upper atmosphere and their manufacture has been stopped in most countries. By the mid-1990s their concentration in the atmosphere had stabilized or begun to decrease (Houghton *et al.* 1996, Fig. 2.10). However, other halocarbons are being manufactured to replace them as refrigerants and aerosols, and the

Increase in the greenhouse effect

manufacture of halocarbons for other uses has continued. These are increasing in the atmosphere, and may in time become abundant enough to have a significant greenhouse effect.

The effect of each of these gases on global temperature depends on their abundance and also on their greenhouse warming effect per molecule. Water vapour is by far the most abundant of the greenhouse gases, but its effect is often ignored in calculations because it varies so much from place to place and from day to day. However, it should not be ignored, because future climate change may increase the average water vapour content of the atmosphere, thereby causing a feedback effect on warming. Among the other greenhouse gases, CO_2 is estimated to have caused about two-thirds of the increase in greenhouse effect since 1800, the remainder being due mainly to methane, nitrous oxide and CFCs.

Aerosols can cause cooling

In order to predict how climate will change in the future we need to consider not only greenhouse gases but also *aerosols*, solid particles and droplets so fine that they remain suspended in the air almost indefinitely. These increase the reflection of short-wave radiation and so have a cooling effect on climate. One source of aerosols has increased substantially during the last 200 years: SO_2 from burning of fossil fuels (especially coal) forms sulphate aerosols (see Box 9.4, p. 258), so the increased cooling effect from them may have partially offset increased warming from greenhouse gases. Since about 1980 the production of SO_2 has decreased in North America and much of Europe, but it is probably still increasing elsewhere (OECD 1997; Houghton 1997), so it is difficult to predict how world SO_2 production will change in the future.

Predicting future climate change

Predicting how the world's temperature will change in future involves predicting how the concentrations of greenhouse gases and aerosols will change, and then how temperature will respond. Because most of the increase in greenhouse gases is caused by people, how much these gases increase in future is (at least in theory) up to us. Even for an agreed projection of future greenhouse gas and aerosol abundance, predicting climate is very difficult. This is partly because there are lots of potential feedbacks: climate change may alter cloud cover, ice cover, ocean currents, plant biomass and various other things that can themselves influence climate. Since this book is primarily about biological aspects of environmental problems, I do not dwell here on the difficulties of long-term climate prediction but instead present a 'central' prediction for temperature rise up to 2100, and then move on to considering how living things would respond to it.

Figure 2.3(a) shows the predicted CO_2 concentration up to 2100 under the 'business-as-usual' scenario, more formally known as IS92a. This assumes no major changes in people's attitudes and priorities towards energy consumption, with continuing increases in the world's population and energy consumption per person up to 2100. IS92a also predicted, on this basis, increases in other greenhouse gases (Houghton *et al.* 1996). Figure 2.3(b) shows estimates of how much the temperature near the

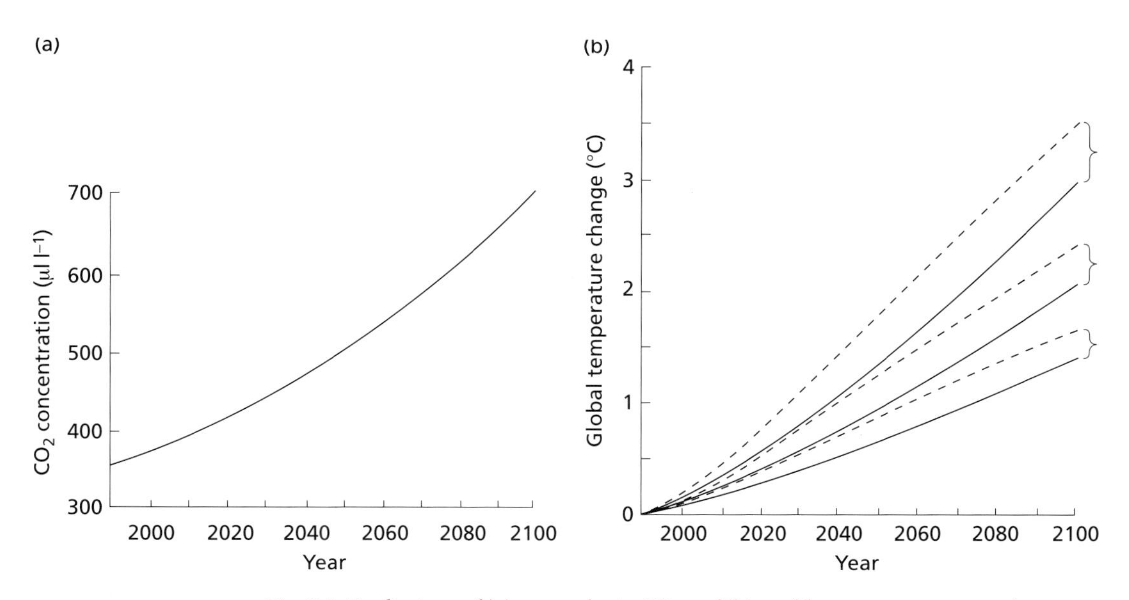

Fig. 2.3 Predictions of (a) atmospheric CO_2 and (b) world mean temperature change 1990–2100, under 'business-as-usual' scenario. The lines in (b) all assume the same increase in CO_2 and other greenhouse gases. - - - aerosols remain constant after 1990; —— aerosols increase. The three pairs of lines reflect uncertainty in how temperature will respond to the changes in gases and aerosols. From Houghton *et al.* (1996), Houghton (1997).

Earth's surface, averaged over the whole world, will increase if these 'business-as-usual' changes in gases occur. It shows alternative predictions depending on whether or not the amount of aerosols in the atmosphere increases after 1990. All the predictions in the graph are for the same increase in greenhouse gases; the three alternative pairs of lines arise from the difficulties of modelling climate change. The temperature rise from 1990 to 2100 is likely to be within the range 1–3.5°C. The 'best estimate' is a 2° warming between 1990 and 2100.

One way to check the accuracy of the climate prediction models is to use them to predict backwards how the temperature changed during the last 100 years or so, based on the concentrations of greenhouse gases known from bubbles in ice, and compare that with the actual mean temperatures. Taking aerosols into account as well, the models mostly predict that world mean temperature should have risen 0.3–0.5° from 1880 to 1990 (Houghton *et al.* 1996, p. 424). The real rise was within that range (Fig. 2.7(c)), though not steady throughout the period.

There are lots of other things we would like to know about future climate besides mean world temperature. How much will the temperature change at different times of year? Will there be changes in rainfall, cloudiness, wind? Will there be more catastrophic events such as hurricanes? Will climate change be greater in some parts of the world than in others? Answers to all these questions have been published, but at the moment they carry considerable uncertainty. Chapter 3 comments further on uncertainties about future changes in rainfall in different parts of the world.

Will sea level rise?

Another concern is rise in sea level. This could happen for two reasons: (1) expansion of the water in the oceans as it gets warmer; (2) melting of ice in glaciers and the polar ice caps. Under the 'business-as-usual' scenario (Fig. 2.3), sea level is projected to rise by 20–90 cm up to AD 2100, with a best estimate of 50 cm (Houghton *et al.* 1996, Chapter 7). Expansion of water will contribute more than half of this. Contrary to some people's expectation, increased melting of the Antarctic ice-cap will contribute little or nothing: Antarctica is so cold that a few degrees' warming will cause little increase in melting. A rise of 50 cm may not sound much, compared with daily tidal ranges or even waves. The crucial events for coastal regions may be occasional flooding caused by exceptionally high tides.

Can CO_2 concentration be stabilized?

Before considering in detail the possible effects of CO_2 increase and climate change on living things, we can say something about the very long-term prospects. Figure 2.2 shows us that there is enough fossil fuel, if we burn it all, to increase the atmospheric CO_2 and temperature far above the predictions for 2100 in Fig. 2.3. Clearly this would be far too disruptive to life on Earth, and the human race must find some way of stopping the increase in CO_2 long before that. Calculations have been carried out to show how CO_2 emissions would have to change to stabilize CO_2 at various concentrations and various times. Figure 2.4 shows one of them, to give stabilization of CO_2 at 550 $\mu l\ l^{-1}$ (about twice the preindustrial concentration) in about 2150. The emissions in Fig. 2.4(a) are from fossil fuels plus land use changes. This shows that, to achieve stabilization, these emissions need to be reduced far below present levels. However, perhaps surprisingly, CO_2 concentration stabilizes long before emissions reach zero. This is because the oceans and land would continue to act as sinks for CO_2. For example, it will take centuries for the increased bicarbonate in the surface waters of the oceans to become mixed into deeper layers. Predictions several centuries ahead, as in Fig. 2.4, are obviously not expected to be very accurate, but they do give us some basis for discussing how much CO_2 emissions need to be reduced.

Can living things be used to reduce atmospheric CO_2?

As Table 2.4 shows, the way people alter vegetation, e.g. by felling forests, can substantially affect global carbon sources and sinks. There are basically three ways in which people's use or manipulation of living things might alter the amount of CO_2 in the future atmosphere.

1 We may be able to reduce the rate at which forests are cut down and converted to other, less bulky vegetation types. Chapter 7 considers rates and causes of deforestation. The key questions are whether we can obtain our timber needs in a sustainable way, without reducing the total area of

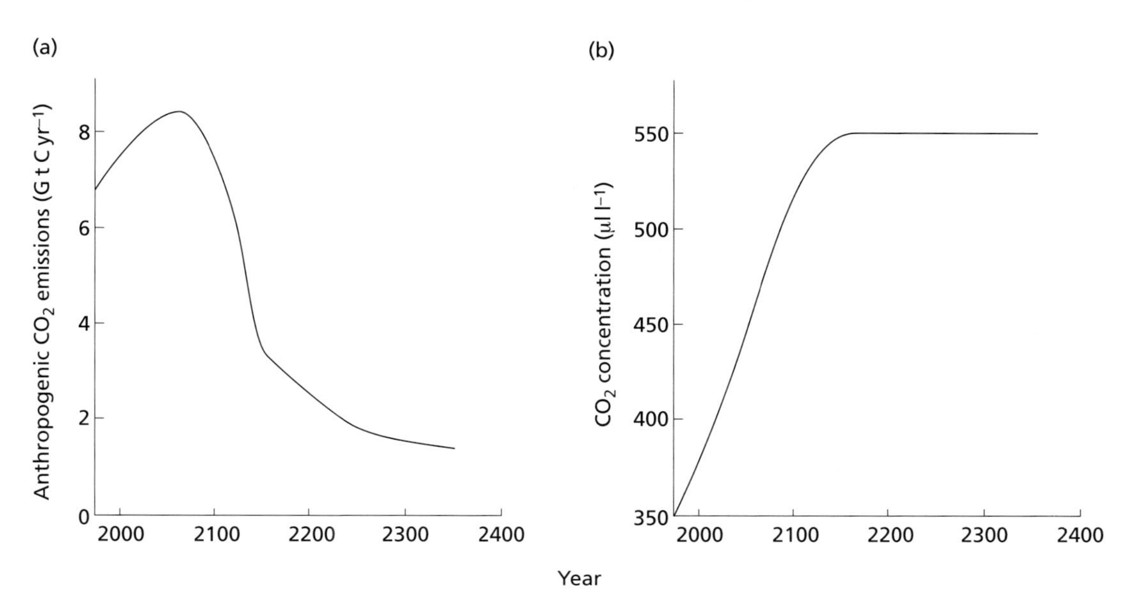

Fig. 2.4 (a) How CO_2 emissions caused by people would need to peak and then decline in order to achieve an atmospheric CO_2 concentration stable at 550 μl l^{-1} by the year 2150. (b) Time-course of atmospheric CO_2, if emissions follow (a). From Houghton *et al.* (1996).

forests; and whether other needs for land, e.g. for farming, will lead to deforestation.

2 We may be able to increase the amount of CO_2 that biological sinks absorb each year.

3 Can we reduce the amount of fossil fuel burnt, by using more biomass as fuel instead? This section considers possibility (3) first, then goes back to (2).

More use of biomass fuels?

Much attention is being given to ways of generating energy that do not create CO_2. Nuclear power stations are one type of non-CO_2 producer, but enthusiasm for these has waned in many countries because of concerns about their safety and how to dispose of the radioactive waste they produce. Box 2.3 lists other energy sources that do not create CO_2. All of them are renewable: in other words, they do not involve depleting a finite resource. Burning biomass (which in practice means plants or plant products) does, of course, produce CO_2. However, it is returning CO_2 that was taken from the air by the plant a few years or a few decades ago, so, on that timescale it is not contributing to CO_2 increase.

Present use of wood as fuel

Until a few hundred years ago plants, especially wood, provided most of the fuel for people throughout the world; in addition, a little energy was provided by animals, wind and water. Today wood forms less than 10% of

Box 2.3. Sources of renewable energy.

Direct solar	H S
heating water	S
photovoltaic cells	S
Wind. Turbines can be on land or off-shore.	S
Waves.	S
Tides.	
Flow of fresh water: hydroelectricity.	S
Geothermal: heat from deep in the Earth.	H
Biomass	H S
purposely grown trees or crops (e.g. sugar cane).	
farm wastes or municipal refuse.	

H: generation of heat, which can be used directly or to generate electricity. The others generate electricity.
S: energy derived from solar radiation.

Some methods of storing electrical energy

As heat.
Pumping up water to a higher reservoir; its later flow down again can generate electricity.
Make H_2 gas by electrolysing water; later combine $H_2 + O_2$ to generate heat.
(Possibly) improved storage batteries.

Further information: Boyle (1996).

total world energy consumption (Table 2.2). However, the percentage of total energy that comes from biomass varies greatly between countries. Table 2.5 (column 7) gives figures for six contrasting countries. In some, such as Congo Democratic Republic, much of the fuel used is wood collected by individuals for use in their own homes. It must therefore be available within walking distance. There is concern that as populations increase the wood-fuel supply will become inadequate (Leach & Mearns 1988); this is discussed in Chapter 7. Here we consider whether more developed countries could reduce their fossil fuel use by increased use of biomass fuels. Table 2.2 shows that to replace all of our present fossil fuel by biomass would require only about 10% of the world's net primary productivity. Viewed like that, biomass for fuel does not look a totally ridiculous solution to our CO_2 problem. But we need to consider what is realistic.

Fuels made from biomass

The fuel produced from biomass can be gas, liquid or solid. Methane has so far been produced mainly from waste, e.g. cattle dung, sewage sludge and domestic refuse, but it can be made from any plant material containing cellulose. Ethanol from plants such as sugar cane has been produced as a petrol substitute. This was favoured in Brazil in the 1970s: production rose greatly up to 1985, but levelled off after that (Goldenberg 1996). One problem is that the fermenter microorganisms can produce only a dilute solution of ethanol, and distilling off the water then requires energy equal to more than half the energy content of the final ethanol product. However, if the waste fibrous material from sugar cane can be used to provide the heat for the distillation, the net energy gain is more favourable (da Silva *et al.* 1978; Hopkinson & Day 1980). Another possible liquid fuel is oil from seeds: this can be separated without distillation. If solid fuel is required wood is the one commonly used, although others such as baled straw are also possible.

If biomass fuel is to increase substantially it seems likely that trees will need to play a major part. Table 7.1 gives examples of yields of wood that can be obtained. Judging from these, and many other yield figures, we cannot expect large-scale, low-input forestry to produce an average of more than 5 tonnes ha^{-1} $year^{-1}$ in temperate regions. In energy terms this is about 10 TJ km^{-2} $year^{-1}$, since the energy content of tree stem material is about 20 kJ/g (1 TJ = 10^{12} J). In the tropics we can expect about 10 tonnes ha^{-1} $year^{-1}$, equivalent to 20 TJ km^{-2} $year^{-1}$.

Which countries could get all their energy needs from biomass?

Table 2.5 provides data on energy consumption for six countries, three in temperate regions, three tropical. Energy includes fossil fuels, plus wood and other biomass, plus electricity generated by other means (e.g. nuclear, hydro). The countries are chosen to provide a wide range of population densities and energy consumption per person, which results in a wide range of energy consumption per area of the country (column 6). Comparing the left-hand and right-hand columns shows that the Netherlands and the UK could not possibly generate all their energy needs from home-grown biomass, because they would need far more than their total land area for biomass forests. The USA, with its lower mean population density, could produce most of its energy needs only by covering almost the whole land area (including prairies, deserts and Alaskan tundra) with biomass forests. In contrast, the three tropical countries could, using part of their area, grow enough biomass to provide their present energy usage. Bangladesh already gets about half of its fuel energy from biomass, but to increase this substantially would be difficult because the country has a very high population density, and must grow most of its own food. The other two countries, Peru and Congo DR, have large areas of forest and are estimated to require only about 2% and 1% respectively of their total area to obtain all their energy needs from biomass forests. In fact, in 1995 Congo DR was getting more than 90% of its fuel energy from biomass, whereas Peru got only 25%. Thus of the six countries in the table, Peru is the only one where substantial replacement of fossil fuels by biomass might be possible. However, there are difficult problems involved in the sustainable exploitation of tropical forests: these are discussed in Chapter 7.

Table 2.5 aims to show whether each of these six countries could get *all* of its present energy needs from biomass. It may, of course, be worthwhile to replace only some of the fossil fuel use by biomass: Table 2.5 helps to show the limitations on this, for countries with various population densities and energy uses per person. Presenting figures country by country implies that each would need to be self-sufficient for biomass

Table 2.5 Area, human population and energy consumption data for selected countries

Country	Area* ($km^2 \times 10^3$)	People total (m)	People per area (km^{-2})	Energy consumption total (10^{15} J yr^{-1})	Energy consumption per person (10^9 J yr^{-1})	Energy consumption per area (10^{12} J km^{-2} yr^{-1})	Energy consumption % from biomass	Area to provide all energy needs from biomass† ($km^2 \times 10^3$)
Netherlands	34	15.5	454	3381	219	99	< 1	338
UK	242	58.3	241	9185	158	38	1	919
USA	9167	263	29	96128	366	10	4	9613
Bangladesh	130	120	926	685	6	5.3	49	34
Peru	1280	23.5	18	504	21	0.39	25	25
Congo DR (formerly Zaire)	2267	43.9	19	492	11	0.22	91	25

* Excludes major inland waters. Data from UNEP (1991).
† Assumes biomass productivity 10×10^{12} J km^{-2} $year^{-1}$ in temperate countries, 20×10^{12} in tropics.
Population and energy data for 1995, from UN Statistics Yearbook 1995, World Resources 1998/9.

energy. Wood is less dense than coal or oil, so contains fewer GJ per m^3 and so is liable to cost more per GJ (in energy and money terms) to transport. At present fuel biomass is rarely transported between countries, but whether this always needs to be so remains to be seen. Even transport within one country might be a major obstacle, for example to the greater use of biomass for energy in Peru.

Are there more efficient ways of capturing solar energy?

Another way to evaluate biomass as an energy source is to ask whether it is the most efficient and satisfactory method of converting energy from sunlight into a usable form. Of the renewable energy sources listed in Box 2.3, five (marked S) in addition to biomass are derived from solar energy. Photovoltaic cells generate electricity when short-wave radiation falls on them. Some are 16% efficient, i.e. the electricity produced contains 16% of the energy in the impinging solar radiation. So they are about two orders of magnitude more efficient than plants at capturing solar energy, and a further increase in their efficiency may be possible. Large-scale use of photovoltaics is at present limited by the cost of manufacture. In contrast, the technology for harnessing wind energy is well developed and the installation of large wind turbines for generating electricity proceeded rapidly in the 1980s and 1990s. The world's installed capacity for wind energy generation increased 10-fold between 1984 and 1996 (Brown *et al.* 1997).

Table 2.6 sets out a comparison of energy that could be provided by trees, wind and photovoltaic cells in Britain. Britain is well endowed with wind compared to many other countries, but less so with solar radiation. A government committee estimated the amount of electricity that could be generated by wind turbines at suitable sites totalling 4000 km^2, so I have based calculations for the other two energy sources on 4000 km^2 also. Wind appears to be three times as efficient as trees at energy capture.

Table 2.6 Estimates of energy that would be provided in Britain on 4000 km^2, by three alternative renewable methods

Source	Form of energy	Amount ($J \times 10^{15} yr^{-1}$)	Notes
Trees	Combustible mass	40	1
Wind	Electricity	120	2
Photovoltaic cells	Electricity	700	3

Notes: 1. Assumes productivity of usable biomass 5 tonnes ha^{-1} $year^{-1}$ (see Table 7.1); 2. Per area of wind farms. Suitable sites identified by government report (1992); see Boyle (1996, p. 309); 3. Per area in which cell arrays mounted, not just area of cells themselves. Data from government report (1989); see Boyle (1996, p. 131).

However, most of the area between the turbine bases—more than 90% of the wind farm area—would still be available for another use, such as farming. So the true efficiency of land use is at least 10 times higher. It may also be possible to site wind turbines offshore. There is also space between arrays of photovoltaic cells, so their true efficiency is several times higher than shown.

A key weakness of wind and photovoltaic cells is that their output varies from hour to hour and from day to day. They generate electricity; Box 2.3 lists some methods of storing electrical energy, but each has limitations at present. Wood, on the other hand, can be stored until needed.

There seem to be many more opportunities for physicists and engineers to increase our renewable energy sources than there are for biologists. Viewpoints on whether renewable energy can replace much of our present fossil fuel use during the 21st century range from optimistic (e.g. Lenssen & Flavin 1996) to pessimistic (e.g. Trainer 1995).

Increasing the sinks for CO_2

C-sink forests?

A growing tree takes up CO_2 and sequesters C in its biomass. It has been suggested that we should dedicate forests to acting as C sinks. Increasing C sinks in forests is in some ways similar to providing more wood as fuel. Both involve CO_2 being removed from the atmosphere. If the wood is used as fuel the CO_2 is returned to the atmosphere; the saving of CO_2 is because less fossil fuel is burnt. If forests are used as C sinks the wood is not burnt but stored, thereby removing permanently from the atmosphere C from burnt fossil fuel. An advantage of C-sink forests is that they can be far away from the C source: CO_2 produced in an industrial zone can be absorbed by a forest thousands of kilometres away in a sparsely populated area. One piece of evidence for this is the relatively small difference in atmospheric CO_2 concentrations between the northern hemisphere (where most burning of fossil fuels occurs) and the southern hemisphere (Fan *et al.* 1998).

Vitousek (1991) has discussed whether C-sink forests could provide a significant contribution to controlling CO_2 increase. A key point to emphasize is that steady-state forests are not net absorbers of C. If C is to be removed from the atmosphere year by year on a long-term basis, we should need to establish C-sink plantations, harvest them while they are still growing actively, replant the site, and store the harvested wood permanently so that it does not rot (as that would return CO_2 to the atmosphere). This storage would be a formidable activity. It has been estimated (Vitousek 1991) that the total amount of C stored at present in all cut timber, worldwide, in use in houses, furniture, fences etc., plus wood products such as paper, is 4–5 Gt. This is about equal to one year's release of C from burning fossil fuels (Table 2.4). Therefore *each year* we would need to add to the world's C-sink store an amount of wood about equal to the present total; this would be far too much to be useful. The volume of timber would be about 30×10^9 m^3, so if it is piled 50 m high it will require an extra 500–1000 km^2 each year for storage. And this wood has to be stored *for ever*: if it is ever allowed to rot, the C will be returned to the atmosphere. This seems a serious responsibility to place on future generations. The area required for growing these C-sink forests would also be formidable. If we assume the higher of the two wood production rates given earlier, 10 tonnes ha^{-1} $year^{-1}$ as an average, to absorb 6 Gt C per year would require 15 million km^2 to be permanently dedicated to C-sink forests, i.e. about one-third of the world's present forested area (Table 1.1). If the wood were to be used for something that might be acceptable, but if it is just going to be stored it may be hard to justify. These calculations are based on the aim of absorbing all the CO_2 generated from burning fossil fuels each year, but they serve to show why C-sink forests can at best make only a limited contribution to slowing the increase of CO_2 in the future.

An ocean C sink?

Because the total amount of C within living things in the oceans is low compared to the amount in the atmosphere (Fig. 2.2), there seems little scope for their tissues providing an increased C sink. However, the pool of dissolved and suspended dead organic matter is much larger, and some organic matter is lost each year by particles sinking down to the deep ocean. If primary productivity in the oceans could be increased, there might be an increased removal of C in that way. It has been suggested that fertilizing parts of the ocean with iron salts would have this effect. There are large areas of the ocean where concentrations of nitrate, phosphate and silicate are too high for them to be the major limiting factors on photosynthesis and the growth of phytoplankton. The addition of iron salts to these waters has been shown to increase the primary productivity substantially; this was shown in laboratory experiments and also when 450 kg of iron (as Fe^{2+}) was added to a patch of the Pacific Ocean (Martin *et al.* 1994). Joos *et al.* (1991) made predictions of the effect of adding iron salt to 16% of the world's ocean for 100 years, assuming that all that area is at present Fe deficient. The amount of Fe needed would be 1 million tons per year. They predicted that if CO_2 production continues to rise under the 'business-as-usual' scenario, then this Fe fertilization would result in atmospheric CO_2 at the end of 100 years being 720 $\mu l\ l^{-1}$, as against 830 $\mu l\ l^{-1}$ without the fertilization. The rise during the 100-year period would thus be slowed by almost one-quarter. A million tons of iron is not an impossibly large amount compared to more than 500 million tons of iron in the iron ore mined each year (World Resources 1998/9). But before embarking on such a wide-scale alteration of the oceans we would need sounder evidence that this much carbon would really be taken out of circulation, not merely respired back to the atmosphere. And we would want to know more about what other effects this fertilization would have on the oceans and their living things, for example the response of animals to this extra algal biomass and production.

Response of plants to increased atmospheric CO_2

Living things are likely to be influenced by warmer climate, but they may also respond more directly to increased concentrations of CO_2, and I consider this first. Because of the high concentration of HCO_3^- in the oceans, it is often assumed that C supply does not limit the rate of photosynthesis there. However, there is some experimental evidence that, on the contrary, increased atmospheric CO_2 will increase primary production in the ocean (Hein & Sand-Jensen 1997). I do not discuss this further, but concentrate on the response of terrestrial plants.

Carbon dioxide is a rare gas, less than 0.04% of the atmosphere by volume. For the compound that is the ultimate source of the C in every organic compound in living things, this seems a low concentration. Several hundred million years ago CO_2 was much more abundant in the atmosphere than it is now (Moore *et al.* 1996; Berner 1998). This suggests that perhaps the present CO_2 concentration is below the optimum for plants, and they will grow faster if the concentration increases. Box 2.4 summarizes methods that have been used to investigate this hypothesis. Experiments in glasshouses and growth chambers have produced very useful results. However, there is concern about high temperatures in glasshouses and low light intensity in growth rooms, as both of these are known to increase the response of plants to raised CO_2 concentrations (Drake *et al.* 1997; Bazzaz & Miao 1993). Also, because the roots are necessarily confined in fairly small pots, this limits the realism of long-term experiments. Therefore, open-topped chambers and free air enrichment, although more expensive to construct and operate, have clear advantages.

Increasing CO_2 concentration usually speeds up photosynthesis

Table 2.7 summarizes the main effects of increased CO_2 on plants that have been consistently found. It gives two similar figures for the mean effect of doubling CO_2 on photosynthesis, a 58% and a 54% increase. However, the range among species and different environmental conditions is large. Among 36 woody species the response of photosynthesis per unit leaf area varied from an increase of 244% down to numerous non-significant changes and one decrease of 40% (Ceulemans &

Box 2.4. Experimental methods that have been used to investigate the response of plants to increase in atmospheric CO_2 concentration.

Transparent chambers within glasshouse.
 Problems: temperature usually higher than outdoors; roots confined to small pots.
Sealed controlled-environment chambers.
 Problems: light intensity lower than outdoors; roots confined to small pots.
Open-topped chambers.
 Transparent vertical walls, but open at bottom to soil and at top to air and rain.
 Requires sophisticated control of CO_2 supply.
 Problem: temperature, wind and humidity still altered.
Free air CO_2 enrichment. Outdoors, no surrounding wall.
 Problem: requires large supply of CO_2, as well as sophisticated control of its supply rate.

Further information: Ceulemans & Mousseau (1994), McLeod & Long (1999).

Mousseau 1994); see also note 3 of the table. Species with the C4 photosynthetic mechanism have a different initial CO_2 capture step, which makes them more efficient at capturing it from low concentrations. They might therefore be expected to benefit less than C3 species from increased CO_2. Although there is some tendency towards this, there is a wide range of responses in both groups and much overlap between them (Poorter 1993).

Short-term experiments may overestimate the effect of increased CO_2 on photosynthesis and growth, as it has often been found in longer experiments that the effect decreases with time. This is especially the case if the plants are growing in small pots or if nitrogen supply is low, suggesting that other factors become limiting (Ceulemans & Mousseau 1994; Drake *et al.* 1997). However, this is probably not the only reason. Starch concentration in leaves often increases at higher CO_2 (Curtis & Wang 1998), which suggests that translocation may be limiting the use of extra photosynthate for growth. Another question is whether, over many generations, species may adapt genetically to higher CO_2, and the response in photosynthesis and growth rate decrease or disappear. Evidence against this comes from a site in Italy where nearly pure, naturally produced CO_2 emerges from vents in the ground and the concentration downwind is higher than normal. A grass species, *Agrostis vinealis*, collected from the site responded about equally to raised CO_2, in terms of photosynthesis and growth, to the same species from elsewhere (Fordham *et al.* 1997).

It may perhaps seem surprising that, if CO_2 is such a rare commodity in the air, doubling its concentration usually less than doubles the rate of photosynthesis. One reason for this is that stomatal conductance

Table 2.7 Effects of increased atmospheric CO_2 on terrestrial plants

Characteristic	Direction of change	Mean response (%) to doubled CO_2	Notes
Photosynthesis rate	+	58	1
		54	2
Growth rate	+	29	3
Stomatal conductance	–	20	4
Transpiration rate	–		5
Water use efficiency	+		6
Nitrogen concentration	–	15	7
		16	8
Phenolic concentration	+		7

Notes: 1. Mean for 45 species, which had large rooting volume available to them. Drake *et al.* (1997). 2. Mean for woody species in glasshouse and open-top chamber experiments. Curtis & Wang (1998). 3. Among 156 species, effect of doubled CO_2 on final plant weight mostly within range no effect up to 1.9 × as large (Poorter 1993). Among 102 measurements on 59 woody species, effect of doubled CO_2 on weight gain averaged 29%, but ranged from 31% reduction to 284% increase (Curtis & Wang 1998). 4. Mean of 28 species. Drake *et al.* (1997). 5. Drake *et al.* (1997). 6. (Dry weight increase)/(amount of water used). See Chapter 3. 7. In various plant tissues. Bezemer & Jones (1998). 8. In leaves. Curtis & Wang (1998).

decreases (Table 2.7). This happens by the stomata partly closing, and also in some species by a reduction in stomata per mm^2. So the concentration of CO_2 in the photosynthesizing cells is likely to be increased less than the concentration in the air outside. Another outcome is slower transpiration. Because the plant grows faster but uses less water its *water use efficiency* will be higher.

Other effects of increased CO_2

Another consistent response to increased CO_2 is lower nitrogen concentration in plant tissues. Less N means less protein and probably reduced amounts of enzymes, which would be expected to lead to slower metabolism. Lower N% in leaves is generally associated with slower photosynthesis. However, this is more than compensated for by the more direct effect of increased CO_2 (McGuire *et al.* 1995). The most abundant protein in green leaves is Rubisco, and at higher CO_2 plants need less Rubisco.

The concentration of phenolics has often been found to increase in high CO_2. The response of other secondary chemicals is less consistent. Plant material with lower protein concentrations but higher in phenolics would be expected to be less palatable to insects. However, studies of 42 insect herbivore species, mostly Lepidoptera larvae, found a strong tendency for them to eat more if the plants had been grown in higher CO_2; but the insects' growth rate and mortality were not consistently altered (Bezemer & Jones 1998). So the changes in plant composition caused by increased CO_2 seemed to affect the food quality rather than its palatability.

Response of living things to future climate change

As explained earlier, the only result of increasing greenhouse gases that can be predicted with fair confidence is that the average temperature of the world will go up. Other predictions, e.g. of rainfall, wind, local temperature changes, are less certain. So this section is confined to considering how animals and plants are likely to respond to the rise in CO_2 and the average rise in world temperature predicted under the 'business-as-usual' scenario, up to 2100. The prediction (Fig. 2.3) is that, compared with 1990, CO_2 will approximately double to about 700 μl l^{-1}, and temperature will rise about 2°C, i.e. about 0.2° per decade. I consider first how these changes will affect food production, and then how wild animals and plants are likely to respond.

Response of crops

One might perhaps expect that in temperate regions an increase of CO_2 and temperature would always lead to higher crop yields. The reality is likely to be more complicated. Table 2.8 shows results from an experiment with winter wheat, grown in a controlled environment facility but with an incoming light regime closely following that outside throughout the growth period. Air temperature also followed the outside conditions closely, or else was kept 4°C warmer. Atmospheric CO_2 was near ambient or double that. Plant final weight and grain yield were, as expected, higher in the raised CO_2. However, they were lower when the temperature was raised above ambient. This occurred because development during the winter was faster in the warmer conditions; as a result, the wheat plants' leaves senesced sooner and grain formation was completed sooner. The wheat plants at ambient temperature, by maintaining green leaves longer, were able to continue photosynthesis during a time in late

Warmer climate may reduce grain yield

Table 2.8 Summary of effects on wheat of raising air temperature, atmospheric CO_2 or both. Results expressed as change relative to CO_2 350 μl l^{-1} and British ambient temperature

	Atmospheric CO_2 (μl l^{-1}) and temperature (°C)			
Effect on wheat	350 ambient	700 ambient	350 ambient + 4°	700 ambient + 4°
Time to reach stage of development (days)*				
Flower formation	early April	+ 2	– 23	– 23
End of grain fill	early July	0	– 17	– 17
Final whole-plant dry weight (%)	100	125	82	106
Grain yield (%)	100	137	64	91

* + means later; – means earlier. From Mitchell *et al.* (1995).

spring when days were long, and so made more total growth. The result is that the combination of doubled CO_2 and temperature 4°C higher did not increase grain yield but in fact reduced it slightly. However, in most parts of the world doubling of CO_2 will probably be accompanied by a warming of less than 4°C. A combination of experiment and modelling was used by Laurila (1995) to predict yield by a Swedish wheat variety if grown in Finland in the future. He predicted that if doubling of CO_2 is accompanied by a 3°C warming there will be little change in the grain yield, but if the warming is only 2°C yield will increase by about 20%.

Wheat is an annual plant, whose useful product is the seeds. Response to temperature is likely to be different for plants that are perennial or whose useful product is some other part. This was found for the widely used pasture grass perennial ryegrass. As expected, raising CO_2 increased its productivity (Table 2.9). Raising the temperature by 3°C had no additional effect, so the prediction is that if doubled CO_2 is accompanied by a temperature rise of 3°C, herbage production by this grass will increase by about 20%.

Effects on plant water use

Warming may cause increased evaporative power of the air, leading to increased transpiration by plants. However, increased CO_2 will act in the opposite direction by reducing stomatal conductance. In the experiment in Table 2.9, the ryegrass swards in all the treatments received the same amount of irrigation water (and no rain). As expected, increased CO_2 reduced evapotranspiration, but raising the temperature increased it more, and so the combined effect was an increase in evapotranspiration. However, the increase in growth was proportionately greater, so that

Table 2.9 Growth and water use by swards of perennial ryegrass in polyethylene tunnels with controlled CO_2 and temperature. All received the same amount of irrigation water

	Atmospheric CO_2 ($\mu l\ l^{-1}$) and temperature (°C)				
Effect on sward	350 ambient		700 ambient		700 ambient +3°
Above-ground productivity ($t\ ha^{-1}\ year^{-1}$)	5.2	*	6.2		6.2
Evapotranspiration (% of irrigation)	79.5	*	78.0	*	83.5
Water use efficiency ($g\ l^{-1}$)	0.93	*	1.15		1.15
N concentration in leaves (%)	2.07	*	1.60		1.73

Results mean of two years. * difference statistically significant. Water use efficiency = above-ground dry matter production/evapotranspiration. Water use efficiency is equal in columns (2) and (3), although evapotranspiration is not. This occurred because the productivity was measured over the whole season but water use efficiency was for April–October only. Data from Casella *et al.* (1996), Soussana *et al.* (1996).

increasing CO_2 and temperature resulted in a rise in the water use efficiency. So, where the main limiting factor to crop yield is the amount of rainfall or irrigation water, future CO_2 and temperature increase could be beneficial. Water use efficiency is considered in more detail in Chapter 3.

Will less N fertilizer be needed?

Plants grown in higher CO_2 usually have lower concentrations of nitrogen but higher starch. This suggests that in future less N fertilizer may be necessary to obtain the same yield and food energy content. However, a warmer climate may act in the opposite direction. In the experiment with ryegrass (Table 2.9), 3°C warmer partly reversed the effect of doubled CO_2 on N concentration, but the combined effect was still a reduction. All the results in Table 2.9 are for ryegrass grown with a moderate rate of N fertilizer application, 160 kg N ha^{-1} $year^{-1}$. If instead N was supplied at 530 kg ha^{-1} $year^{-1}$, the combined effect of doubled CO_2 and +3°C was little change in N concentration.

Surviving very high temperatures

The combination of high air temperatures and bright sunshine can heat plants and animals, or exposed parts of them, to damaging or lethal temperatures. There is evidence that high-temperature tolerance of plants and animals is increased by *heat-shock proteins*. These proteins are synthesized in all living things (or at least in all species where they have been looked for) in response to a rise in temperature. If the rise is sudden, synthesis can start within a few minutes; but a slow temperature rise can also trigger their synthesis. If the temperature subsequently declines they decrease in abundance over several days, and some of them disappear altogether. There is increasing evidence that some heat-shock proteins reduce heat damage to other proteins by acting as *molecular chaperones*: they bind to other proteins, preventing them from unfolding if some of their cross-bonds break, thereby making it easier for them to reform bonds when the temperature falls again. Heat-shock proteins are not breakdown products of larger proteins: they are synthesized in response to the heat shock, and some of their genes have been identified. So it may be possible in future to increase the high-temperature tolerance of particular animal or plant varieties by gene transfer. For more information on heat-shock proteins and evidence that they increase heat tolerance, see Waters *et al.* (1996) and Park *et al.* (1996).

Future geographical ranges of crops

Climate change will alter the geographical range within which particular crops can grow well. For example, the regions of Europe where maize can be grown for grain should extend northwards. Figure 2.5 shows a prediction of the area within southern Africa which will be favourable for the growth of avocados, in a future of increased CO_2 and temperature. Rainfall is assumed to be unchanged. Research on this species has shown that its range is influenced by rainfall, evaporative demand of the air, and temperatures in the hottest and coldest months (see legend to Fig. 2.5). The predictions do not take into account possible direct effects of CO_2 increase on plant water balance. Nor have they considered possible effects of climate change on soil conditions or pests, two important topics about which we can at present say little. Figure 2.5 shows that some

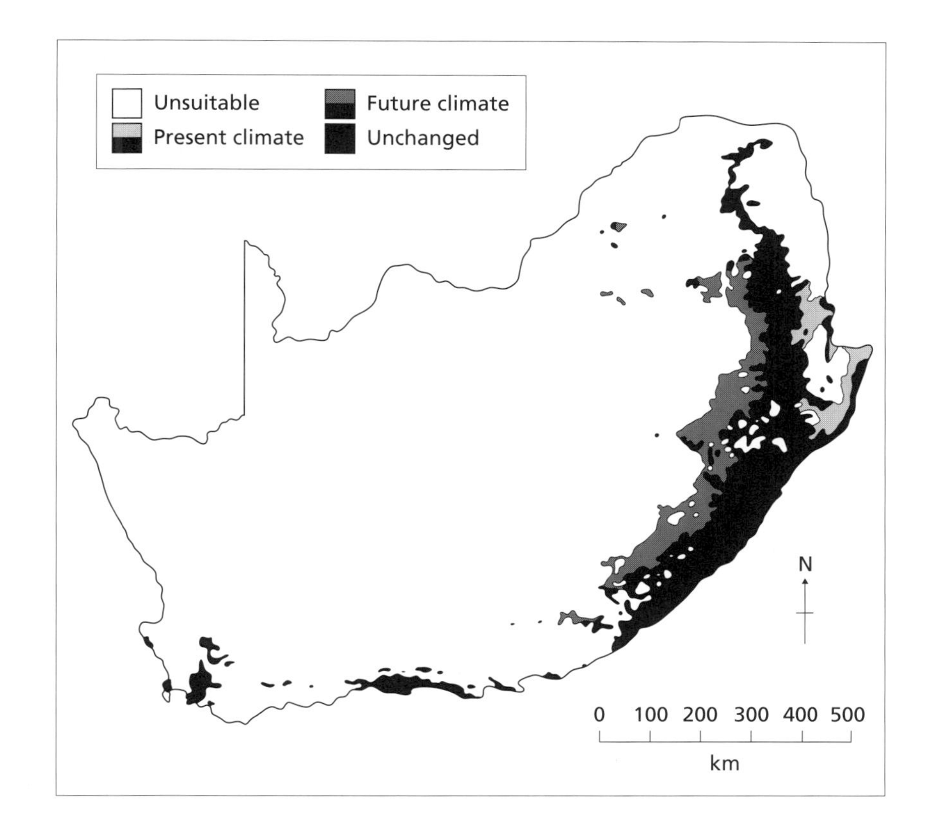

Fig. 2.5 Area in southern Africa with climate allowing good growth of avocado, now, and in future if CO_2 560 μl l^{-1} and temperature 2°C higher. Climate requirements of avocado: mean annual rainfall > 700 mm; mean daily minimum relative humidity ≥ 25%; monthly mean of daily maximum temperature, hottest month ≤ 31°C; monthly mean of daily minimum temperature, coldest month ≥ 4°C. From Schulze & Kunz (1995).

current avocado areas will remain favourable; some will become unsuitable, though a larger area further west will become suitable. Evidently some farmers will have to change the crops they grow. Changing to a new variety of the same species may be satisfactory for some crops. The response of wheat to higher temperatures (see earlier) indicates that farmers in a particular area will need different varieties which can maintain their canopy long enough in the warmer climate. For wheat, existing varieties from further south may well be satisfactory, but for some other species a new variety may need to be bred, since the combination of temperature and day length may never have existed previously.

How wild species will respond to climate change

If the climate gets warmer a wild species may respond in one of three ways.

Box 2.5. Techniques for studying living things and climate of the past 100 000 years.

Distribution of animals and plants

The past distribution of vertebrate animals is indicated by *bones* and of arthropods by *exoskeletons*. Macroscopic remains of plants that can provide information are *seeds, leaves* and *wood. Pollen* grains have the advantage of very large numbers, allowing quantitative assessment of changes in abundance, but the disadvantage that they can be widely dispersed, hence the area of catchment is not well defined. Very useful sources of these remains are sediments at the bottom of lakes, and peat in growing bogs. These combine (1) little physical damage, (2) anaerobic conditions, so slow microbial decomposition, and (3) continued accumulation of the surrounding medium, so the remains are in a vertical time sequence.

Dating the remains of living things

The age of organic materials up to tens of thousands of years old can be measured by *radiocarbon dating*. Among the CO_2 molecules in the atmosphere a small proportion contain the natural radioactive isotope ^{14}C, which has a half-life of 5730 years. The isotope is incorporated into living plants by photosynthesis and from them passes along food chains. The amount of ^{14}C remaining in dead plant or animal material provides a measure of when the plant it originated from was alive. There are technical problems arising from the fact that the $^{14}C/^{12}C$ ratio in the atmosphere has not remained constant throughout the last 100 000 years. This can be corrected for, but sometimes 'uncorrected radiocarbon years BP' are quoted. (BP = before present).

Past temperatures

Temperatures that occurred in the past can be estimated using the stable isotopes ^2H (deuterium) and ^{18}O. These both occur naturally in a small proportion of water molecules. Because they alter the molecular mass, they alter slightly the rate at which the molecules evaporate, condense or freeze. Hence the *isotope ratios*, $^2H/^1H$ and $^{18}O/^{16}O$ in ice indicate the temperature at which snow formed in the air overhead. Oxygen isotope ratios in $CaCO_3$ in skeletons of ocean animals can be used in a similar way to indicate the temperature of the water at the time they were formed. This has been applied particularly to cores from Antarctic and Arctic ice and from ocean sediments containing foraminiferan shells, since both these sources provide long vertical time sequences.

Further information: Moore *et al.* (1996).

1 It may be able to continue in the same area, either because the new climate is within its existing tolerance range or because it adapts (i.e. changes genetically).
2 It may migrate to a new habitat range, so remaining in a favourable climate.
3 If it cannot do either of these it will become extinct.

Fig. 2.6 Past and present distributions of three small mammal species, 13-lined ground squirrel (left-hand picture), long-tailed shrew and Hudson Bay collared lemming (right-hand picture). ● Site at which bones of all three species, dating from 18 to 10 000 BP, occur together. Shading and hatching: present distributions. From Graham (1997).

Past migrations and extinctions

We know of examples of all three occurring as the world warmed after the last Ice Age. Figure 2.6 shows the present distribution of three small mammals which all lived together at a site in southern Pennsylvania

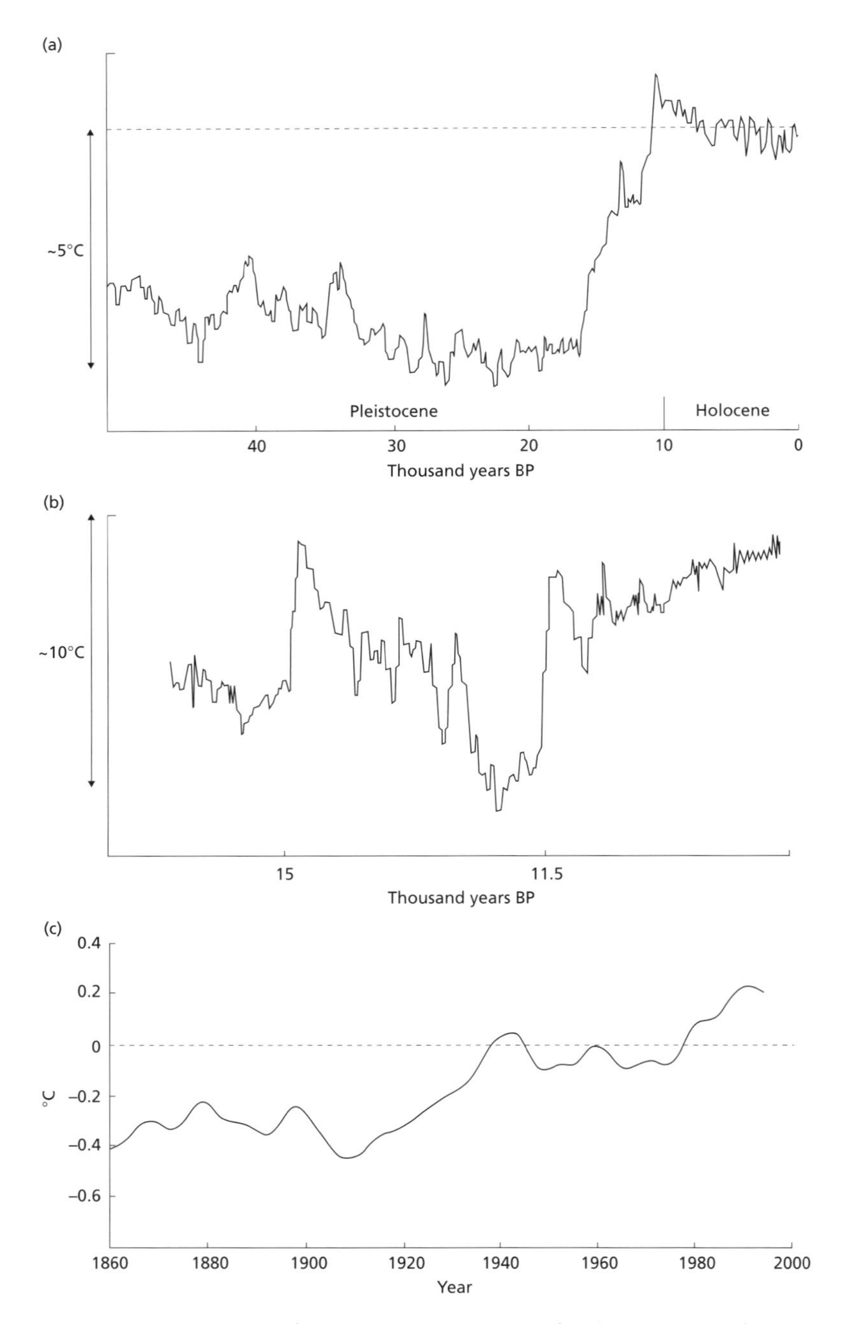

Fig. 2.7 Air temperatures in the past, on various timescales. (a) Antarctica, during past 50 000 years. (b) At a site (Dye 3) in Greenland, from about 15–10 000 BP. (a) and (b) are based on oxygen-isotope determinations on ice from deep cores; they indicate the amount of temperature change, but not the exact temperature at any particular time. (c) World mean air temperature, land and sea sites, 1861–1994. The line has smoothed year-to-year fluctuations, to show trends on a decade-by-decade timescale. Mean 1961–90 taken as 0. Sources: (a) and (c) Houghton *et al.* (1996); (b) Johnsen *et al.* (1992); reprinted with permission, copyright Macmillan Magazines Limited.

Table 2.10 Periods of warming

Period	Place	Length (yrs)	Temperature rise (°C)	Rate of rise (°C per century)	Method of temperature measurement	Source
AD 2000–2100	World	100	2	2	Predicted	1
AD 1910–40	World	30	0.5	1.6	Thermometers	2
BP 11 K	Greenland	50	7	14	O isotopes	2
BP 12.5 K–9 K	Greenland	3500	10	0.3	O isotopes	3
BP 16 K–11 K	Antarctic	5000	6	0.1	O isotopes	2
BP 13 K–7 K	Tropical Indian Ocean	6000	2.5	0.04	O isotopes	4

Data sources: (1) see text; (2) Fig. 2.7; (3) Houghton *et al.* (1996); (4) Van Campo *et al.* (1990).

about 18–10 000 years ago. The shrew still lives in that area, but the lemming migrated northwards and the ground squirrel westwards. Among large mammals of Eurasia at the end of the Ice Age, reindeer and musk-ox are examples of species that survived within their former range, whereas woolly mammoth, mastodon and woolly rhinoceros became extinct (Sher 1997). However, extinctions at that time were not all due to climate change: hunting by people was probably also involved (Stuart 1991).

These examples show that when we want to predict future responses of species to climate change we can usefully learn from the past, in particular how species responded to the warming at the end of the last Ice Age.

Temperature change in the past

Box 2.5 summarizes methods that can be used to find out about species in the past and the temperatures at the time. Figure 2.7(a) shows temperatures in Antarctica over the last 50 000 years, determined by oxygen isotope measurements on ice from layers in a deep ice core. Temperatures had been colder than present since 120 000 years ago ('the last Ice Age'). From 16 000 BP warming began (BP = before present), and continued until 10 000 BP, since when it has varied much less. 10 000 BP is taken as the start of the *Holocene*, or postglacial, period, the previous 2 million years being the *Pleistocene*. Table 2.10 shows that the average long-term rate of warming between 16K and 7K years BP was at least an order of magnitude slower than the rate predicted for the coming century. However, more detailed records indicate that the warming was not at a steady rate throughout the period. As shown in Fig. 2.7(b), the temperature warmed, cooled again and then warmed again very suddenly. This second warming was substantially faster than is expected in the coming century (Table 2.10). It is shown in several Greenland ice cores, and there is evidence for a sudden warming at the same time in Switzerland (Dansgaard *et al.* 1989). There were probably some periods when warming by 7°C or more occurred in even less than 50 years (GRIP 1993), so there have been short periods when warming was as fast as expected for the coming century, or faster, but none is known to have gone on for as long as a century.

Will species be able to stay put?

Some species should be able to remain in much of their present range if the climate warms by 2°C. Sykes (1997) predicted ranges suitable for some European tree species in a doubled CO_2 climate. Species that could remain in most of their present range, but also have the opportunity to expand, include beech (*Fagus sylvativa*), sycamore (*Acer pseudoplatanus*) and fir (*Abies alba*), whereas spruce (*Picea abies*) is predicted to disappear from much of its present range in central Europe and Scandinavia. Whether invertebrates can remain will depend much on their local microclimate. Experiments on soil microarthropods in the Arctic indicate that they can survive a warming of several degrees C (Hodkinson *et al.* 1996). Among the ant species in Spain studied by Cerdà *et al.* (1998), depending on what time of day and what season of the year they were active, some were foraging in sites close to their high-temperature tolerance, others far from it. So some could only survive a climate warming by changing their foraging habits.

Will species adapt genetically?

Species can adapt genetically to differences in climate, as shown by the fact that relevant ecotypic variation occurs within species. Plants or seeds of the same species collected from different altitudes in mountains have been found to differ genetically in ways that adapt them to the climate of their particular altitude (Clausen *et al.* 1948; Slatyer 1977). Butterfield and Coulson (1997) gave examples of genetic variations within insect species that were evidently adaptive to their local climate. These examples show that adaptation to higher temperature can occur within an animal or plant species, but can it occur fast enough to allow species to remain in the same range if the climate warms at the rate of 2°C per century? Some examples of rapid evolution of relevant characters are known. Body size in mammals can evolve rapidly. Many mammal species decreased in body size at the end of the Ice Age (Lister 1997). This could assist in heat regulation by altering surface:volume ratio, but it could also be related to changes in diet. One species which got smaller then was red deer. We know its size can change rapidly, because after British red deer were introduced into New Zealand their progeny were 2–3 times heavier within 20 years.

An experiment was provided by a nuclear reactor in South Carolina whose cooling water, for 13 years, went into a long artificial pond, which was thus very hot at one end but cooling down to near normal at the other. Bluegill fish from the hot end could survive a temperature about 2°C higher than could those from the cool end (Table 2.11). When the fruit fly *Drosophila* was reared at different temperatures for 60 generations or more (this took 4 years), there was a statistically significant increase in its tolerance to high temperatures, although the change was small (Table 2.11(b)).

Thus we have evidence that species can sometimes change their tolerance of high temperatures quite rapidly. However, at the end of the last Ice Age many species migrated as the world warmed, so they evidently did not adapt fast enough to allow them to stay put. The best information we have on the rates of spread at the end of the Ice Age is for plants. Pollen sequences, dated by radiocarbon (see Box 2.5), have been studied at

Table 2.11 Effect of long-term higher temperature regime on subsequent survival of very high temperature

(a) Bluegill fish living in water heated by nuclear reactors, S. Carolina.		
Conditions of pond where fish lived	Temperature that killed fish (°C).	
Parts often above 50°C, seldom below 30°C	40.9	
Near normal for S. Carolina	39.0 *	
(b) Fruit-fly, *Drosophila melanogaster*, reared for about 60–100 generations.		
	Percent surviving 39.5°C for 30 min	
Reared at (°C)	Female	Male
16.5	36.4	7.1
25.0	40.4*	9.9*

* Difference statistically significant ($P < 0.05$).
Data sources: (a) Holland *et al.* (1974); (b) Huey, Partridge & Fowler (1991).

Migrations at the end of the Ice Age

enough sites in eastern North America and western Europe to allow maps of spread of individual species to be made (Davis 1981; Huntley & Birks 1983). Figure 2.8 shows two examples among North American tree species. They were both responding to climatic warming, but they did not spread at the same rate or in exactly the same direction. Table 2.12 shows rates of spread for members of major tree genera in Europe and North America. Obtaining these rates from pollen data is not entirely straightforward (Delcourt & Delcourt 1991) and the figures in Table 2.12 may need some revision, but there is no serious doubt that rates of spread often averaged several hundred metres per year. Most tree species do not produce seed in their first year, so it is informative to work out the average distance each species must have spread per 'generation', i.e. per length of time from seed germination to seed production by the resulting tree. Table 2.12 shows such calculations for North American trees. These indicate that the seed of some species must have dispersed several kilometres. One might expect species with winged, wind-dispersed seeds to have spread more rapidly than those with larger seeds which have no obvious means of dispersal, but this was not consistently the case.

Were these rates of spread keeping pace with climate change, or were they limited by the ability of the species to spread? Today a distance of 110–170 km north–south corresponds to a 1°C difference in mean annual temperature. As the long-term average rate of warming was 0.1–0.3°C per century (Table 2.10), species would need to have migrated at about 0.1–0.5 km year^{-1} to keep pace, so the observed rates were of the right order. If the rise of 7°C that occurred in Greenland about 11 000 BP (Fig. 2.7(b), Table 2.10) also occurred across North America, species responding to it quickly would have moved northwards much faster. It is not clear that the pollen records show species migrating much faster at that time.

Table 2.12 Rates of spread of tree genera during the postglacial period

		Europe*	North America		
	Type of seed	Rate of spread (km yr^{-1})	Rate of spread (km yr^{-1})	Age (yr) at first seed production	Rate of spread (km per generation)
Birch (*Betula*)	winged	> 2			
Pine (*Pinus*)	various	1.5	0.3–0.4	3–5	1–2
Hazel (*Corylus*)	large, hard coat	1.5			
Oak (*Quercus*)	large, hard coat	0.15–0.5	0.35	20	7
Elm (*Ulmus*)	winged	0.5–1.0	0.25	15	4
Spruce (*Picea*)	winged		0.07–0.3(– 2)	4	0.3–1(– 8)
Beech (*Fagus*)	large, hard coat	0.2–0.3	0.08–0.3	40	3–12
Hemlock (*Tsuga*)	winged		0.03–0.2	15	0.5–3
Chestnut (*Castanea*)	large, hard coat		0.1	12	1

* Mainland Europe.
Data from Davis (1981), Webb (1986), Ritchie & MacDonald (1986), Birks (1989), King & Herstrom (1997).

How did plants manage to spread so rapidly?

When trying to predict how fast plant species will be able to move in the future, one problem is that we do not understand how many of them managed to spread as fast as they did in the past. In modern times many species seem to spread much more slowly than the rates in Table 2.12. Chapter 11, which is about restoration of communities, expresses concern about how slow species can be to recolonize apparently suitable habitats, even over distances of only a few hundred metres. It gives examples for trees, herbaceous plants, lichens and several groups of insects. The time available for colonization was often only decades, but sometimes several hundred years. Figure 10.9 (p. 307) shows that even after a century, one herbaceous woodland species has rarely colonized another woodland if there is a gap of more than a few hundred metres. These gaps were usually farmland. Such non-natural habitat can certainly be a barrier to spread, a barrier which was not present at the end of the Ice Age but will be present during the warming of the future.

Range extension by some species in response to warming of the 20th century has sometimes been slower than would keep pace with temperature change. Grabherr, Gottfried and Pauli (1994) compared the distribution of plant species on high mountains in Austria and Switzerland with precise records made 70–90 years earlier. Disturbance by people was slight. Mean annual temperature rose 0.7°C during the period, and if that is the determining factor the plants should have extended their altitude upwards by 16–20 m per decade. In fact, most had extended by less than 1 m per decade. On mountains in northern Canada white spruce trees 100–150 years old were found close to the tree line, indicating little or no advance of the boundary during that time, although the density of trees within the stands had increased (Szeicz & MacDonald 1995).

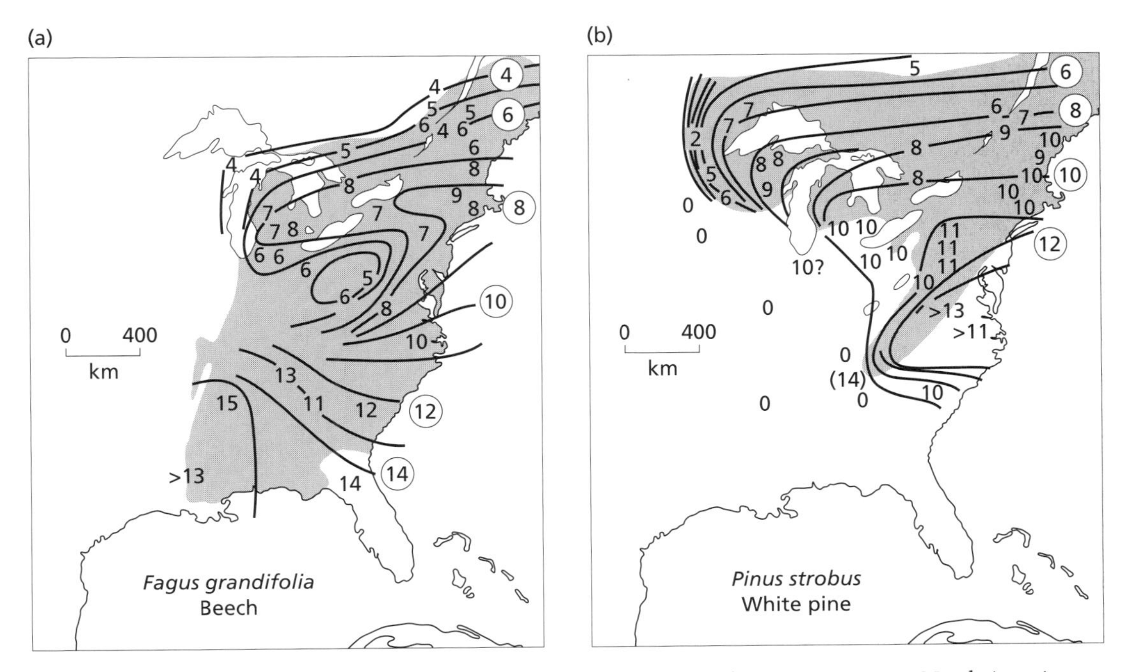

Fig. 2.8 Change in distribution of beech and white pine in eastern North America during the last 15 000 years. Uncircled figures show the date of first arrival at a site (in thousands of years BP). Circled figures apply to the thick lines, which approximately join points of equal arrival date. 0: never present. Shaded areas are the present distribution. From Davis (1981).

Research by Cain, Damman & Muir (1998) on a herbaceous species of northeastern North America, *Asarum canadense*, makes clear the difficulty of explaining rapid spread in the past. This has a seed weighing 14 mg which is usually dispersed by ants. Although the species' past distribution is not known, it must have extended at least 450 km since the end of the Ice Age. Even if that took 15 000 years the average was about 30 m year^{-1}, or 300 m per generation, since it takes 10 years to produce seed. Cain *et al.* recorded the distance seeds moved in the natural habitat. Most moved less than 1 m, though two individual seeds moved 24 and 35 m. Even 35 m per generation is an order of magnitude too slow to account for the actual long-term range extension in the past. This problem applies to many other species: Cain *et al.* gave an extensive list of the maximum seed dispersal of other species, reported in the literature, which shows that seeds of most herbaceous species travel less than 100 m, though a few can sometimes travel several kilometres. Table 2.13 shows the maximum distance that seeds of some common tree species have been reported to travel from the tree. These may be compared with the rate of spread per generation that occurred in the past, given in Table 2.12. Movement of oak and beech seeds by small mammals is clearly far too limited, and among the wind-dispersed species only hemlock is known

Table 2.13 Maximum distance (km) that seeds of some North American tree species are known to travel

	Dispersal agent		
	Mammal*	Wind	Bird
Pine (*Pinus*)		0.04–0.5	4–22
Oak (*Quercus*)	0.05		2
Elm (*Ulmus*)		0.3	
Spruce (*Picea*)		0.2–0.5	
Beech (*Fagus*)	0.01		4
Hemlock (*Tsuga*)		1.6	

* Excluding bats.
Picture: bluejay carrying acorn. Datafrom Cain *et al.* (1998).

Dispersal by birds

to travel far enough to account for its past spread. However, dispersal by birds could give about the right rate of spread for oak and beech, and amply so for pine. The observed bird-dispersal of pine seed was by nutcrackers; acorns and beech mast was by bluejays (Vander Wall & Balda 1977; Hutchins & Lanner 1982; Johnson & Webb 1989). These birds store seeds in caches for future use but do not always use them all. The caches may be some distance from the seed-bearing trees, sometimes under trees but sometimes in the open. So this could be a way that a forest extends beyond its existing boundary.

We may wonder whether migrating birds can carry seeds much longer distances, in their digestive systems or stuck to their outsides. Apart from the question of whether seeds can ever remain viable after passing through a bird, there is the problem that birds migrate in the wrong direction—when the seeds are ripe in autumn they fly from cooler to warmer latitudes. Another possibility is that seeds were carried by rivers. There are major north-flowing rivers in Europe and Asia, but in North America all the major rivers flow southwards, eastwards or westwards. Some species could have been carried by the ocean. It is known that the hard-coated seeds of hazel can germinate after floating in seawater for some days. It is likely that it invaded Britain in this way, since the pollen records show that it first occurred along the west coast of Wales, northern England and Scotland but along the east coast of Ireland, and then spread inland (Birks 1989).

Thus it seems likely that at the end of the Ice Age the observed rates of spread were near the maximum that plant species could attain. The evidence includes the facts that (1) different species extended at different rates, suggesting that properties of the plants and their dispersers, rather than temperature change, were controlling the pace; and (2) study of present-day plants does not suggest an ability to migrate faster than they did in the past.

Past range changes of animals

This section on dispersal rates has been all about plants. We have no similar data on past rates of dispersal for any animal groups. Some mammal groups are now far from where they were in the late Pleistocene. Figure 2.6 shows that the Hudson Bay collared lemming's present range is separated by about 1600 km from where it occurred in Pennsylvania some 10 000 years ago. Graham (1997) lists other North American mammals grouped according to how their range has changed since 18 000 BP. Some moved northwards about as fast as the collared lemming, some more slowly. Some moved westwards, others eastwards; others have changed their range little. Some insects have moved by several thousand kilometres. For example, several beetle species known to have lived near the Great Lakes 13–10 000 years ago are now found only in northern Canada or Alaska (Morgan *et al.*, in Porter 1983). A dung beetle species that occurred in Britain 25 000 years ago is now found no nearer than western China (Ashworth 1997).

Detailed studies of the sequences of beetle exoskeletons preserved at some sites in Britain show changes that occurred over periods of 500 years. Figure 2.9 shows the abundance of species which today have distributions either to the north or to the south of Britain. Following 13.5 K BP and again following 10.5 K BP there was a sudden near-disappearance of northerly species. Southerly species appeared, but their arrival was slower, especially after 10.5 K. This second change in species could be a response to the sudden warming about 11 K BP (Fig. 2.7(b)).

Will wild species be able to move fast enough?

The key question we want to answer is: if temperatures increase by 2° per century, over a century or more, will wild species be able to move their range fast enough? The answer for animals is that we do not know. We tend to assume that most birds and large mammals will not have a problem; for smaller animals we cannot say. The answer for plants seems to be that most species will not be able to move fast enough. The rates required will be much faster than at the end of the Ice Age, yet their rates then were probably at or near their maxima.

A clear message from the records of the past is that whole communities did not move as a unit: sometimes species moved in the same direction but at different rates, sometimes they moved in different directions. This was true for both plants (Table 2.12) and animals (Fig. 2.6). In North America today the northern parts of the ranges of beech and white pine overlap substantially, yet before 12 000 years ago their ranges did not overlap at all (Fig. 2.8). The distribution maps of Huntley and Birks (1983) show similar examples for European species (e.g. beech and deciduous oaks) that grow together today over wide areas but which were formerly only found far apart. Animals did not necessarily keep in step with plants. The changes in British beetle species indicated in Fig. 2.9 were faster than plant changes at that time: about 10 000 BP a diverse suite of insects from more southerly regions moved into Britain when the only established trees were birches, and it is not clear how there were food chains to support them all (Coope 1987). If species behave in such an opportunistic and individualistic way in the future, it is not possible to predict what sort of species combinations and interactions will arise.

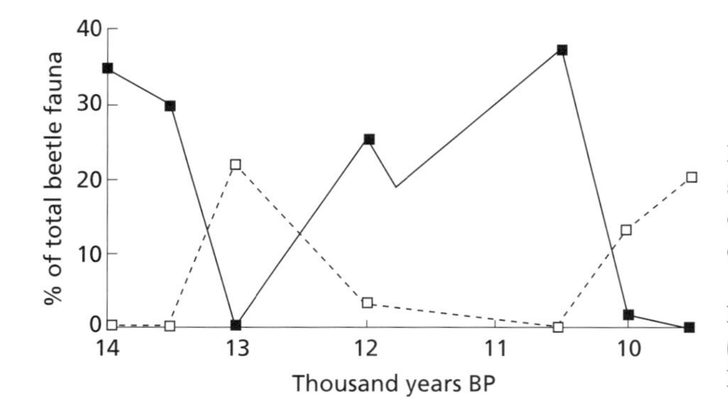

Fig. 2.9 Diagram summarizing changes in species composition of beetle fauna of Britain during the period 14–9.5 K BP. Species classified according to their present distribution: —— mainly northwards from Britain, – – – mainly southwards from Britain. (Species with wider distribution not shown.) From Ashworth (1997).

Climate change was disruptive to communities in the past, and will probably be disruptive in the future.

It is not easy to decide how people can best help to minimize the effects of climate change and help species to survive. Should we try to maintain existing communities? What if this means trying to prevent the establishment of invaders responding to climate change? Should we try to translocate whole communities? Or should we accept that individualistic species response to warming has happened in the past, and try to promote that? But might we then be trying to set up new communities that are in fact not ecologically viable? Information that may help towards answering these questions comes in Chapters 10 and 11. Chapter 10 discusses keystone species—examples where the presence of one species is essential for others to survive. Chapter 11 is about the recreation of communities within their former ranges, but some of the topics it considers are also relevant to helping communities to migrate.

Conclusions

- Food production systems, past and present, have varied greatly in their energy production per hectare. Without modern high-input farming systems the present farmed area could not feed the world's present population.
- Most of the net CO_2 transfer to the atmosphere each year comes from burning fossil fuels. About half this net transfer ends up in sinks on land and in the oceans; half remains in the atmosphere, increasing CO_2 concentration.
- Carbon dioxide, methane and nitrous oxide are likely to increase in the atmosphere during the 21st century, causing global warming, probably within the range 1–3°C.
- In theory the CO_2 increase could be greatly retarded by using biomass fuel instead of fossil fuels, or by growing C-sink forests; but for practical reasons their contribution can only be small.
- Increased CO_2 concentration usually increases plant growth.

However, a 1–3° increase in temperature can be either harmful or beneficial to animals and plants.
- Probably many wild species will be unable to move their range fast enough to keep pace with a temperature rise of 2° in a century.

Further reading

Climate, energy balance:
Robinson & Henderson-Sellers (1999)

Photosynthesis, productivity:
Larcher (1995)

The carbon cycle and climate change:
Houghton *et al.* (1996)
Moore, Chaloner & Stott (1996)
Houghton (1997)

Climate and living things in the past:
Moore, Chaloner & Stott (1996)
Delcourt & Delcourt (1991)

Renewable energy sources:
Boyle (1996)

Chapter 3: Water

Questions

- Could cutting down tropical forest result in reduced rainfall?
- In the semi-arid Sahel region of Africa, has the vegetation cover decreased in recent decades? Has the rainfall decreased? If so, what are the causes? Are people to blame?
- Are large herbivorous mammals limited by water supply in some areas, even where vegetation is abundant?
- Can crop plants produce more food for the same amount of water used? How?
- Can crop plants be developed which can use seawater, or other saline water, as their only source of water?

Background science

- Relationships between water use and growth, for plants and animals.
- What features of vegetation affect (1) energy balance, and (2) transpiration and evaporation.
- How rain forms.
- Distribution of large herbivores in relation to drinking water.
- Limitations to irrigation farming.
- Ways that NaCl in soil can harm plants. Strategies of plants that minimize damage by NaCl.

Plants: CO_2 gain and water loss

Green plants need to take up carbon dioxide in order to photosynthesize. The CO_2 has to enter plant cells through a wet surface, so any surface that is taking up CO_2 is also losing water. Because the concentration of water in plants is high but the concentration of CO_2 in the atmosphere is low, a plant loses many molecules of water for every molecule of CO_2 it gains: commonly plants operating the C3 or C4 photosynthetic system lose about 500–2000 mol of H_2O per mole of CO_2 gained, though the ratio is generally lower for plants with crassulacean acid metabolism (CAM) (Stanhill 1986; Nobel 1991b). Most vascular plants have the ability to reduce transpiration greatly by closing their stomata if soil water supply becomes deficient; but this is at the expense of also stopping photosynthesis.

Animals: O_2 gain and water loss

Land animals are substantially different in their water balance, because they take in oxygen instead of CO_2. In air oxygen is about 600 times more concentrated than CO_2, so we might expect that instead of the unfavourable ratio (H_2O lost/CO_2 gained) of 500–2000 suffered by plants, the H_2O lost/O_2 gained by animals would be about 1–3. However, most animals lose water through their skins and in urine, as well as through their respiratory surfaces. This varies a lot between species: some insects and mammals lose scarcely any water by those two routes. Another important difference between animals and plants is that animals create water from the organic matter of their food by respiration, in contrast to plants, which use water in photosynthesis. If the animal is respiring sugar the ratio of moles of water produced by metabolism to moles of oxygen used is 1:

$$C_6H_{12}O_6 + 6O_2 = 6H_2O + 6CO_2$$

These features make it possible for some animals, notably many adult insects and some mammals, to survive without drinking liquid water (Schmidt-Nielsen 1997). Some mammals of arid habitats obtain all the water they need from metabolic water plus the liquid water in their food (Nagy, Bradley & Morris 1990; Nagy 1994). Some can even survive for much of the year on seeds, which contain little liquid water (Nagy & Gruchacz 1994). However, many animals do require a drink at regular intervals. Information on how this affects their distribution in semi-arid rangeland, and hence their ability to make use of the available vegetation, is given later in this chapter.

The previous two paragraphs were a gross oversimplification of the water relations of living things, but they serve to emphasize an important difference between plants and animals. Animals which have efficient control of their water loss can continue active for long periods without taking in any liquid water, apart from what is in their food. Plants can survive long periods without water if they are in a dormant state, for example as seeds, but they can only photosythesize and grow when they are able freely to take in and transpire water.

Plant growth and transpiration are correlated

Photosynthesis and transpiration are both strongly affected by stomatal aperture, and also by the amount of short-wave radiation falling on the leaf surface. This is the basic reason why photosynthesis and water loss tend to be closely correlated. Figure 3.1 shows that there is a clear relationship, worldwide, between the net primary productivity of natural vegetation and the precipitation (rainfall + snow water) at the site. It might seem surprising that over the great range of vegetation types such a relationship holds, but all land plants are constrained by this key fact that when CO_2 is gained water is lost. Figure 3.2 shows the growth of a single species—wheat—in relation to water supply. It uses the inelegant term 'evapotranspiration', which means evaporation plus transpiration, i.e. the total amount converted from liquid water to vapour. Evaporation can be rainwater caught on leaf surfaces and evaporated before it

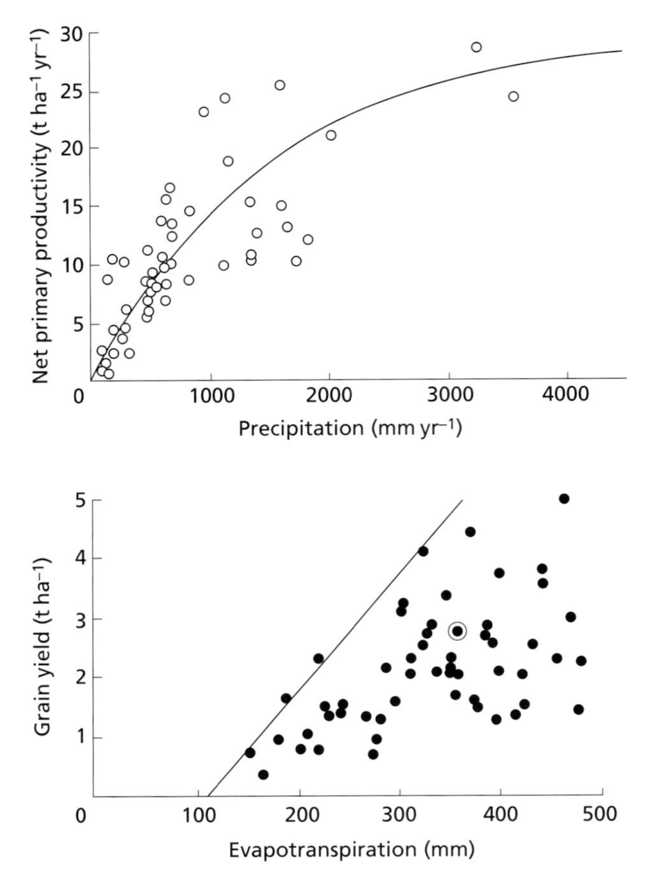

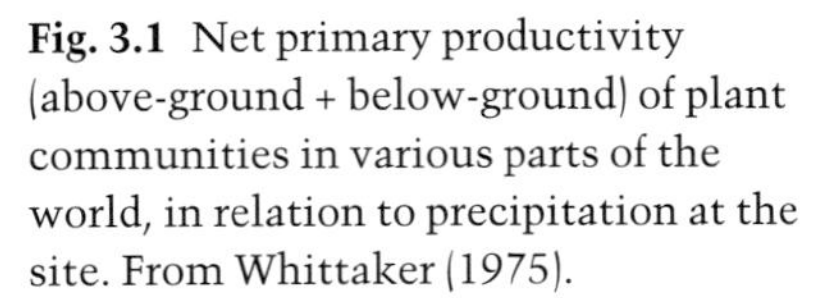

Fig. 3.1 Net primary productivity (above-ground + below-ground) of plant communities in various parts of the world, in relation to precipitation at the site. From Whittaker (1975).

Fig. 3.2 Relationship between grain yield of wheat and evaporation + transpiration at sites in South Australia. Each point is for one site in 1 year; many sites supplied data for several years (shown as several points). The boundary line indicates the approximate upper limit for grain yield at any particular evapotranspiration. From French & Schultz (1984).

reaches the ground, as well as evaporation of water from the soil. In this area of low rainfall there was little run-off into rivers, and evapotranspiration was approximately equal to rainfall during the growing season. About 110 mm was lost as evaporation from soil; the rest was transpired by the wheat plants. The amount of water available for uptake and transpiration evidently set an upper limit for growth and hence for grain yield. However, the yield was often lower than this upper limit, owing to other unfavourable factors such as nutrient deficiency, pests or weeds.

If people alter vegetation, can that result in reduced rainfall?

Case 1: Amazonia

Tropical rainforest is being cut down and some of the area converted to grassland or farmland. Chapter 7 (Forests) considers this: how much forest is being lost, in which regions, and what is the ecological significance. Here I consider the question: If a large proportion of the present Amazon rainforest is converted to grassland or open savanna woodland, would that alter the rainfall?

At first sight the answer may seem to be obviously yes. Surely if forest is replaced by smaller plants there will be less transpiration, hence less moisture in the air and so less rain? The real situation is, however, more complex. We need to consider the following questions:

1 Does forest actually put more water vapour per year into the atmosphere than grassland does? We should bear in mind that the conversion of water from liquid to vapour requires energy; the energy must come from the sun; the amount of short-wave radiation falling on a hectare is the same whatever the vegetation.

2 Can vegetation affect rainfall in other ways than via the water vapour content of the air?

3 Does most of the rain falling on Amazonia come from local evaporation and transpiration? or does most of it come from the oceans?

Evapotranspiration: more from forest than grassland?

The first question can be answered by measurement and experiment. The amount of water vapour lost from a whole valley can be calculated from the amount of rain falling (measured by rain-collecting gauges) and the amount of water lost in the outflowing river. Over a whole year:

Evapotranspiration = rainfall – runoff

This equation assumes the valley has an impermeable base (clay or rock) so that there is no loss by deep seepage. A forested and an unforested valley can be compared, but they could differ in other ways, e.g. exposure to wind. A better way is to first measure the water balance of two forested valleys, then to find out the effect of clear-felling one of them while keeping the other as a control. Such experiments have been carried out at several sites in the USA, also in Kenya. Part of the mixed deciduous forest at Hubbard Brook, New Hampshire, was clear-felled; the felled trees were left and herbicide was applied for 3 years to prevent any regrowth (see Chapter 4). During those 3 years the annual evapotranspiration was reduced by 58% (Bormann & Likens 1979). At other sites, where forest was replaced by vegetation of lower stature, evapotranspiration was usually reduced by about 10–40% (Penman 1963; Lewis 1968). So evapotranspiration is affected by the vegetation but is not proportional to the biomass or leaf area of the plants.

What causes the difference?

How is this difference in evapotranspiration possible, if the amount of energy received by both vegetation types is the same? Box 2.1 summarizes the fate of the energy in the short-wave radiation that impinges on plants and soil. One common difference between forest and grassland is that grassland reflects more short-wave radiation (its *albedo* is greater). Therefore, it absorbs less radiant energy, so less is available to power evapotranspiration. A second difference is the canopy structure: forests tend to have their leaves more spaced out, allowing wind to blow through the canopy more freely; also the top of the canopy is more irregular (called *greater aerodynamic roughness*), which promotes wind turbulence above the canopy. These differences in structure tend to increase water loss from forest. There may also be differences in stomatal closure, for example if grassland has shallower roots and exhausts the available soil water more quickly during rainless periods. Higher transpiration by forest owing to canopy structure and stomata results in more energy (latent heat of evaporation) being used in transpiration. The canopy therefore becomes cooler, so less energy will be lost through long-wave radiation and heating the air.

Thus conversion of the Amazonian forest to grassland probably would reduce the amount of water vapour going into the air, but would also affect the energy balance. Would this affect the amount of rain falling? It is clear that not all the water that falls on the Amazon basin comes from evapotranspiration there. The evidence is the existence of the Amazon River. About half the rain that falls on Amazonia flows out to sea; it follows that, to maintain the water balance of the continent, about half the rain that falls must be water that evaporated from the oceans (Salati & Vose 1984). The Amazon Basin is bordered by rising ground to the north and south, and by the high Andes to the west. The prevailing winds are from the east, from the Atlantic Ocean, and carry moisture across Amazonia. If half the rain comes from the ocean, the other half comes from evapotranspiration within Amazonia. So this evapotranspiration could be important. But we have not yet directly tackled the question, would removing the forest alter the rainfall? Would it increase it or decrease it? By how much?

Does Amazonian rainwater all arise locally?

The average length of time a water molecule spends in the atmosphere before it falls as rain is about 10 days; the average distance it travels during this time is about 1000 km (Berner & Berner 1996). So we need to consider these questions on a very large scale—hundreds of kilometres, or perhaps a whole continent. Do we have any experimental evidence on that scale?

Changes in land use and rainfall in Israel

The nearest to a suitable experiment was on the land at the southeast corner of the Mediterranean Sea, the semi-arid to arid region of the Negev and adjacent Sinai (Otterman *et al.* 1990). Following the founding of the state of Israel in 1948, grazing was reduced on the Israeli side of the border. Irrigated farming started in the late 1950s and afforestation in the early 1960s. By 1972 the straight boundary between Israel and Egypt was clearly visible from outer space, darker on the Israeli side because of more vegetation (Otterman 1974). Comparing the annual rainfall for 1942–62 vs. 1963–88, the mean rainfall was 16% higher after 1962. In percentage terms the increase was greatest in October, the start of the rainy season. This could have been caused by the increased vegetation cover (Otterman *et al.* 1990), but this is not certain because there were no comparable rainfall data from the Egyptian 'control' area.

As we cannot carry out adequate experiments to investigate the effect of vegetation on rain, we have to work from a basic understanding of how rain is formed, and then construct mathematical models.

How rain forms

For rain to fall, the water vapour must coalesce into drops large enough

Box 3.1. Summary of processes by which a reduction in the amount of vegetation (tallness, structural complexity or amount of ground covered) can affect rainfall.

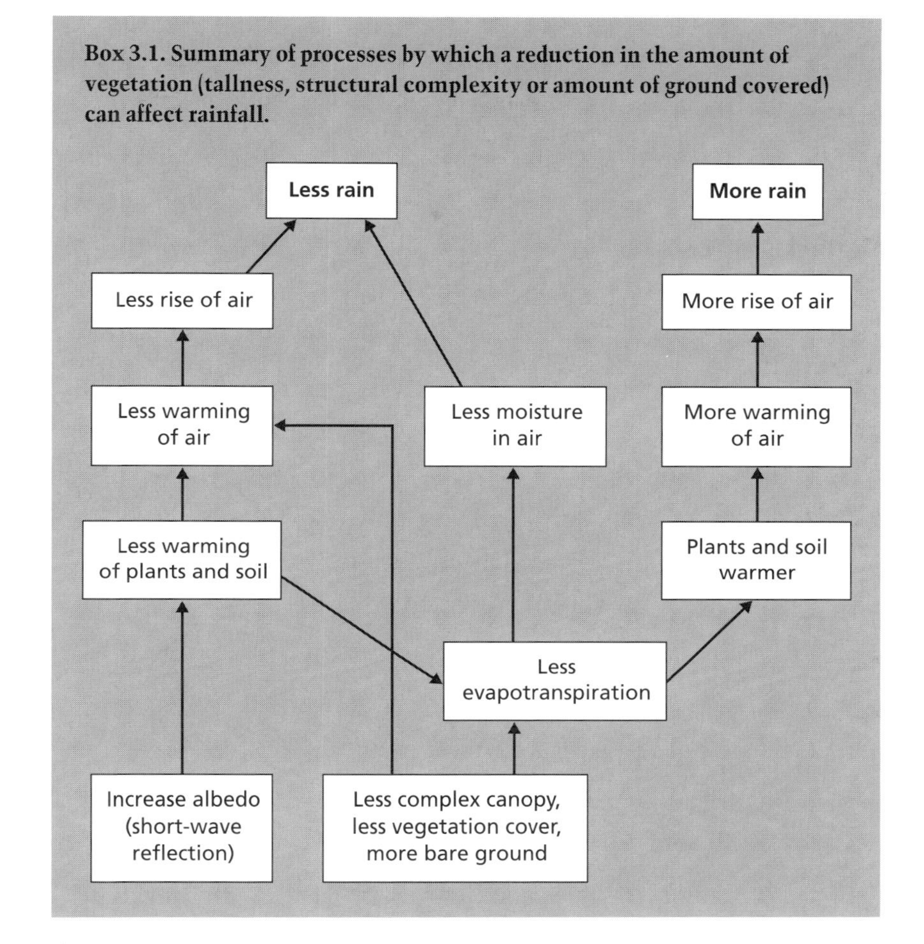

to fall. Nuclei—minute solid particles (e.g. NaCl crystals or fine dust from soil)—can help this drop formation, but the key requirement is for the air to cool so that the amount of water vapour it can hold decreases. Over land the normal way that air cools is by rising upwards. If air blowing sideways meets mountains it will rise, but that is not going to happen in the low-lying Amazon Basin: there air rises because it is heated near the ground. It may seem curious that air has to be warmed before it can cool, but that is an important feature of climate. It means that the ability of forest or grassland to warm the air above it can have an effect on how much rain falls. Box 3.1 shows the ways in which conversion of forest to grassland, by altering the albedo and canopy structure, could influence rainfall. The arrows in the centre and left lead to 'less rain', but those on the right lead to 'more rain'. In other words, there are counteracting effects, and it is not obvious what the net effect on rainfall will be. What is needed is a model.

Deforesting Amazonia: models

The effect of deforesting Amazonia has been modelled by several groups of scientists, using general circulation models: in other words, the models are primarily atmospheric physics, though with substantial attention to what happens in the soil-vegetation–atmosphere interface. One team has published several papers, in which their modelling developed and hopefully became more realistic (Dickinson & Henderson-Sellers 1988; Henderson-Sellers *et al.* 1993; McGuffie *et al.* 1995). They also extended from deforesting only the Amazon Basin in the first paper, to deforesting the Amazon plus southeast Asia in the second, and finally tropical Africa as well in the third. Here I summarize mainly the results for Amazonia. These models predict energy balance, wind, evapotranspiration and rainfall: not just means for the whole year, but values for each month and how they vary across South America. The Amazon Basin is assumed to be deforested instantly: all forest one day, the next day all grassland or open shrubland. Principal changes assumed to occur are:

a decrease in:

1 the percentage of ground covered by vegetation;

2 leaf area index;

3 vegetation tallness and structural complexity;

4 depth of rooting.

an increase in:

1 albedo.

Reduced rainfall predicted

Averaged over the whole year and the whole Amazon Basin, the prediction by McGuffie *et al.* (1995) is that deforestation would cause a decrease in both evapotranspiration and rainfall (Table 3.1(a)). The average rainfall over the Amazon Basin at present is about 2000 mm year^{-1}, so the predicted change is substantial. Reduced rainfall is predicted for every month. The decrease in rainfall is greater than the decrease in

Table 3.1 Predicted effect of deforestation on the water balance of tropical areas. –means that deforestation causes a decrease, + that it causes an increase. Units mm year^{-1}

(a) Predicted effect on Amazon Basin of converting all tropical forest to savanna woodland or grassland. First column: predictions by model of McGuffie *et al.* (1995). Right-hand: their summary of range of previous predictions by other models.

	McG *et al.*	Others
Change in evapotranspiration	–231	–164 to –985
Change in rainfall	–437	–640 to +394

(b) Averages for eight areas in the Amazon Basin each about 400 × 500 km, if only those areas deforested. Predictions by Sud *et al.* (1996).

Change in evapotranspiration	–420
Change in rainfall	–111

(c) Averages for southeast Asia after tropical deforestation. Predictions of McGuffie *et al.* (1995).

Change in evapotranspiration	–128
Change in rainfall	–48

evapotranspiration. Rainfall decreases not only because there is less moisture in the atmosphere but also because the convective rise of air is less. There are changes in wind, and less moist air is brought in from the Atlantic.

But not all models agree

Table 3.1 shows that there is considerable disagreement between the models of Amazon deforestation. The models differed in their assumptions about what characteristics of the vegetation changed with deforestation, and by how much. These differences depended partly on whether the new vegetation was taken to be grassland, shrubland or woodland. The length of time the model was run after deforestation also varied. The models did all agree that evapotranspiration would decrease. Most predicted that rainfall would also decrease. However, a model by Polcher and Laval (1994) disagreed and predicted that deforestation would be followed by an increase in rainfall. One weakness of that model is that it assumed that canopy roughness would be unchanged: this is clearly unrealistic, and would have a substantial effect on rainfall. Also the Polcher and Laval model was run for only 1 year after deforestation, which may be too short for the climate change to approach equilibrium. Most other models predicted a decrease of rainfall within the range of about 300–600 mm, i.e. of the order of one-quarter, and most agreed that rainfall would decrease more than evapotranspiration.

These predictions are for deforestation of the whole Amazon Basin, an extreme scenario. Sud, Yang and Walker (1996) modelled the effect of converting eight areas, each about 400 × 500 km, to savanna, leaving the remainder as forest. Table 3.1(b) summarizes the predicted effects on the water balance of the deforested areas. Rainfall and evapotranspiration were both predicted to decrease, but unlike most of the predictions for whole-Amazon deforestation, rainfall decreased less than evapotranspiration: rainfall was influenced by still-forested areas nearby. In reality the areas of Amazon rainforest that have been cut down are each much smaller than 400 × 500 km, and global circulation models are not at present capable of predicting whether their rainfall will be significantly altered.

Effects of deforestation elsewhere

It should not be assumed that converting all tropical forest into grassland or woodland will have the same effect everywhere as in South America. Table 3.1(c) shows predictions for southeast Asia. Although the reduction in evapotranspiration would be quite substantial, the effect on rainfall would be much less than most models predict for the Amazon. The basic reason for this is that in southeast Asia the land masses are much smaller, and so all rainforest is closer to ocean and the climate and moisture supply much more dominated by the oceans. So we can make no global generalization about the likely effect of deforestation on climate: each area needs to be considered separately, in the light of its size, its closeness to an ocean and its prevailing winds.

Although they do not fully agree with each other, the models of deforestation of Amazonia do give us a perspective on the likely effects. Deforestation of the Amazon Basin will not change it to desert, but the rainfall could decrease enough to have an effect on the vegetation. In regions now forest but close to the edge of the savanna zone, deforestation might push down the rainfall enough to make it difficult or impossible for forest to return later. Destruction of the forest would then be effectively permanent: the boundary between forest and savanna would be permanently shifted.

Case 2: Semi-arid North Africa, the Sahel

Substantial areas of the tropics and subtropics are semi-arid, meaning that although they are not so dry as to be full desert, yet water supply is seriously limiting to plant and animal life. The natural vegetation is mostly savanna, i.e. grassland, shrubland or open woodland. People living in these areas may be able to grow crops in the moister part of the year, or may depend on grazing animals.

The word *desertification* has been used in relation to these areas. A United Nations conference on desertification was held in Nairobi, Kenya, in 1977, which helped to put the subject on the political agenda. So what does the word mean? It involves a decrease in vegetation and degradation of the soil, but there still seems to be some uncertainty about its precise meaning. One commentator on the 1977 conference said: 'Practically everyone knew intuitively that desertification was bad, irrespective of what it referred to' (quoted by Thomas & Middleton 1994). So I will avoid the term.

Sahel region

The Sahel is a semi-arid region of Africa bordering the south side of the Sahara Desert. Figure 3.3 shows a map of the annual rainfall in West Africa (though the Sahara desert and the Sahel also extend further east). Starting at the southern coast of West Africa and going northwards, the rainfall decreases gradually until you reach full desert. So the Sahel does not have clear natural boundaries. Hulme (1992) defines it as the region receiving between 100 and 600 mm year^{-1}, but others (e.g. the *UNEP Atlas of Desertification*) regard it as wider, with the southern boundary about the 1000 mm year^{-1} isohyet.

Here we consider four questions about the Sahel region:

1 Has the rainfall in this region decreased in recent decades?

2 Has the vegetation changed in recent decades?

3 If the vegetation has changed, could this be due solely to lower rainfall, or may land use by people also be involved?

4 Could reduced rainfall be partly caused by a change in vegetation?

Lower rainfall after 1970

Figure 3.4 shows how rainfall in the Sahel varied during the 20th century. The rainfall from the late 1960s onwards was on average lower than during any previous part of the 20th century. This was not worldwide: a world map (Houghton *et al.* 1996, Fig. 3.9) comparing the rainfall 1975–94 with that in 1955–74 shows that tropical Africa north of the equator was the largest area of land where there was a decrease. Most subtropical and temperate areas had increased rainfall.

Has the vegetation of North Africa changed during recent decades?

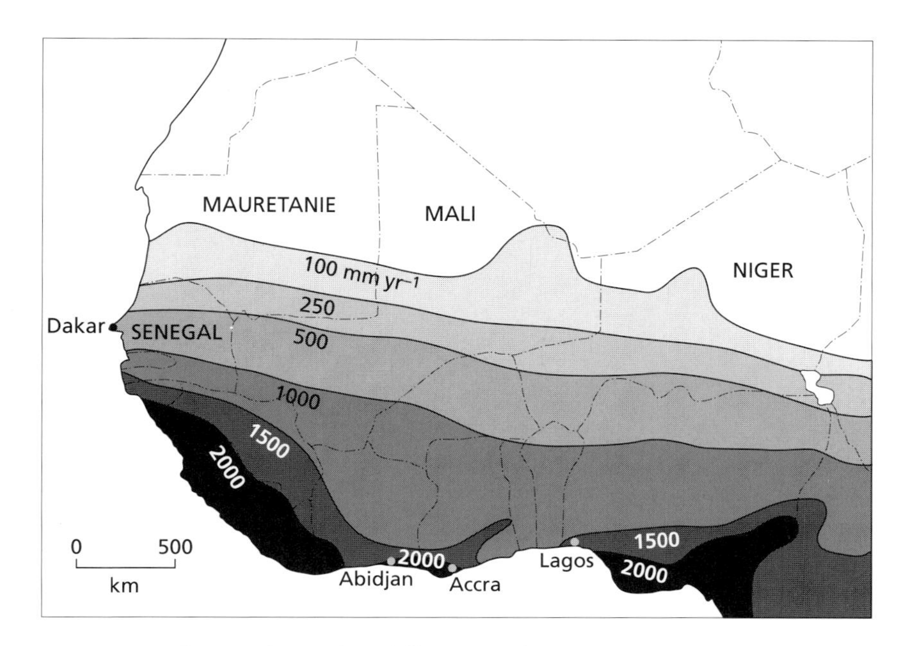

Fig. 3.3 Distribution of annual rainfall in West Africa. From Savenije (1995). Reprinted with permission from Elsevier Science.

Is the Sahara Desert expanding?

There have been numerous statements in the non-scientific literature about the Sahara desert expanding. Thomas and Middleton (1994) quoted politicians at different times giving the rate of southward expansion as 5, 9 or 17 km year^{-1}, and the UN Environment Programme estimating '27 million hectares lost a year to the desert or to zero economic productivity'. There are problems with measuring vegetation change in the Sahel:

1 Like the rainfall, vegetation changes gradually as one moves southwards across the region; it is difficult to define boundaries on the ground.

2 Vegetation change in time is also likely to be gradual, when considered over a large area. So, ground-based estimates of long-term vegetation change are unlikely to be reliable.

More reliable information on vegetation change can be made by remote sensing from satellites. Box 3.2 summarizes some of the types of information that can be obtained by remote sensing. The box explains briefly a 'greenness index', which is a measure of the amount of green plant cover. Figure 3.5 shows the greenness of the Sahel yearly from 1980 to 1992, determined from satellite recordings in the region, with long-term mean annual rainfall of 200–400 mm. The graph shows a marked loss of vegetation from 1980 to 1984. This is equivalent to a southwards movement of the boundary of the desert by 250 km. But after 1984 there was a partial recovery, followed by no marked trend up to 1992. Comparing this graph with the rainfall in the corresponding years (Fig. 3.4), there is a close correspondence: rainfall also declined from 1980 to 1984, then

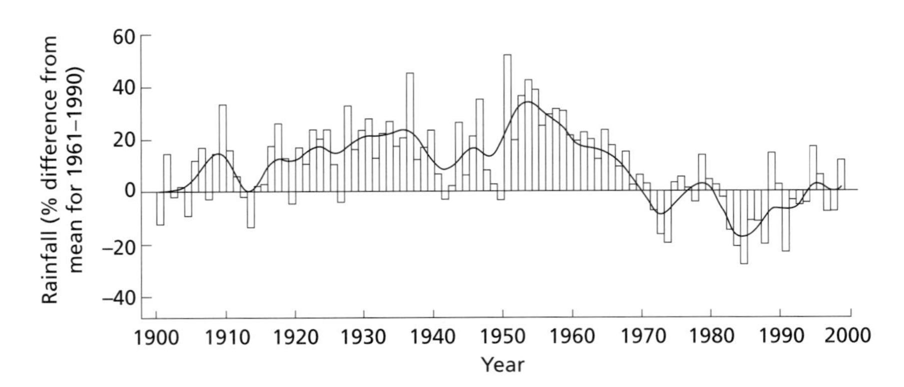

Fig. 3.4 Annual rainfall, 1900–98, at sites in Sahel region of West Africa, expressed as percentage deviation from the mean for 1961–90. The bars show data for individual years, the curved line smoothes the data over 10 years. Figure provided by Dr Mike Hulme, University of East Anglia.

Vegetation and rainfall changed in parallel

recovered to the relatively wet year of 1988, then fell back. So the satellite records over this fairly short period do not indicate a steady loss of vegetation, but rather change from year to year in parallel with rainfall. However, we should not assume that over the timescale of the whole 20th century (or longer) vegetation change in the Sahel has been due solely to rainfall. There is no doubt that people and their land-use practices can alter vegetation greatly. Chapter 6 gives information on how different grazing regimes can affect vegetation, though few of the examples there are from semi-arid regions. Here I ask what may have been the cause of the prolonged drought in the Sahel from the late 1960s onwards. Possibilities are:

What caused the long drought?

1 It was a natural fluctuation in rainfall.
2 It was caused by the increase in greenhouse gases described in Chapter 2.
3 It was caused by changes in vegetation brought about partly by people.

Natural climate fluctuations?

The most direct evidence relating to possibility (1) would be that equally severe droughts have occurred in the region before. There were no continuous rainfall records before the 1890s, but some indications can be obtained from written reports, for example of lake levels and of where crops were being grown (Nicholson 1989). These indicate a low rainfall during the first half of the 19th century and especially from about 1820–40, but a period of higher rainfall from about 1870–95. Lake Chad, for example, had lower water levels from 1800 to 1860 than for the following 100 years. There is also some evidence of earlier drought periods, e.g. about 1740–60. So, droughts lasting about two decades have probably happened before.

Greenhouse effect?

Could the increase in CO_2 and other greenhouse gases be causing a decrease in rainfall? Chapter 2 had much to say about global warming,

Box 3.2. Information from satellite remote sensing that can be useful to ecologists.

Artificial satellites now in orbit provide images of the Earth's surface, at various resolutions from about 10 m upwards. Most older satellites record reflected radiation in several wavebands; some newer ones record in up to 200 very narrow wavebands. Comparison of the intensity of different wavebands can provide information of interest to ecologists. Some satellites use radar or laser beams, which can provide additional information.

Well established methods are already in use for observing and measuring:

clouds, wind, rain, water vapour, ozone;
surface temperature of vegetation and exposed soil;
ocean surface temperature, which can indicate regions of upwelling;
fires, their frequency and area;
distinguishing different types of vegetation, hence measuring areas of different types, and changes (e.g. deforestation);
amount of chlorophyll, hence amount of phytoplankton in water or green leaf area on land.

Methods are under development for measuring:

atmospheric aerosols;
water content of vegetation;
transpiration rate;
stem volume per hectare in forests;
concentration of constituents in plants, e.g. N, cellulose, lignin;
area and abundance of individual species of arable crops and forest trees;
net primary productivity;
damage to crops by pests and diseases.

An example of the use of two wavebands

A 'greenness index' (normalized difference vegetation index, NDVI) is given by $(IR - R)/(IR + R)$ where IR and R are the amounts of radiation measured by the satellite in the near infrared and red wavebands, respectively. This method is based on the fact that chlorophyll reflects more near infrared (about 1 μm wave-length) than red. The difference between the amount reflected in these two wavebands is much greater for green leaves than it is for dead leaves, stems or soil, so this greenness index is related to the amount of chlorophyll or green leaf biomass per unit ground area.

Further information: Campbell (1996); Drury (1998); Danson & Plummer (1995).

but said that the effects of increased greenhouse gases on rainfall are still uncertain. As already mentioned, the drought from the 1970s onwards shown in Fig. 3.4 is a regional, not a worldwide phenomenon. Could increased greenhouse gases cause local or regional changes in rainfall, for example by changed ocean or wind currents? The short answer is that we do not know. Climate models have been used to predict future changes in

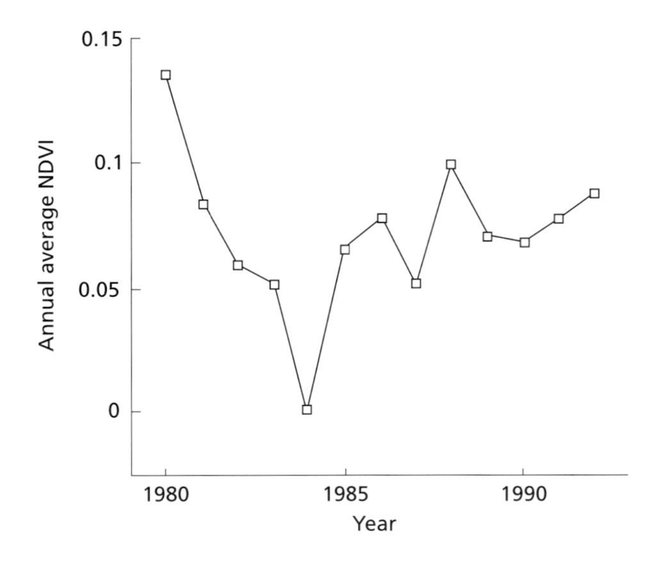

Fig. 3.5 Average normalized difference vegetation index (NDVI, 'greenness index') in the Sahel region of West Africa for each year from 1980 to 1992, measured by remote sensing from a satellite. For further explanation of NDVI see Box 3.2. From Tucker, Newcomb & Dregne (1994).

regional rainfall, and maps have been published, but the different models' predictions do not agree with each other. Houghton *et al.* (1996, Fig. 6.9) summarized predictions by four models of the rainfall change between the 19th century and the mid-21st century in five regions of the world. For the Sahel two of the models predict little change and two predict increased rain. For the other four regions (central North America, central Europe, Australia, southeast Asia) the models cannot even agree whether the rainfall will increase or decrease, let alone by how much. So we must await better models based on a better understanding of what controls rainfall.

Could the drought be caused by changed vegetation?

Could decreased rain in the Sahel be caused by changed vegetation? This depends on how much the rain there is determined by local conditions. Earlier I gave an approximate rule that a water molecule travels about 1000 km in the atmosphere before it falls as rain. That was meant to be *very* approximate. Nevertheless, Savenije (1995), using more sophisticated methods, has estimated that in West Africa more than 1000 km from the southern coast almost all the rain comes from local evapotranspiration, and scarcely any from far-away ocean or forest. So this would apply to much or all of the Sahel, depending on the definition of its southern boundary.

Each year an atmospheric feature called the *intertropical convergence zone* (ITCZ) moves northwards across the Sahel and then southwards again. North of the ITCZ there is virtually no rain; south of it rain occurs; so the movement of the ITCZ determines the length of the rainy season at each point (Le Barbé & Lebel 1997). If reduced rainfall in the Sahel is strongly influenced by conditions in the oceans or long-distance air movements, we should expect this to have its effect by a change in the movement of the ITCZ and a shortening of the rainy season. Le Barbé & Lebel (1997) made a detailed study of the pattern of rainfall in southwest

Niger, comparing 1950–69 with 1970–89. There was no consistent change in the length of the rainy season. The decline in rainfall was primarily in the central months of the rainy season, July and August, and was due to fewer rainfall events (i.e. showers and storms), not to less rain per event. The drought is thus not linked to changed behaviour of the ITCZ, but is more likely to be related to changed local convection. This makes it worth looking further at how changed vegetation could influence the rainfall.

Fewer rainfall events

Dirmeyer and Shukla (1996) used a general circulation model to carry out a 'desertification experiment'. The model predicted the effects of semi-desert becoming full desert and savanna becoming semi-desert, not just in North Africa but in all the dry lands of the world. This would result in a reduction in vegetation cover and an increase of bare ground, increased albedo, reduced tallness and structural complexity of the vegetation, and reduced rooting depth. So, in spite of the very different vegetation from the Amazon Basin, the changes would be broadly in the same direction as caused by deforestation there, and the mechanisms by which rain could be affected are basically those set out in Box 3.1.

A model predicting effects of reduced vegetation

The change in vegetation was assumed to occur instantly across the world, and the model predicts daily energy and water balances after that. It predicts that after 2 years a new climate equilibrium would have been reached. Table 3.2 summarizes the water balance after that, for the areas of North Africa in which the vegetation had been altered; this means approximately the Sahara plus the Sahel regions. It predicts that evapotranspiration and rainfall would decrease by about one-third. In view of the approximations in the model we cannot rely on this prediction to be exactly correct, but it does at least indicate that reducing the vegetation of North Africa could reduce its rainfall.

However, the rainfall changes predicted by Dirmeyer and Shukla were not the same in all the areas: North and South Africa were predicted to suffer large reductions in rainfall, Australia and central Asia small reductions, and the semi-arid areas of North and South America little change. Table 3.2 shows the predictions for Asia: the semi-arid region here stretches from the Caspian Sea eastwards across to western and northern China. The reduction in evapotranspiration and rainfall caused by reduced vegetation would be much less than in North Africa, whether expressed in millimetres or as a percentage. One reason for this may be that the Asian region already contains more full desert and less semi-desert than North Africa, so the potential decrease in vegetation is less. But each area can be affected by its surroundings. The Asian region receives more of its rain from elsewhere than does North Africa (compare ET and rain figures in Table 3.2). Dirmeyer and Shukla suggested that in Africa the deserts of the north and south are reinforcing each other, whereas Asia (and Australia) have no corresponding desert on the other side of the equator. The main message is that the effect of reduced vegetation can be different in each semi-arid area of the world.

Table 3.2 Predicted effects on evapotranspiration (ET) and rainfall in desert and semi-arid regions of North Africa and central Asia if their vegetation is reduced

	North Africa		Asia	
	ET	Rain	ET	Rain
Present rate (mm $year^{-1}$)	505	579	679	916
Predicted reduction				
mm $year^{-1}$	178	198	52	50
%	35	34	8	5

Data from Dirmeyer & Shukla (1996).

This section has been much dependent on models. Models are inevitably a simplification of reality, and the conclusions should be viewed in that light. The predictions for the effects of deforestation of Amazonia and reducing the vegetation of North Africa have in common that they indicate the possibility of a positive feedback: reducing the vegetation could reduce the rainfall, which could further reduce the vegetation. In the Sahel, as in Amazonia, reductions in the vegetation might be irreversible if rainfall decreases. We have no evidence that such irreversible change has already happened, but this is a possibility to bear in mind for the future.

Food production where water supplies are limited

There are substantial areas of the world where temperature and incoming radiation are favourable for plants and animals but rainfall is inadequate. This section considers ways in which food is grown in these areas, and asks whether there are ways by which food production there can be increased in the future.

Farming: rainfed or irrigated

In these areas farming may be either *rainfed* or *irrigated*. Rainfed means that the crop or pasture plants are dependent for water on the rain that falls on their field. Water may be supplied from wells or reservoirs for animals to drink, but not to water the plants they eat. In most semi-arid areas there is each year a dry season when little or no rain falls, as well as a rainy (or at any rate more rainy) season. There will thus be a part of the year when the soil contains available water, plants can open their stomata, transpire, photosynthesize and grow; but another season when available soil water is exhausted, the plants cannot photosynthesize and grow, but nevertheless have to survive. Successful growth of a crop will thus depend on (1) its having highly efficient water use during the growing season, and (2) its being able to survive the dry season.

Many crop plants are annuals, which can survive the dry season as seeds. Annual crop species have been selected and bred to adjust the time and the amount of water required by the plant between germination and seed maturation, so that the crop grows throughout the time that water is available but completes seed swelling before the water supply runs out. Among perennials in savanna some woody species shed their leaves in the dry season, and the leaves of many grasses die. This affects the food supply for herbivorous mammals.

Drinking water for animals . . .

Becoming dormant during dry weather is not an option open to grazing mammals: they continue to require food, and most require regular access to drinking water. How often they need to drink depends on their metabolic state, what sort of food they are eating and how hot and dry the weather is. For example, in temperate climates beef cattle may drink only every alternate day, whereas lactating cows may drink as often as eight times a day (Church 1988). Western (1975) studied the distribution of herbivorous mammals in a 3000-km^2 area of savanna, the Amboseli region in southern Kenya. Within this area, 600 km^2 has natural supplies of drinking water all the year round, because of run-off from Mount Kilimanjaro; the remainder has drinking water only in the rainy season (October–April). Counts of large mammals, in relation to distance from the nearest watering point, were made from aircraft and from the ground. The species differed considerably in how far from water they were willing to go. Figure 3.6 shows results for three species in the dry months. Thomson's gazelle was rarely more than 4 km from water, whereas Grant's gazelle was about equally abundant up to 12 km from water, and in a later study was observed more than 30 km from water. Zebra was intermediate between these two in its distribution.

. . . native mammals in Africa . . .

. . . cattle in central Australia

During 1970–75 the distribution of individual cattle in a large paddock near Alice Springs, central Australia, was recorded from an airplane on many occasions. In this region rainfall averages 275 mm year^{-1}, but occurs irregularly. The paddock was 170 km^2 in area and contained five sources of water, from dams and boreholes, where the cattle could drink. Figure 3.7(a) shows the number of cattle per unit ground area, during 2 years of about average rainfall. Beyond the immediate environs of the water point the density of cattle decreased with distance. This suggests that grazing of the vegetation will be uneven: there may be overgrazing near the water but underuse far from it. There is also the danger of soil erosion by the cattle's hooves close to the water point. Figure 3.7(b) provides evidence that there was indeed uneven use of the vegetation in this paddock. Vegetation cover was estimated by the difference between reflection of green and red, recorded by a satellite (similar, though not identical to the greenness index in Box 3.2). The upper line was from measurements made soon after very heavy rain. It shows vegetation cover increasing with distance from the water points. The lower line shows that after about a year with low rainfall the vegetation cover had decreased, especially near to the water points. Evidently widely spaced supplies of drinking water present a limitation to efficient use of grazing land.

Herbivores also need a regular supply of food, and this can be a problem if the climate has an annual rainy/dry cycle. In northern Queensland,

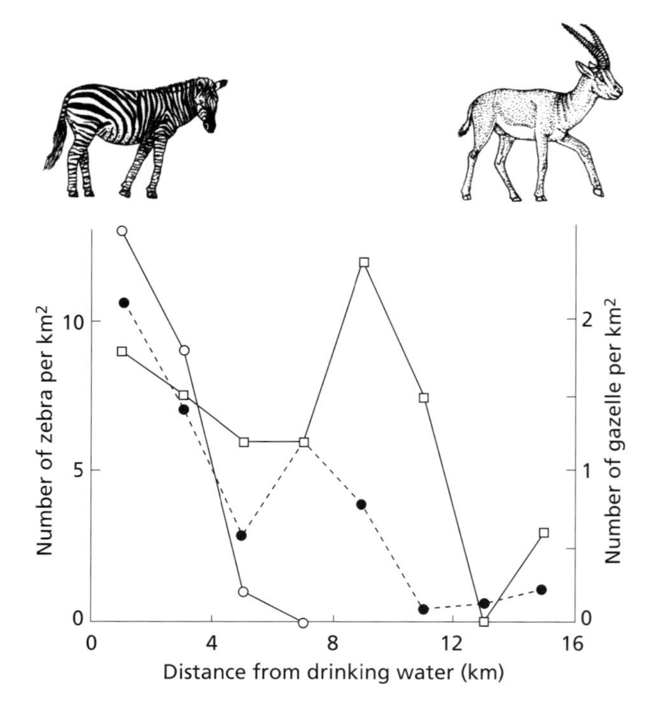

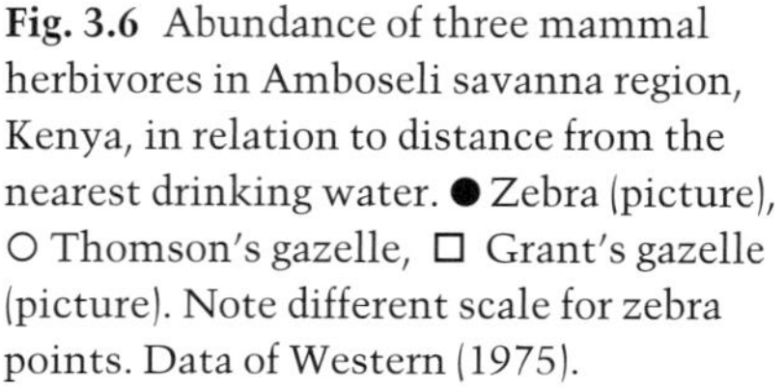

Fig. 3.6 Abundance of three mammal herbivores in Amboseli savanna region, Kenya, in relation to distance from the nearest drinking water. ● Zebra (picture), ○ Thomson's gazelle, □ Grant's gazelle (picture). Note different scale for zebra points. Data of Western (1975).

cattle feeding on native plants gain weight during the rainy season but lose it during the dry season, when leaves of the native grasses die back (McCown & Williams 1990). This is not because of lack of forage, since there is dead foliage available: rather, it is because the old, dead grass leaves have a high-fibre low-protein content, and hence low digestibility. Feeding non-protein nitrogen supplements to the cattle in the dry season results in their eating more of the dead grass and reduces cattle mortality; however, it does not have much effect on the weight loss, perhaps because of the low digestibility of the dead grass. Another approach is to grow a legume, stylo (*Stylosanthes* spp.), as forage. The leaves die in the dry season but retain more nutritional quality than the dead leaves of native grasses, and cattle will eat them. For wild grazers in East African savanna also, during the dry season the shortage of digestible forage can lead to their energy intake being less than their maintenance requirement (Hodgson & Illius 1996, especially Fig. 9.3).

Efficiency of water use by plants

In semi-arid regions a key aim in crop growth is to maximize the efficiency of water use by the crop during the growing season. Efficiency of water use means:

amount of growth ÷ amount of water used.

Growth can mean dry weight increase of either the whole plant or of its useful parts only (e.g. seeds). Amount of water used can mean (1) amount

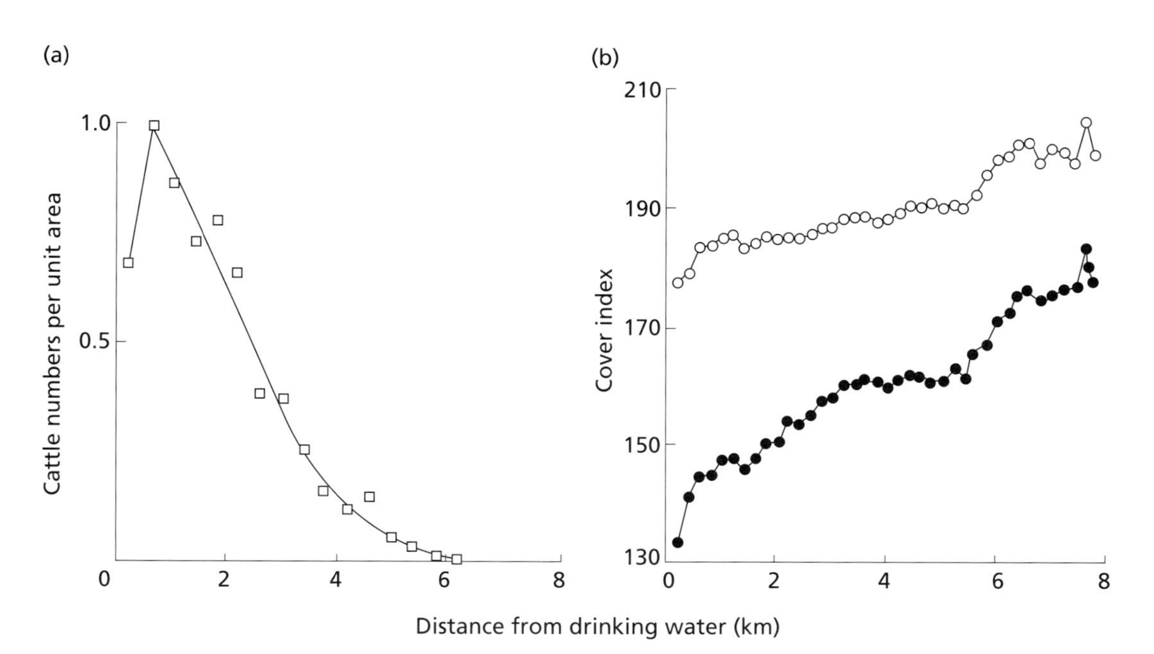

Fig. 3.7 Distribution of cattle and vegetation in a large paddock in central Australia, in relation to distance from nearest source of drinking water. (a) Number of cattle per unit ground area (arbitrary scale), during 1972 and 1973. (b) Vegetation cover, estimated by reflected radiation recorded by satellite: ○ Early 1983, soon after heavy rain; ● about 1 year later, during which little rain fell. Data from Pickup & Chewings (1988), Pickup (1994).

of water transpired or (2) amount transpired + evaporated, or (3) amount of rain, which equals transpiration + evaporation + run-off. It was explained earlier in this chapter that for any individual plant water loss and CO_2 uptake tend to be closely correlated. The ratios quoted earlier of moles water transpired:moles CO_2 taken in correspond to efficiency of water use of about 1–4 kg dry weight increase per tonne water transpired. This is whole plant growth per water transpired, whereas a farmer is more likely to be interested in the production of useful parts per amount of rainfall. In Fig. 3.2 the encircled point near the middle can be taken as an example. This gives an efficiency of water use of 0.8 kg grain yield per tonne of rain. (Because there was very little run-off at this site, rainfall and evapotranspiration were approximately equal.) The sloping boundary line in the figure gives the maximum efficiency expressed as grain yield/transpiration, after allowing for 110 mm lost by evaporation; this gives 2 kg grain per tonne of water.

How efficiency of water use might be increased

This suggests several ways that the efficiency of water use, if expressed as (amount of useful product/amount of rain) might be increased:

1 Increase the proportion of rain that is taken up by the plant and transpired, and reduce the proportion 'wasted' by evaporation or run-off. Ways of reducing evaporation from the soil include arranging for the crop to germinate as soon as there is adequate soil moisture, and having a crop variety which expands its leaf canopy quickly.

2 Avoid crop yield being reduced by limitations other than water. Most of the points in Fig. 3.2 lie below the upper boundary line because of limitations such as nutrient deficiency, pests and weeds. Evidently these reduced grain production more than they reduced water use (French & Schultz 1984).

3 Crop breeding may increase the proportion of total dry weight growth of the plant that ends up in useful parts.

The importance of some crop characteristics is illustrated by an experiment that compared 10 varieties of wheat first released for use at various times between 1860 and 1986. They were all grown in 1987 in the wheat belt of Western Australia, where water supply is limiting for grain production. Figure 3.8 shows that in the more modern varieties water use during the season was slightly lower. This was primarily due to the leaf canopy expanding more rapidly, and hence lower evaporation from the soil early in the season; whole-season transpiration did not differ much between varieties. Total above-ground plant growth over the season differed little between varieties, but the grain yield increased substantially in more recent varieties (see Fig. 3.8). If water use efficiency is calculated by grain yield/evapotranspiration it increased nearly twofold between the older and the more recent varieties, but this was mostly due to the increased harvest index (the proportion of growth going into grain). It was due much less to any increased efficiency of water use, in the sense of whole plant growth per gram water taken up.

CAM plants have higher water use efficiency

So far we have considered crop plants with the C3 or C4 photosynthetic mechanism, whose efficiency of water use (calculated as plant dry weight increase/evapotranspiration) is usually below 4 kg tonne^{-1}. However, plants with the crassulacean acid metabolism (CAM) type of photosynthesis are able to take in CO_2 at night, when transpiration is slow, and can keep their stomata closed by day. They can therefore have a substantially higher efficiency of water use than plants with C3 or C4 photosynthesis, often in the range 6–15 kg tonne^{-1} (Larcher 1995). Although there are thousands of species with CAM photosynthesis, only a few at present have useful products. Examples are pineapple, agaves for fibre, and a few species used for animal forage. CAM photosynthesis is biochemically less efficient than C3 and C4 (Nobel 1991b), and growth rates may be further reduced by the leaves often being thick and succulent. However, Nobel (1991b) gives figures for above-ground dry weight production by CAM plants in favourable conditions of 18–47 tonnes ha^{-1} year^{-1}, which are within striking distance of the highest productivities reported for C3 and C4 crop plants, up to 70 tonnes ha^{-1} year^{-1}. So, attempts to breed more CAM species with useful products might be worthwhile.

Increasing CO_2 will raise water use efficiency

One way to increase efficiency of water use is to increase the concentration of CO_2 around the leaves of the plant. This causes the stomata to partially close, so that transpiration is reduced. But in spite of the reduced

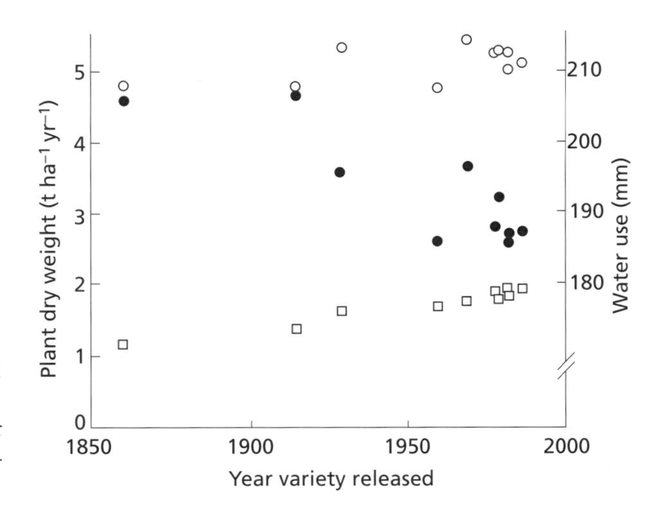

Fig. 3.8 Comparison of 10 wheat varieties all grown in the same experimental area in the same year. ○ Total above-ground dry matter at harvest; □ grain yield; ● water use (evapotranspiration). Data of Siddique *et al.* (1990).

stomatal aperture, the rate of photosynthesis usually increases (see Tables 2.7, 2.9). Table 3.3 shows the effect of doubled ambient CO_2 on some crop species. They were grown in pots of soil in temperature-controlled glasshouses. The soil was fully watered at the start but no more water was added after that, so the plants grew on the stored water. The first three species were grown at higher temperatures than the others, as they normally occur in warmer climates. Doubled CO_2 increased the growth of all the species, but had a variable effect on their transpiration per unit leaf area. Water use efficiency increased for all species.

Results from such pot experiments may not apply exactly to whole fields of a crop outdoors. Effects of reduced stomatal aperture on transpiration can be partly offset by higher leaf temperature and drier air near the leaf, both caused by reduced transpiration (Eamus 1991). These may be different in the field. Nevertheless, the expected increase in CO_2

Table 3.3 Effect of doubled atmospheric CO_2 concentration on growth and transpiration of crop plants. Results expressed as ratio of (value in doubled CO_2)/(value in present-day normal CO_2)

Crop	Whole-plant dry weight	Transpiration per unit leaf area	Water use efficiency (wt gain/transpiration)*
Maize	1.36	0.67	1.60
Rice	1.52	(1.0)†	1.53
Cotton	1.19	0.85	1.40
Wheat	1.83	0.61	1.87
Broad bean	1.61	0.82	1.59
Oilseed rape	1.48	(1.0)†	1.50

* These figures were not obtained by simply dividing column (1) by column (2). They used transpiration per plant, which was in some species affected by increased leaf area in the high-CO_2 treatment.
† No significant effect of increased CO_2. Data from Morison & Gifford (1984).

concentration of the world's atmosphere may well result in substantial increases in crop production in regions where water is the primary limitation, without requiring increased irrigation. This is not to claim that the CO_2 increase will be beneficial in all respects; its effects were discussed at length in Chapter 2.

Dry weight increase is not the only sort of growth. Cell expansion, by water intake to the cell, controls stem and leaf extension, which in turn influences light capture; root extension influences soil exploitation for water and nutrients. However, in the longer term cell expansion requires metabolic energy to maintain osmotic concentrations, either by uptake of ions or by synthesis of organic solutes; so photosynthesis is essential for growth. Osmotic adjustment in the plant's tissues can make an important contribution to its continued ability to grow when water supply is deficient, e.g. in wheat (Morgan 1983).

Genes for drought tolerance?

The successes of genetic engineering could lead us to ask, can we find genes for drought tolerance? There are proteins which are synthesized in response to water deficits: these are known as LEAs—*late embryogenesis abundant proteins*—because they were first discovered at late stages of seed maturation and drying. However, they have also been found in the vegetative parts of plants after water stress. Transfer of a gene for an LEA into rice was found to increase its drought tolerance (Xu *et al.* 1996), but some other LEAs have not increased drought tolerance (Bray 1993; Turner 1997). LEAs may act in several different ways: some may be molecular chaperones (see p. 34) or protease inhibitors, which could reduce the damaging effects of water deficit on other proteins. The genetic engineering approach to increasing drought tolerance deserves further research. However, drought tolerance involves many characteristics of the plant, and therefore many genes. So, the genetic manipulation of plants to increase drought tolerance is bound to be complex.

Irrigation

The alternative to depending on local rainfall is to irrigate. This is not a new idea: some early farming communities depended on complex irrigation systems to produce their food. Examples include ancient civilizations on the Tigris and Euphrates rivers (present-day Iraq), the Indus valley (present Pakistan), the irrigated rice region of southern China, and parts of Mexico and nearby Central America. The dependence today of agriculture on irrigation varies greatly from country to country: the percentage of farmland irrigated ranges from 100% in Egypt to less than 1% in Belgium (World Resources 1998/9).

Sources of water for irrigation

Many dry areas draw their irrigation waters from rivers which are fed by rain far away in another country. In modern times, when large dams can be built, there can be serious disagreements about how much water each country takes from a river system. For example, Iraq is much dependent on the Tigris and Euphrates for irrigation water, but about two-thirds of their flow is from rain that falls in Turkey. The Euphrates also flows through Syria. Each of the three countries has built dams and irrigation networks, and has plans for more, but there is no unified plan for use of the water in this catchment (Hillel 1994).

The other main source of water for irrigation is *aquifers*, underground reservoirs in porous rocks. This water may have accumulated over centuries or even thousands of years. There is water under the Sinai Desert which has been there 20 000–30 000 years, according to radiocarbon dating: so it accumulated when the climate was much different from today (Issar 1985). The danger is that this 'fossil water' may be used faster than it is arriving, so the aquifer is depleted and eventually exhausted. The great Ogallala aquifer, which stretches from South Dakota to Texas, has been depleted in its southern part, and as a result the irrigated area has had to be reduced (Gleick 1993).

Worldwide, the total irrigated area increased fairly steadily between 1961 and 1994, from 139 to 249 million ha (Brown *et al.* 1997). Falkenmark (1997) made predictions of whether water supplies will be adequate for irrigation in 2025 in each major area of the world, taking into account expected increases in human population and assuming that each area aims to be self-supporting in food production. His predictions particularly identify the Middle East and the Indian subcontinent as regions which by 2025 will not have adequate water supply for irrigation.

Can we irrigate with salty water?

This section concentrates on the question of whether crop plants can be grown with a saline water supply. This is important for two reasons. First, there are large areas where fresh water is in short supply but where the sea is close by. Examples are Pakistan, Arabia, North Africa, Western Australia, and some western regions of North and South America. Crop production there could be substantial if the crops could be irrigated with seawater. Although seawater can be desalinated this uses a lot of energy, so it is better if we can use it pure or partly diluted.

The second reason is that when crops are irrigated with fresh water the concentration of salts can, over many years, build up in the soil, to the point where crop production is reduced or even prevented altogether (White 1997, Chapter 13). This is because even fresh water contains some dissolved salts, and if these are not all taken up by the crop or flushed away they will accumulate. This is not a new problem: salinity has been suggested as the main cause of the demise of some past civilizations. An example is the great Sumerian civilization, which flourished in the southern part of present Iraq in the third millennium BC, and especially from about 2400–1700 BC. They had a great city, Ur, which had temples, palaces and schools. They had writing, art, a complex social organization and an army. All this required efficient food production, which in that low-rainfall area was dependent on irrigation from the River Euphrates. We know from their written records that during the period 2400–1700 BC their cereal yields declined, and that salt appeared on fields which had previously been salt free. The Babylonian civilization that followed was

Box 3.3. Some properties of typical seawater.

The concentration of dissolved solutes is typically about 35 g l^{-1}, often expressed as 35 parts per thousand, 35‰.

The most abundant solutes

	g l^{-1}	mol l^{-1}
Cl^-	19.5	0.55
Na^+	10.8	0.47
SO_4^{2-}	2.7	0.028
Mg^{2+}	1.29	0.053
Ca^{2+}	0.41	0.010
K^+	0.38	0.010

Osmotic potential –2.3 MPa

Source of data: Gleick (1993).

in new lands further north. We cannot say definitely that salination on its own ended the Sumerian civilization, but it may have played an important part (Jacobsen & Adams 1958).

Growing crops in saline water

Box 3.3 summarizes some properties of seawater. Although Na^+ and Cl^- are the most abundant ions, other elements are also present, in lower concentrations, for example inorganic N and P.

Salt-tolerant varieties

Some plants can grow rapidly with their roots in saline soil: these are called *halophytes*. For, example, some species grow well in salt marshes; some native plants in salt marshes have net primary productivities as high as farmers obtain from their crops on fertile soils inland, using modern farming methods (Long & Mason 1983). This suggests that it may be possible to develop crop plants that grow well on saline soil. One approach has been to test many varieties of a crop species for ability to grow in saline water. This has had some success in identifying more salt-tolerant varieties in a few species (see Fig. 3.9, Table 3.6), but in general there has been little increase in salt tolerance produced by such screening (Noble & Rogers 1992). It appears that among existing varieties of many species the range of salt tolerance is small.

. . . or hybrids

Another possibility is to cross-breed crop species with wild plants that grow in saline areas. King *et al.* (1997) crossed several wheat varieties with a grass, *Thinopyrum bessarabicum*, which grows on the coast of Crimea. By chromosome doubling fertile hybrids were produced. Table 3.4 shows results for one of the hybrids, which had some salt tolerance. The hybrids had various characters from the grass parent, and would not themselves be suitable as a crop, but they might be a starting point for further breeding. A better approach may be to insert individual genes for salt tolerance into crop plants. The experiment by Xu *et al.* (1996) described

Table 3.4 Survival and seed production by wheat variety Chinese Spring and by the fertile hybrid produced by crossing it with the salt marsh grass *Thinopyrum bessarabicum*. The rooting solution contained NaCl at 0.2 mol l^{-1} (about half the concentration in seawater), or none.

	NaCl	Survival (%)	Grains per plant
Wheat	–	94	121
Wheat	+	0	–
Wheat x *T. bessarabicum*	+	39	23

Data from King *et al.* (1997).

earlier, in which a gene for production of an LEA protein was inserted into rice, resulted in increased survival and growth under saline conditions.

The basis of salt tolerance in plants

The basic problem, as with drought tolerance, is that tolerance of salinity is unlikely to be controlled by a single gene, but rather depends on a whole suite of interacting characters. Therefore, a better way forward may be to try to understand the basis of salt tolerance in plants, so that key characteristics can be identified. Plants growing with their upper parts in air and their roots in saline soil have two sorts of problem:

1 related to the osmotic effect of the salt and resultant water deficits;
2 toxic effects of Na and Cl.

Box 3.4 summarizes harmful effects that can occur.

Control of Na^+ and Cl^- entry

Plants have the ability to control the entry of Na^+, Cl^- and other ions into their roots, and therefore to alter how salinity affects them. Table 3.5 summarizes results from an experiment with oranges in which the effects of genetic differences in the roots can be seen: the shoots were all of the same variety, but the rootstocks on to which they were grafted were of two alternative varieties. Troyer roots took up more Cl than Cleopatra roots, resulting in higher concentrations in roots and leaves. However, with Na the main difference was in how much was sequestered in the root tissue: Troyer retained more Na and passed on less to the leaves than did Cleopatra.

The exclusion of NaCl is one way in which plants increase their

Table 3.5 Concentration of Na^+ and Cl^- (mmol g^{-1} dry wt) in roots and leaves of young orange trees. All shoots were of the same variety but grafted on to two alternative rootstocks. Grown in sand culture with full nutrient solution including NaCl at 45 mmol l^{-1}

	Na^+		Cl^-	
Rootstock	roots	leaves	roots	leaves
Troyer	0.34	0.53	0.51	0.76
Cleopatra	0.10	0.66	0.25	0.52

Data from Bañuls *et al.* (1991).

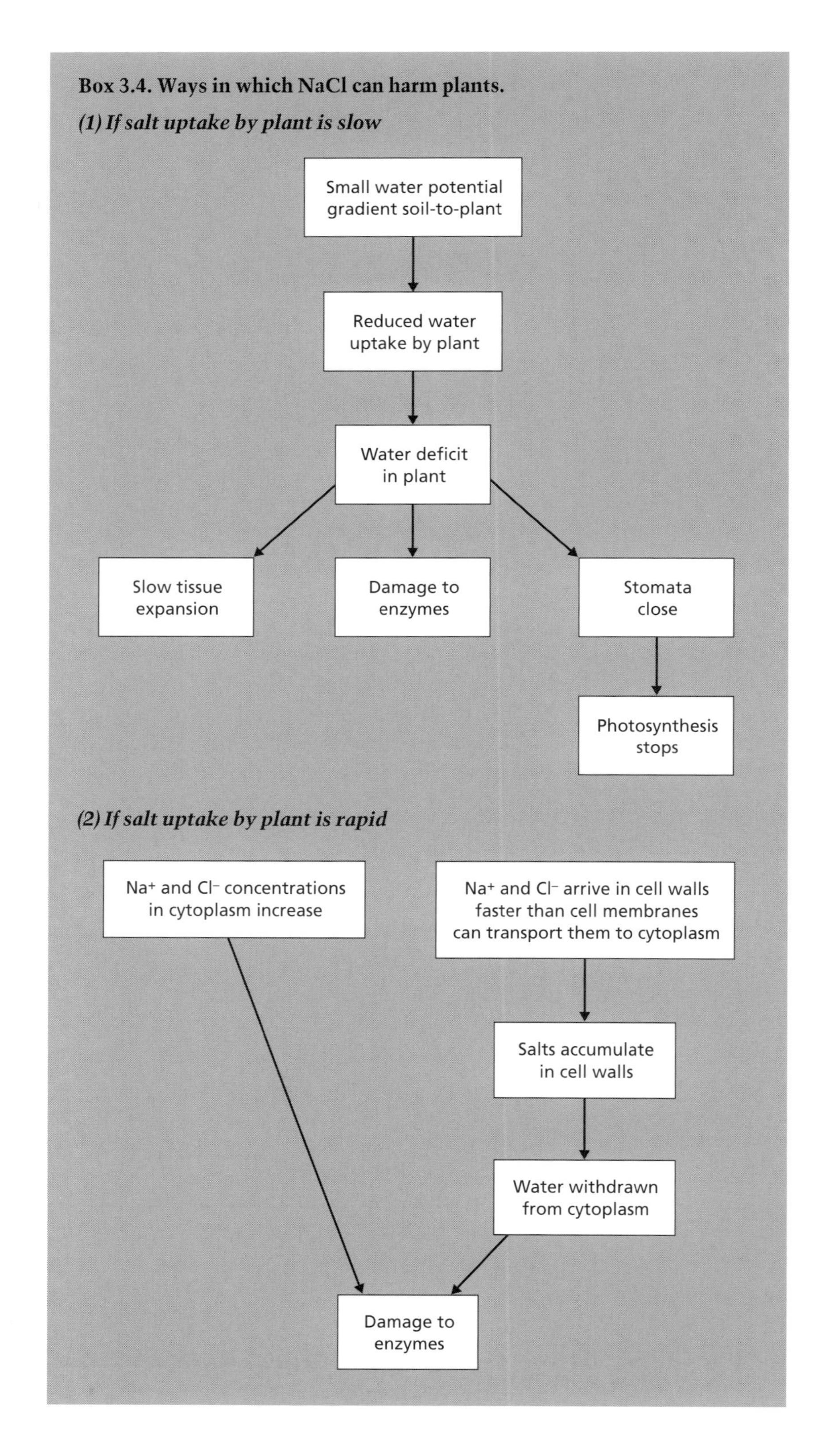

Box 3.4. Ways in which NaCl can harm plants.

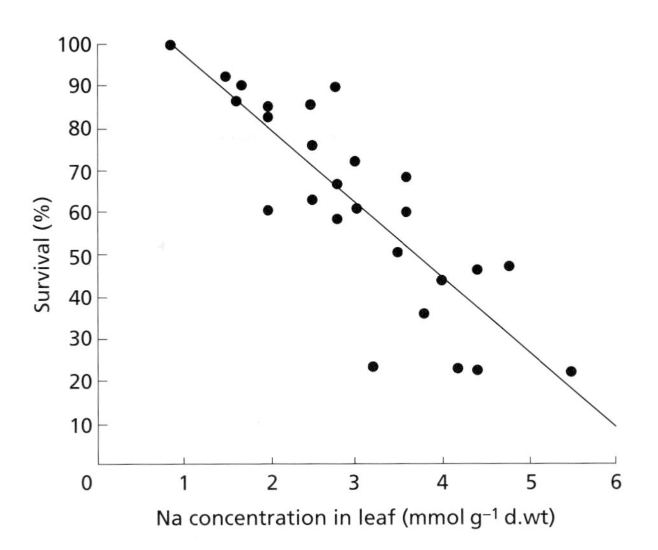

Fig. 3.9 Survival of 26 varieties of maize after the roots had been subjected to NaCl solution for 30 days, in relation to the Na concentration in the third leaf (measured after 15 days). From Hajibagheri, Harvey & Flowers (1987).

tolerance of NaCl in the rooting medium. Figure 3.9 shows results from an experiment in which NaCl was added to the solution around the roots of 26 maize varieties: 30 days later the percentage of plants still alive varied greatly between varieties. This survival showed a strong tendency to be higher the lower the concentration of Na that had built up in the leaves. These plants merely survived; but NaCl uptake can also affect growth. Table 3.6 shows results from an experiment in which lucerne plants were graded for salt tolerance by the amount of necrosis in the leaves when grown with saline solution around the roots. There was little difference between the three groups in concentration of Na or Cl in the roots, but a substantial difference in the shoots, with salt-sensitive plants containing the highest concentrations. Tolerant plants grew the fastest in high-NaCl solution and sensitive the slowest. It seems likely

Distribution of Na^+ and Cl^- within the plant

Table 3.6 Growth and NaCl concentration of lucerne plants when there was 0.25 mol l^{-1} NaCl in the rooting medium. The strains had been grouped into three salt-tolerance classes. Dry weight (shoot + root) and NaCl concentration measured at age 75 days

		Concentration in tissue (% dry wt)			
		Na^+		Cl^-	
Group	Dry wt (g)	roots	shoots	roots	shoots
Tolerant	1.24	2.61	3.13	2.81	3.88
Moderate	0.99	2.84	4.21	2.99	5.97
Sensitive	0.59	2.96	5.01	2.91	6.74
Stat sig*	sig	not sig	sig	not sig	sig

* sig means there is a significant difference within the column ($P < 0.05$). Data from Noble *et al.* (1984).

Table 3.7 Chloride concentrations (mmol g^{-1} dry weight) in parts of white clover plants. The plants had previously been selected for high or low average Cl^- concentration in the whole shoot. Plants grown with 40 mmol l^{-1} NaCl

	Plant group	
Tissue	High-Cl	Low-Cl
Expanding leaves	0.17	0.17
Fully expanded leaves	0.69	0.37
Petioles	0.63	0.43

Data of Rogers *et al.* (1997).

that this was related to the difference in Na and Cl concentrations in their shoots. So, in this species the amount of these ions passed on from the roots to the shoots may be important in determining how much growth the plants can make if the medium is saline.

The distribution of Na and Cl within the shoot may also be important. Rogers *et al.* (1997) grew white clover plants of a single variety in nutrient solution with added NaCl, and then measured the Cl concentration in the whole, macerated shoot. There were differences between the individuals in Cl concentration, and on the basis of this they were grouped into high-Cl accumulator and low-Cl accumulator plants. Table 3.7 shows that the difference in Cl concentration was greatest in the fully expanded leaves, less in the petioles, with no difference in the young, still-expanding leaves. One might argue that differences in Cl would be most important in the young leaves, where growth is rapid. However, the expanded leaves provide the major photosynthetic area of the plant, so Cl toxicity there could be important. This experiment also showed that a single round of selection within a single variety can separate out substantial differences in ion transport ability.

To reach the shoot, Na^+ and Cl^- must enter the xylem. Some of the ions probably reach the xylem through 'leaks', i.e. pathways which do not involve crossing a membrane. Such pathways occur where young lateral roots pass out through the endodermis and cortex. But probably other Na^+ and Cl^- ions pass through membranes during their passage across the root. So, selectivity of the membranes would be important. It would also be important when ions are reabsorbed from the xylem by living root cells: this is thought to be a way in which some varieties retain more Na and Cl in root tissue and pass less to the shoots (Yeo *et al.* 1977). If any treatment can increase the ability of roots to reduce the Na and Cl reaching the xylem, this could be beneficial to the plant. Calcium ions are known to enhance the selective abilities of membranes, and adding a low concentration of calcium salt to NaCl around roots can reduce the amount of Na^+ reaching the xylem (Gorham *et al.* 1985). Figure 3.10 shows how adding a Ca salt can reduce the harmful effect of NaCl. Melon was grown in a wide range of NaCl and $CaCl_2$ concentrations. The high-

Ca can reduce harm by Na

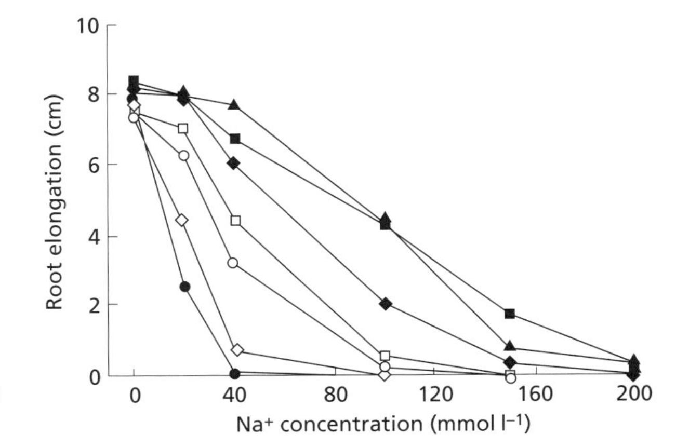

Fig. 3.10 Root extension during 3 days by melon, with roots in different concentrations of NaCl and $CaCl_2$. $CaCl_2$ (mmol l^{-1}): ● 0.05, ◇ 0.10, ○ 0.25, □ 0.50, ◆ 1.0, ▲ 5.0, ■ 10.0. From Yermiyahu *et al.* (1997). Reproduced with kind permission from Kluwer Academic Publishers.

est Ca concentration is similar to that in seawater, the highest Na about half that in seawater (see Box 3.3). When no NaCl was in the root medium any $CaCl_2$ concentration within the range tried was about equally favourable for root extension, but when NaCl was in the range 20–150 mmol l^{-1}, $CaCl_2$ had a substantial beneficial effect. The top lines ($CaCl_2$ 5 or 10 mmol l^{-1}) represent a purely osmotic effect, as the non-penetrating solute mannitol produced a closely similar result. But when $CaCl_2$ was 1 mmol l^{-1} or less, the NaCl was evidently having an additional harmful effect. One message is that experiments that treat the plant with pure NaCl solution, without other salts, may be showing more harmful effects than would real seawater, which contains some Ca^{2+} and other ions (see Box 3.3). Also, some saline irrigation water may have an unfavourable Na/Ca ratio, which could be improved by mixing with more Ca-rich water or adding lime.

So far we have considered partial exclusion of Na^+ and Cl^- as a strategy for the plant. Its limitation (Box 3.4) is that it may leave the plant with a water deficit, which will reduce growth in various ways. Are there any strategies adopted by plants that allow more Na^+ and Cl^- to be taken up but minimize their toxic effects within the tissues? Many species native to salt-marshes take up substantial amounts of Na^+ and Cl^- yet grow fast. Maybe we can learn from them.

Table 3.8 Concentrations of solutes (mmol l^{-1}) in leaf tissue of *Suaeda maritima* grown with NaCl around the roots

Solute	Cytoplasm	Vacuole
Glycinebetaine	830	0
Na^+	109	565
K^+	16	24
Cl^-	21	388
Sum	976	977

From Harvey *et al.* (1981).

In crop plants that take up Na^- and Cl^-, their continued transport to leaves may result in the concentration building up as the leaves get older. Table 3.7 illustrates this. Many salt-marsh species are succulent: it has been suggested that this is a dilution strategy (Flowers & Yeo 1986): as salts are drawn into the leaf in the xylem flow some of the water is also retained there and dilutes the salt to some extent. However, it is doubtful whether converting our common crop plants to succulent varieties is a realistic way forward.

Putting much NaCl into vacuoles . . .

A more relevant strategy concerns the location of the excess salt within each cell. Some salt-marsh plants have high concentrations of NaCl in their vacuoles but much lower concentrations in the cytoplasm. Thus most enzymes are separated from the higher, and potentially damaging, concentrations of Na^+ and Cl^-. However, the plant needs to maintain an equal osmotic potential in the cytoplasm and the vacuole. This is done by increasing the concentration in the cytoplasm of a low-molecular weight, soluble non-toxic organic compound. Because osmotic effect is generated by each dissolved molecule, the lower the molecular weight the more osmotic potential can be generated per gram of organic compound. In response to drying soil the solute most commonly produced in large amounts in plants is the amino acid proline; but in response to salinity plants most often increase glycinebetaine, $(CH_3)_3$–N–CH_2–COO^-. Salinity causes this to increase not only in halophytic species but in some common crop plants, including wheat, barley, oats and maize (Paleg & Aspinall 1981, pp. 171–204). The location of Na^+ and Cl^- mainly in the vacuole and glycinebetaine mainly in the cytoplasm has been demonstrated on very thin sections of leaves of the halophyte *Suaeda maritima* (Harvey *et al.* 1981; Hall *et al.* 1978); concentrations of ions can be shown by X-ray scans, glycinebetaine by staining with iodoplatinate. Table 3.8 shows how the high concentration of glycinebetaine in the cytoplasm allows the osmotic potential to be the same as in the vacuole, yet the concentration of Na^+ and Cl^- much less. Synthesis of glycinebetaine has a cost in energy and carbon, which may slow down growth, but it is kept to a minimum by (1) using a low-molecular weight compound to generate the osmotic potential, and (2) using glycinebetaine only in the cytoplasm, not in the much larger volume of the vacuole.

. . . and a non-toxic osmoticum in the cytoplasm

Using native salt-tolerant species as crops

An alternative approach, instead of trying to make our existing crop species more salt tolerant, would be to use native species from high-salt habitats such as salt marshes. O'Leary, Glenn and Watson (1985) experimented with growing halophytes in nearly pure sand in a desert area of northern Mexico, irrigated with seawater to which some nutrients were added. They obtained rapid growth: some species gave productivities of 12 tonnes ha^{-1} $year^{-1}$ or more. A problem was the high salt concentration in the tissues. For example, one species, *Salicornia europaea*, had an ash content (a measure of salts) of 41% of dry weight; in non-halophytes grown on non-saline soil the ash content is usually less than 5%. If the salt content is high, animals are likely to refuse to eat the plant.

However, these plants could be useful as seed crops. The ash content of seeds of *Salicornia europaea* was only 7.5%, and they contained 28% oil. This compares respectably with 36% oil in sunflower seeds, which are often grown as a source of oil for human food.

Conclusions

- The abundance, structure and colour of vegetation can affect the water vapour content and the temperature of the air above it, both of which can affect rainfall. So the influence of vegetation on rainfall is complex.
- Changes in vegetation are likely to affect rainfall over scales of hundreds or thousands of kilometres, so they have not been investigated experimentally.
- Models predict that converting the forest of the whole Amazon Basin to open woodland or grassland would reduce the rainfall substantially.
- In the Sahel area of Africa rainfall has been lower since 1970 than it was earlier in that century. Climate measurements and modelling indicate that vegetation change brought about by people could be one cause of this, though not necessarily the only one.
- In semi-arid areas the local abundance of wild and domestic mammals is related to distance from drinking water.
- Future increase in atmospheric CO_2 is likely to increase plant growth per unit amount of water used.
- There are some other possibilities for increasing food production by crop plants per amount of water used, but they are limited by the close correlation between the rates of photosynthesis and transpiration.
- Ability to survive and grow under saline conditions can be influenced by various plant characters, including:
  (a) control of uptake of Na^+ and Cl^- into the root;
  (b) distribution of Na^+ and Cl^- among different parts of the plant;
  (c) distribution of Na^+ and Cl^- between cytoplasm and vacuole;
  (d) ability to synthesize large amounts of low-molecular weight soluble osmoticum.
- It is probable that crop plants can be made more salt tolerant, thereby allowing irrigation with saline water; but it is unlikely that a single gene will provide all the tolerance we need in any species.

Further reading

Sharing water between countries:
Hillel (1994)

Climate, rain formation:
Lamb (1995)
Berner & Berner (1996)
Robinson & Henderson-Sellers (1999)

Desertification:
Williams & Balling (1996)

Water balance of animals:
Schmidt-Nielsen (1997)

Water balance of plants:
Larcher (1995)
Taiz & Zeiger (1998)

Crop plants in relation to water:
Turner (1997)

Salinity:
White (1997, Chapter 13)

Chapter 4: Soil

Questions

- Does converting natural vegetation to farmland increase soil erosion?
- Are rates of erosion from farmland fast enough to be serious?
- How can erosion rates be reduced?
- What properties of a soil make it less prone to erosion? How can we enhance them?
- Can adequate amounts of food be produced without using inorganic fertilizers?
- What are the advantages and disadvantages of inorganic fertilizers? Can we use organic manure instead? Can we use human excreta?
- Can nitrogen fixation provide enough nitrogen for high crop yields?
- Do some farming or forest management systems result in harmful amounts of nutrients being lost by leaching?

Background science

- Processes and rates of rock weathering.
- Erosion in the past and its relations to human land use.
- Soil aggregates: what holds them together.
- Soil organic matter: what affects its abundance.
- Natural inputs of major essential elements: their sources.
- Amounts of nitrogen and phosphorus removed from farms in food harvested. How this compares with natural and fertilizer inputs.
- Nitrate leaching from farmland and forests.
- Nitrogen fixation: processes and rates.
- Phosphorus balance of farming systems of the past that did not use fertilizers.

Soil is important. If all the world's soil suddenly disappeared, life on land would be reduced to some crusts of algae and lichens on rocks, perhaps supporting some fungi, bacteria and small invertebrate animals. It would lead to an abrupt end for land-living vertebrates such as us.

Sustainable management of soil

The underlying theme of this chapter is sustainable management of soil. Whether we are using an area for growing crops, for grazing animals, for growing timber, or whether it is primarily for wildlife, it is essential that the soil should remain productive, that its ability to support plant growth should not decline. It has sometimes been suggested that whole civilizations have collapsed because their agricultural soil was eroded away or became 'exhausted'—for example the Maya civilization of Central America (Russell 1967). A soil can become less productive for various reasons, for example if it becomes too saline (see Chapter 3) or if toxic chemicals accumulate in it (see Chapters 9 and 11). This chapter concentrates on two aspects, erosion and plant mineral nutrition.

Erosion

Soil erosion can be dramatic and obvious. On 11 May 1934 (according to the *New York Times*) 'New York was obscured in a half-light' by a great cloud of dust blown from the farmlands of the mid-West more than 1000 km away. But slow erosion that is not obvious can also be important. Erosion that is too slow to be noticeable within the lifespan of one farmer could still remove all the topsoil within several thousand years. Erosion has been going on for a long time: much of the sedimentary rock in the Earth's crust was formed from the products of previous erosion. The volume of sediment now on the floors of the oceans is so large that it must be the product of millions of years of erosion (Howell & Murray 1986). So erosion is not always the fault of people, but we have sometimes increased it.

How land use can affect erosion

Erosion when forest or grassland is converted to arable

The rate of erosion may increase when forest is felled and the land converted to farming. Davis (1976) provided evidence on how erosion in a small area of Michigan changed when the first European settlers cleared forest and replaced it with farmland. Before Europeans settled in the area the vegetation was deciduous forest with some pine, as shown by pollen records. After the settlers arrived, about AD 1830, the forest was quickly cleared for farming. In 1976 it was mostly meadows and cornfields. The area drains into a small lake which has no outflow, so the sediment in the bottom of the lake can be used as a measure of erosion. The layers in the sediment could be dated by pollen and ^{14}C (see Box 2.5). Figure 4.1 shows the rate at which sediment reached the lake, from 1800 to 1970. Before 1830 erosion averaged 9 tonnes km^{-2} $year^{-1}$. At about the time of clearance there was a very large peak of erosion, and then it settled to a steadier rate with some fluctuations and short bursts. So there is no indication that changes in farming practice between 1880 and 1970 increased the erosion rate. The average rate from 1900 onwards was about 90 tonnes km^{-2} $year^{-1}$, about 10 times the preclearance rate.

Converting long-established grassland to arable can also increase erosion. Table 4.1 shows rates of soil loss caused by rain in two areas of Oklahoma where the natural vegetation is prairie. The rate of erosion was measured by catching surface run-off water in flumes on the soil

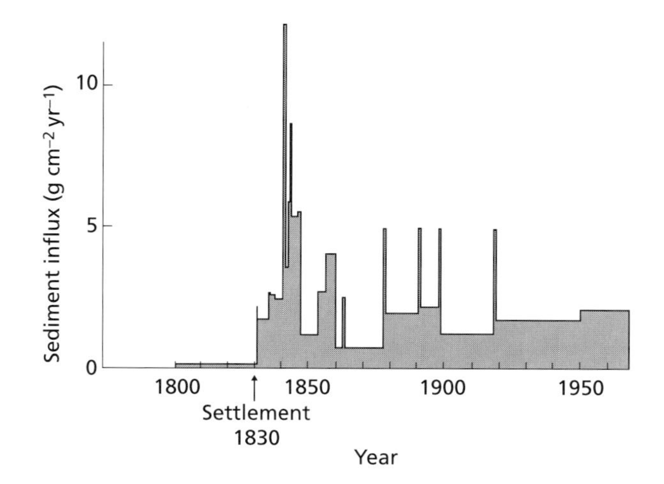

Fig. 4.1 Soil erosion from a catchment in Michigan, USA between 1800 and 1970, measured by the rate of sediment deposition in a lake. From Davis (1976).

surface and determining the concentration of suspended sediment. Erosion was more than 100 times faster in the arable areas than in prairie. This is the total soil washed off through the whole year, including the period when the cereal fields were bare of vegetation.

Similar observations—that converting forest or grassland to arable increases erosion—have also been made elsewhere. However, this is not universally true. For example, in low-rainfall areas of Kenya the rates of soil loss by water erosion, determined from the amount of suspended sediment in rivers, were in the order: grazing land > agricultural land > forest (Morgan 1986).

Soil erosion by wind is more difficult to measure, but this can be done, for example by careful positioning of air samplers. Fryrear (1995) determined soil loss from fields in the US mid-West which were either bare or partly covered by crop residues. During 10 episodes (lasting up to 35 h) of very high wind speed the soil loss in a single episode ranged from 50 to 7000 tonnes km^{-2}.

What depth of soil is lost?

The rates quoted, ranging from a few tonnes up to thousands of tonnes, may sound like rapid rates of loss but they are spread over a square kilometre. It may be helpful instead to view them as loss of soil depth. Because the bulk density of soils usually averages not far from 1 tonne m^{-3},

Table 4.1 Rates of soil erosion (tonnes km^{-2} $year^{-1}$) in two areas of Oklahoma, USA

	El Reno		Woodward	
Land type	Individual fields	**Mean**	Individual fields	**Mean**
Native prairie	3.2, 3.3, 3.7, 4.0	**3.6**	1.1, 10.6	**6**
Cultivated cereal fields	269, 613, 1284	**722**	3960	**3960**

Data from Sharpley *et al.* (1992).

1 tonne km^{-2} corresponds to a thickness of about 1 μm. So, for example, 100 tonnes km^{-2} year^{-1} would correspond to a loss of about 1 cm in a century.

Rates of erosion vary greatly from one part of the world to another, depending on such things as how friable the rock is, how steep the ground, how heavy the rainstorms. Parts of the Himalayas, for example, suffer very rapid erosion. This can remove soil, some of it from areas being used for farming (Ives & Messerli 1989); or erosion can be so fast that there is very little soil and the bedrock is being eaten into (Burbank *et al.* 1996). Of this material carried by rivers from the Himalayas, some is deposited in Pakistan, northern India and Bangladesh, forming fertile floodplains and allowing high densities of farming populations (see Table 2.5). So one person's loss by erosion can be another person's gain. Later in this chapter I describe how soil eroded from Ethiopian mountains helped to maintain the food production of Egypt for thousands of years. However, not all soil eroded by wind or water is deposited elsewhere on land: a substantial percentage ends up in the oceans (Milliman & Meade 1983).

Past erosion in Greece

I suggested earlier that erosion over thousands of years could lead to complete loss of the topsoil. Is there any evidence that this has in fact happened in areas where farming has been going on for several thousand years? In Greece farming started about 8000–9000 years ago. Today much of the countryside is rocky, with soil in shallow layers or pockets, supporting a vegetation of shrubs and annual herbaceous plants. A visitor who has come to see the land where democracy was born, where Plato and Aristotle wrote their great philosophies, where the first Olympic Games were held, may well wonder whether the limited areas of fertile soil could really have supported such a civilization. Or has much of the former soil been eroded away? Part of the answer is that the ancient Greeks were great traders and obtained some of their food from elsewhere in the Mediterranean. But there is also good evidence that soil has been lost from the hillsides.

The main evidence comes from studying soil profiles in valleys, where some of the material eroded from hillsides would be deposited. Layers of coarse material have been found which must have been formed by relatively rapid loss of soil from surrounding hillsides (Van Andel *et al.* 1990; Zangger 1992). Sometimes these contain lumps of rock so large that they must represent rock-slides; others are finer and represent soil erosion by water. These erosion events probably lasted between a few decades and a few centuries. In between were long periods when fine material arrived slowly and soil could form. One might wonder whether these erosion events were triggered by climate conditions. However, they occurred at different times in different parts of Greece, which has led researchers to look more carefully at whether people might be involved.

Van Andel, Runnels and Pope (1986) made a detailed study of one area in the east of the Peloponnese region of southern Greece. They found evidence of seven major erosion periods. The first three, which between them included more than half of the total erosion, were long before farming started. The dates of the last four are shown in Fig. 4.2, which

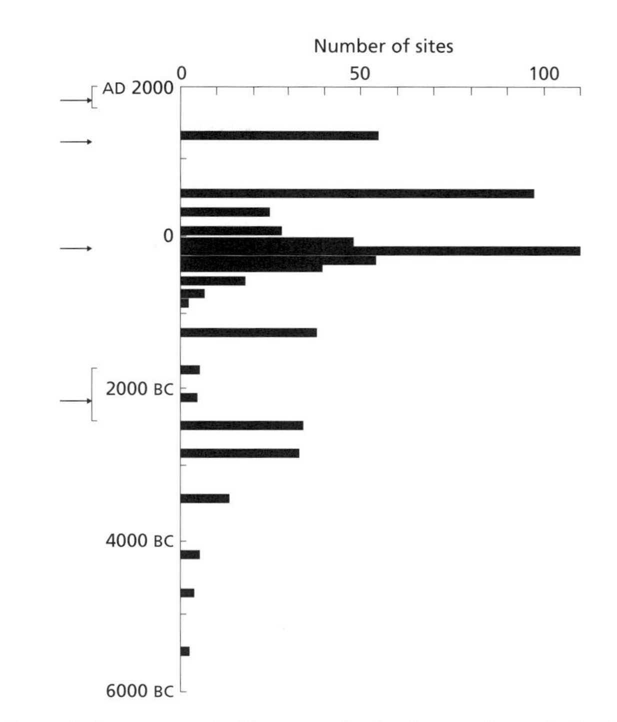

Fig. 4.2 Soil erosion and human occupation in an area of Greece during the last 6000 years. Arrows on left show times of four major erosion periods. Bars show number of sites occupied by people. From van Andel *et al.* (1986).

also shows the number of sites occupied by people during each period of about 100 years. A 'site' could be a single house, a group of houses or a whole village, indicated by buried remains. Settlement and farming in the area started about 6000 BC, but there was no major erosion for about 3000 years after that. So farming did not inevitably cause erosion. Each of the four erosion periods was associated with a period of high population, and therefore presumably of farming activity. The most recent extends into the 20th century. The first two erosion events shown in Fig. 4.2 were evidently towards the end or just after the end of a high population period, the second of them being the great classical period of philosophy, plays and democracy. Rapid soil erosion also occurred in other parts of Greece at that time (Van Andel *et al.* 1990). There are two possible explanations for erosion occurring near the end of a high population period:

1 The high human population led to use of more intensive farming systems, which in turn led to soil erosion, a decline in food production and collapse of the population. So, did the great classical age of Greece end because of soil erosion?

2 The population declined for some other reason (such as war or disease), farms were abandoned, terracing on slopes was not maintained, and so erosion occurred.

We do not have a sound basis for deciding between these two explanations. Whichever is true, it is clear from Fig. 4.2 that soil erosion has been associated with human activity in Greece over more than 4000 years. However, even larger erosion events occurred long before the first farmers, so the lack of soil on Greek hillsides is not entirely the fault of people.

Is new soil formed fast enough to balance loss?

Over timescales of millions of years much soil has been lost to the oceans and ended up on ocean floors. Yet most of the land surface is still covered by soil. Clearly, the formation of new soil worldwide must approximately have balanced loss by erosion. So, how fast is soil formation? and could it balance erosion losses quoted earlier?

Rates of rock weathering

The key initial process in soil formation is physical weathering of rocks. Rock material is broken down to fine particles: sand (a few millimetres across or less), or smaller silt or clay-size particles. The main processes that can lead to this physical weathering are listed in Box 4.1. Rates of physical weathering are difficult to measure: it often occurs out of sight, at interfaces between soil and rock, so the available figures can only be considered approximate and only moderately reliable. Information from several sources drawn together by Morgan (1995) and Bland and Rolls (1998) gives rates mostly in the range 6–100 tonnes km^{-2} $year^{-1}$ (6–100 μm $year^{-1}$). These may be compared with erosion rates quoted by Morgan, Morgan and Finney (1984) from various parts of the world with various types of ground cover, including forest, grassland, crops and bare soil. These range from 0.5 to 44 000 tonnes km^{-2} $year^{-1}$ (about 0.5 μm–4 cm $year^{-1}$). Clearly some erosion rates are slow enough to be balanced by soil formation, but many are much faster than soil formation. This comparison assumes that the soil mineral particles have all formed *in situ*, but this need not be so: in many temperate and northern regions they were formed and brought to the site by glacial action, or they may have been brought more recently by water or wind. Many sites are net receivers of soil.

Ways of reducing erosion

So far, this section has shown that the conversion of forest or grassland to

Box 4.1. Processes that cause weathering of rock to form fine mineral particles.

Temperature extremes. Includes: freeze–thaw; growth of ice crystals; heating by sunshine or fire.
Wet–dry alternation. Water getting into cracks makes them swell, then drying shrinks them again.
Salt crystals expand when hydrated.
Roots of trees grow into cracks, then grow fatter.
Glaciers grind the underlying rock. They can later move the fine material and deposit it elsewhere.
Chemical action. Organic acids (from plants and dead organic matter) and CO_2 dissolve in water to form dilute acid, which can get into cracks and dissolve the adjacent rock.

Further information: Bland & Rolls (1998); Butcher *et al.* (1992)

arable can greatly increase the rate of soil erosion, and that these enhanced rates can be serious, at least in the long term. So what can we do about it? Giving up crop production worldwide is not a desirable option. Can farming methods be modified to reduce erosion?

Box 4.2 summarizes the main factors that influence the rate of erosion. Something can be done to influence each of these. Wind can be reduced by belts of trees, hedges or fences. Its effects can also be reduced by keeping a vegetation cover over the soil for as much of the year as possible, e.g. by having a *cover crop* that covers the ground after the main crop has been harvested. The vegetation canopy can also reduce the erosive effect of heavy rain. On the other hand, light rain which would have scarcely any erosive effect if it fell directly on the soil surface can accumulate on leaf surfaces and then fall as large drops. So, over a whole year the canopy can either increase or decrease erosion by rain (Morgan 1995). Leaving crop residues (a mulch) on the soil surface can be more effective against rain. Slopes can be levelled by terracing, and this is widely done in parts of Asia. It used to be widespread in Europe, too: ancient Greece has already been mentioned as an example. Remains of these terraces can still be seen in European countries. Terracing on steep slopes is not easily compatible with modern mechanized farming, but on less steep slopes erosion may be reduced by ploughing along contours (rather than up and down the slope) or by planting bands of grass along them.

Table 4.2 provides an example of the substantial effects that farming methods can have on erosion rate. It summarizes results from an experiment in which grain sorghum was grown with three sorts of management contrasts. Some plots had conventional tillage (which involves the use of ploughs and harrows) and no other plants present in winter after the sorghum was harvested; this was compared with 'no-till', where the only disturbance to the soil is making slits in which to sow the seed, and in these plots clover was grown in winter to cover the ground between sorghum crops. The crops were grown in this way for 4 years; then

Box 4.2. Factors affecting the rate of soil erosion.

Wind and rain. A few short periods of exceptionally high wind or heavy rainfall can cause much of the total erosion during a year; so total annual rainfall or mean wind speed are often poor predictors of erosion rate.
Landform. Steepness is particularly important; the length of a slope and its shape can also have an effect.
Soil structure. In many soils the abundance, size and strength of aggregates (crumbs) is particularly important, since small particles are more easily washed or blown away. Large pores between aggregates allow rapid infiltration of rain water; if rainfall exceeds infiltration, surface run-off occurs, which is when most water erosion takes place.
Vegetation. Above-ground canopy, roots, microorganisms associated with roots, and dead plant material on the soil surface can all reduce erosion.

Table 4.2 Abundance of water-stable aggregates, and soil loss by erosion during a simulated rain storm, in plots in Georgia, USA, on which grain sorghum was grown with conventional tillage, or grain sorghum was grown in summer and clover in winter with no tillage

	Sorghum, conventional tillage	Sorghum–clover, no-till	Statistical significance*
Water-stable aggregates (% of soil dry weight)	50	87	sig.
Soil loss in 1 h (t km^{-2})			
If residues of sorghum left on soil surface	213	41	sig.
If residues removed	450	133	sig.

* sig. indicates a statistically significant difference ($P < 0.05$) between the figures in the two columns. Data from West *et al.* (1991).

erosion was measured when there was no cover of living sorghum or clover, by a device that simulated an extremely heavy downpour of rain for 1 h. During this 'rainstorm' some plots had sorghum residues on the ground. Table 4.2 shows that the three management treatments combined could reduce erosion by a factor of 10 (from 450 to 41 tonnes km^{-2}). Leaving the crop residues on the soil surface made a substantial contribution to reducing erosion. However, erosion was slower in the no-till plots even when the residue treatment was the same. In these plots the soil was found to have more water-stable aggregates. This leads us on to consider what water-stable aggregates are, why they are important and what holds them together.

Soil aggregates

How easily soil can be blown away by wind depends much on the particle size. Clay particles (about 2 μm diameter or less) tend to stick together, forming a sticky mass when moist and setting hard when dry. So they are not easily blown away. Particles larger than this can be blown away, but the larger they are the faster they fall (through air) and so the less far they are likely to be blown. Above about 20–50 μm diameter they fall quite rapidly. So the particles most prone to long-distance wind erosion are in the silt size range, about 2–20(– 50) μm. If silt particles are bonded together into aggregates (crumbs) which are larger than this, they are less likely to be blown away. If these aggregates are easily broken down by rain they will not last long. Aggregates which retain their structure even when sprayed with water or shaken in water are called *water stable*.

Aggregates reduce erosion

Aggregates also reduce water erosion, because the large pores between them allow water to percolate freely downwards after rain. If, instead, the soil is poorly aggregated the pores will be much smaller and water is likely to run off across the surface after heavy rain: this is how most soil loss by water erosion occurs. A well-aggregated soil also promotes growth and functioning of plant roots because it allows every root to be close to both water and air: water is held in the fine pores within each aggregate, and air in the larger pores between them.

What binds particles together to form aggregates

Box 4.3 lists the main agents that contribute to holding soil particles together to form water-stable aggregates. The influential review paper by Tisdall and Oades (1982) suggested that roots and hyphae are primarily important in holding together aggregates about 0.2 mm in diameter or larger; aggregates smaller than this are much dependent on cementing agents. Evidence that both organic and inorganic cementing agents contribute was provided by Wierzchos *et al.* (1992). They treated soils with either hydrogen peroxide, which oxidizes and removes organic materials, or with acetylacetone, which removes organically bound aluminium and iron. Examination by electron microscope found that both these treatments resulted in collapse of pores and loss of structure within aggregates, at sizes of a few micrometres or less. Another oxidizing agent, periodate, also causes loss of structure in small aggregates (Tisdall & Oades 1982): periodate selectively breaks down polysaccharides and is evidence for their importance as gums. Metal ions and organic gums may well enhance each other's effects. Because clay particles are negatively charged, and so are many organic substances in soil, they tend to repel each other electrically; positively charged ions, especially Al^{3+} and Fe^{3+}, can link them together.

Earthworms deposit conspicuous casts—soil and litter that has passed through their guts. We can ask whether these contribute to aggregate formation. At some sites worm casts are substantially higher in organic

Box 4.3. Agents that contribute to holding soil aggregates (crumbs) together.

Adhesion between clay particles after a succession of wet-dry cycles

Cementing agents

- Inorganic
  - iron oxides
  - aluminium oxides
  - calcium carbonate
- Organic
  - polysaccharide gums and mucilages
  - other humic materials

Living things

- Fungal hyphae
- Roots

Further information: Tisdall & Oades (1982); Oades (1993)

matter than the rest of the soil and also substantially more resistant to erosion by rain (De Vleeschauwer & Lal 1981). However, this is not consistently true: it varies between earthworm species and between soil types (Schrader & Zhang 1997). Earthworms also increase the abundance of large pores (Lachnicht *et al.* 1997), which can promote water percolation and so reduce erosion.

Binding by fungal hyphae

Evidence that fungal hyphae can contribute to binding larger aggregates has been provided by some pot experiments. Table 4.3 shows results from one experiment in which a treatment allowed a comparison of ryegrass with higher or lower amounts of mycorrhizal infection, with corresponding differences in amount of hyphae in the soil. (Some of the soil hyphae presumably belonged to non-mycorrhizal saprophytic fungi.) The treatment did not alter the amount of root. The treatment with less fungal hyphae also had fewer large water-stable aggregates. In another experiment Thomas, Franson and Bethlenfalvay (1993) arranged to have soil compartments containing (1) soybean roots plus associated mycorrhizal fungi; (2) mycorrhizal fungi but no roots; (3) non-mycorrhizal roots; or (4) neither roots nor mycorrhiza. This was done with the aid of fine-mesh gauze which allowed hyphae but not roots to pass through. The amount of water-stable aggregates was greatest in treatment (1) and least in (4); evidently roots and hyphae were both contributing to aggregate stability.

Aggregates in prairie and cornfield

Evidence of the importance of roots and fungi also comes from a field experiment described by Jastrow (1987, 1996). In 1969–71 near Chicago nuclear physicists built a large particle accelerator ring. The 314 hectares of land thus enclosed had originally been prairie, but had been under cultivation for at least 100 years. From 1975 onwards some plots were planted with species from native prairie (grasses and forbs) with the aim of reconstructing vegetation similar to that of the original prairie. (For a summary of the development of prairie vegetation there over 11 years, see Fig. 11.4, p. 335.) Figure 4.3 compares the soil of some of these plots,

Table 4.3 Root and fungal development and abundance of water-stable aggregates in an experiment in which ryegrass (*Lolium perenne*) was grown in pots, some of which received treatments to reduce mycorrhizal infection

Characteristic	Control	Mycorrhiza reduced	Statistical significance*
Root length (m per g soil)	0.76	0.77	n.s.
Arbuscular mycorrhizal infection in roots (% of root length infected)	21.1	8.9	sig.
Hyphal length in soil (m g^{-1})	12.5	6.3	sig.
Water-stable aggregates > 2 mm diameter (% of total soil)	14.7	10.3	sig

* sig. indicates that the values differ significantly ($P < 0.05$) between the treatments. n.s. = not significant. Data from Tisdall & Oades (1979).

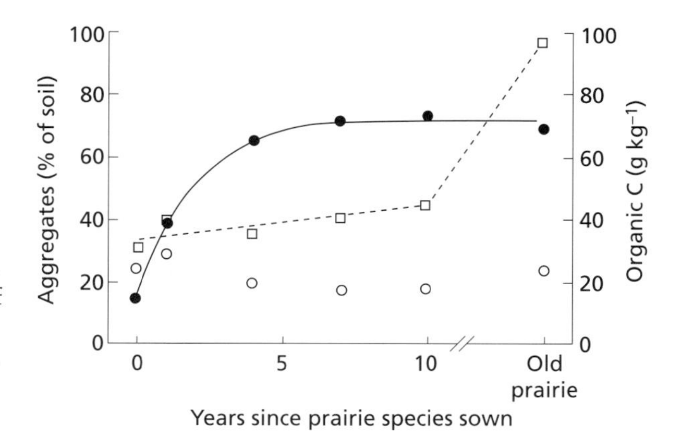

Fig. 4.3 Amount of water-stable aggregates and organic matter (measured as organic carbon) in soil at nearby sites in Illinois which were growing corn (= year 0), had been planted with native prairie species or were remnants of old prairie. ● aggregates >1 mm diameter, ○ aggregates 0.2–1 mm diameter, □ organic C. Data from Jastrow (1987, 1996). Reprinted with permission from Elsevier Science.

which had been under prairie species for 1–10 years, with a field still under corn and a patch of undisturbed prairie nearby. The cornfield and prairie differed little in the amount of water-stable aggregates 0.2–1 mm in diameter, but the prairie had many more of the larger aggregates. After prairie replanting the abundance of these larger aggregates had recovered to old prairie levels within about 5 years. The old prairie soil was also much higher in organic matter than the cornfield, but only a small part of this difference had been made up after 10 years under prairie plants. So, the rapid recovery of crumb structure was apparently not dependent on the recovery of total organic content. It is possible that certain key organic constituents regenerated quickly, but it seems more likely, since only the larger aggregates increased in abundance, that it was due to more direct effects of the roots and their associated fungi. Roots, fungi and bacteria exude a variety of organic chemicals, some of which are mucilages (Watt *et al.* 1993). Some fungi are sticky: soil particles can be seen sticking to them (Tisdall & Oades 1979). So they do not just act as a simple net. Most prairie plants are perennial, so their roots will be present throughout the year, unlike corn roots. They are also likely to form a denser root system than corn roots, and to have more mycorrhizal fungi.

The reverse process, conversion of grassland to arable, has also been studied. When old grassland in England was ploughed up and converted to arable the abundance of large aggregates dropped sharply in the first year after ploughing, though not right down to their abundance in old arable (Low 1972). Soil organic matter (measured here as total nitrogen) fell proportionately much more slowly. So the abundance of large aggregates seemed to be associated with the presence of live grass roots.

Soil organic matter

This evidence for direct effects of roots and hyphae in holding large aggregates together should not lead us to forget about cementing agents. These are crucial in holding together small aggregates (less than about 0.2 mm

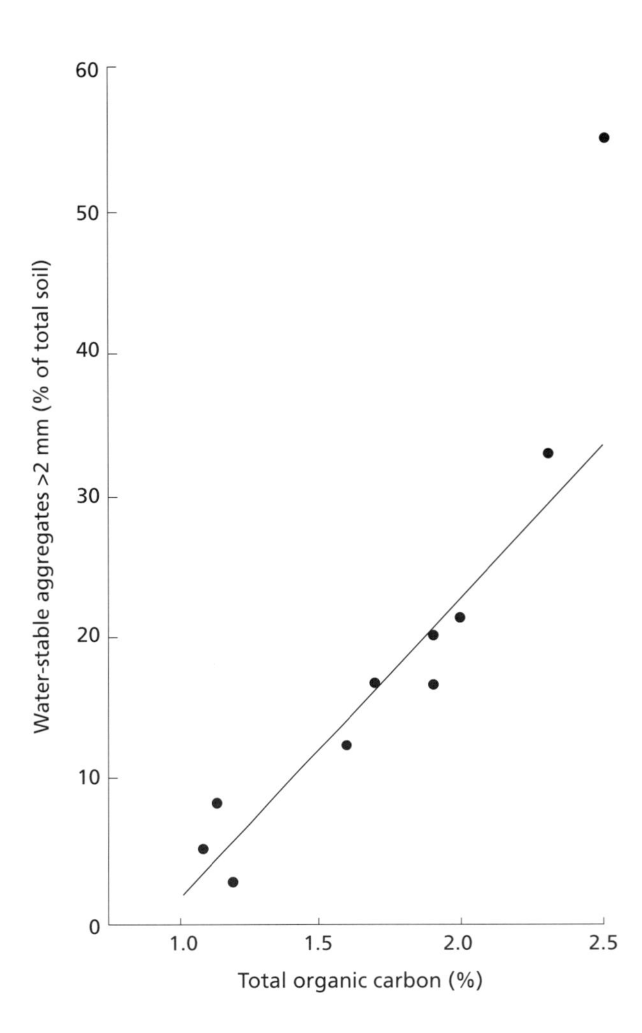

Fig. 4.4 Relationship between organic matter content and abundance of water-stable aggregates more than 2 mm in diameter, in soils from adjacent sites in South Australia that had carried different crops, rotations or pasture for 50 years. From Tisdall & Oades (1980).

diameter), and larger aggregates are likely to be made of these smaller aggregates bound together. It has been found in various parts of the world that higher organic matter content in soil tends to be associated with a greater abundance of water-stable aggregates. Figure 4.4 shows an example of this. Presumably this is partly because some of the organic matter is acting as gums and mucilages. In addition, more organic matter is likely to support more saprophytic fungi. This has led to interest in how farming practices, including inorganic fertilizer use, affect soil organic matter content.

Organic matter and aggregates

What affects the amount of organic matter

In plots at Rothamsted Experimental Station in England on which winter wheat has been grown since 1843, with inorganic fertilizer application or with no fertilizer, the soil organic matter content has remained virtually unchanged, whereas the application of farmyard manure has produced a marked increase (Fig. 4.5(a)). The abundance of water-stable aggregates over 0.5 mm in diameter has increased substantially in the plot given farmyard manure (Wild 1988, p. 437). Very different results

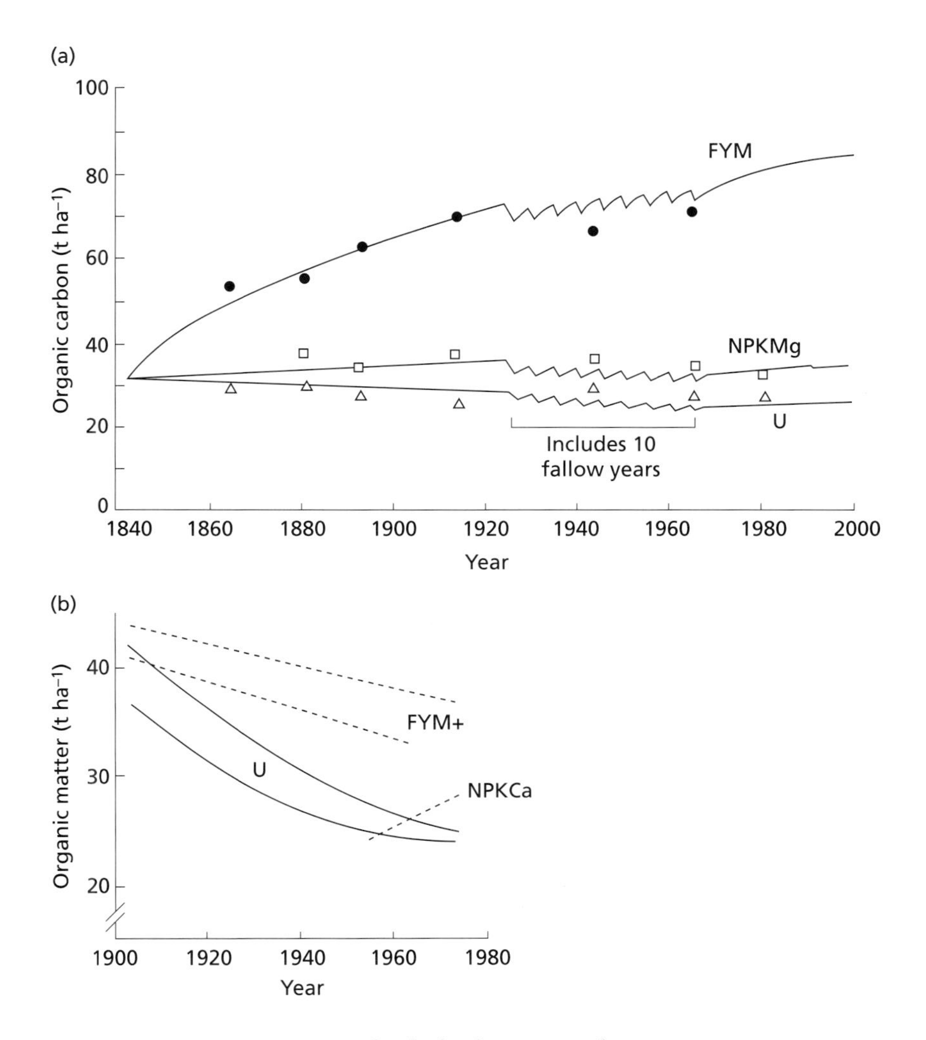

Fig. 4.5 Organic matter content of soils that have received contrasting manure treatments in long-term experiments in the UK and USA. (a) Rothamsted, southern England, top 23 cm of soil. Winter wheat grown since 1843. The plots received annually farmyard manure (FYM), inorganic fertilizer (NPKMg) or no addition (U). The symbols show measured values, the lines are fitted by a model. From Jenkinson (1991). (b) Urbana, Illinois, top 17 cm of soil. Corn grown since 1876. The treatments were: unfertilized throughout (U); unfertilized until 1904, thereafter farmyard manure, lime and phosphate (FYM +); unfertilized until 1955, thereafter inorganic NPK fertilizer and lime (NPKCa). The lines are statistical best fits. From Odell *et al.* (1982, 1984).

have been obtained from plots on the campus of the University of Illinois, on which corn has been grown since 1876 (Fig. 4.5(b)). In the unfertilized plots soil organic matter has decreased substantially since 1904; the addition of farmyard manure has slowed this decline but not reversed it. In contrast, after inorganic fertilizer additions were started to some previously unfertilized plots, their organic matter increased.

It may seem surprising that these two experiments, both growing cereals, should give such different results. What happened before the experiments started is significant. The Illinois corn plots had previously been prairie, and a decline in soil organic matter has frequently been observed after old grassland is converted to arable. In contrast, the Rothamsted plots were already under arable cultivation when the experiment started, and probably had been since Roman times, about 2000 years previously, so the soil organic matter had plenty of time to reach equilibrium. Whether the amount of organic matter increases, decreases or stays constant depends on the balance between inputs and losses. Organic matter is lost by being decomposed by soil microorganisms. If the amount of organic matter is to remain constant this loss must be balanced by inputs. Inputs can be (1) residues of the crop—roots, stems, leaves—left after harvest, and from any cover crops or plants grown in rotation as a green manure; or (2) organic manures added, such as farmyard manure or sewage sludge. In both the experiments in Fig. 4.5 farmyard manure (which is usually cattle faeces and urine mixed with straw) increased the amount of soil organic matter. In the Illinois corn plots inorganic fertilizer increased soil organic matter, presumably because increased crop growth resulted in more crop residues left at the end of each growing season.

An increase in temperature within the range 0–35°C usually increases the metabolic rate of soil microorganisms, and hence the rate at which organic matter is decomposed. Photosynthesis and net primary productivity are less affected by temperature. Therefore, as the temperature rises the balance between organic matter inputs and losses is expected to alter, resulting in less organic matter in soil (Kirschbaum 1995). The tendency for soils in the tropics to have low amounts of organic matter can cause problems for soil structure and erosion, and also in their mineral nutrient status (see later).

Rate of breakdown of organic matter

The organic matter in soil comprises a great diversity of chemicals, ranging from simple sugars and amino acids to complex polymers with molecular mass >1 million (Wild 1988). Some are broken down by microorganisms within a few days; others are extremely resistant to microbial attack. Using radiocarbon dating it has been shown that there is a substantial component of the organic matter in the topsoils of the unfertilized Rothamsted and Illinois plots (Fig. 4.5) that is 2000–3000 years old (Jenkinson & Rayner 1977; Hsieh 1992). Other sites in the US and Canadian mid-West—some still prairie, others farmland—also have soil organic matter that is several thousand years old (Paul *et al.* 1997). Whether these very long-lived components of the organic matter contribute much to holding aggregates together is not known.

Management for improved soil structure and reduced erosion

Does this scientific research suggest ways to manage soil to promote crumb structure, thereby making the soil less prone to erosion? One aim should be to maintain or increase soil organic matter. If we knew more about the chemistry of soil organic matter, in particular about the cementing agents, maybe we would find ways of managing for particular desired chemical compositions. For the moment, the basic recommendation would be to add organic manure, of either animal or plant origin. Crop residues provide one such type of input of organic matter. One way in which plant breeders have increased seed yields of crop plants over the years is by increasing the proportion of plant growth that goes into seeds (see Fig. 3.8), but this tends to reduce the amount of residue.

Another aim could be to promote soil fungi. Generous applications of P fertilizer tend to reduce mycorrhizal infection of crop plants (de Miranda *et al.* 1989; Thomson *et al.* 1992): this could affect decisions about how much inorganic fertilizer to apply.

The experiment summarized in Table 4.2 suggested that reducing tillage promotes aggregate abundance, and this has been confirmed by other experiments (Beare *et al.* 1994). It may seem surprising that implements as large and crude as ploughs and harrows can affect crumbs a few millimetres across. One way they may do this is by damaging fungal hyphae: it has been shown that physical disturbance of soil can reduce the amount of live arbuscular mycorrhizal fungus (Jasper *et al.* 1987; Evans & Miller 1988). Also, some earthworm species are more abundant in no-till (Edwards & Lofty 1982). Possible effects of earthworms on the resistance of soil to erosion were summarized earlier.

Figure 4.3 shows that aggregate abundance increased greatly during the first few years after the conversion of arable to grassland. The reverse happens when grassland is ploughed and converted to arable: aggregate abundance declines, but not all in the first year (Low 1972). So, a few years under grass would be expected to have a beneficial effect on the aggregate structure for several years of arable that followed.

These are some of the options for farmers. Others were mentioned briefly earlier: terracing, contour ploughing, shelter belts, cover crops, surface mulching. Which methods are most effective, and which are worth the time, effort and cost, will depend very much on local conditions, including soil type, the steepness of the land, and climate. In many parts of the world soil changes are slow enough not to be a serious threat within the lifetime of one farmer, yet they might eventually put an end to farming there for ever. This section on erosion started with the *New York Times* describing massive loss of soil from the mid-West in 1934. The trigger for that wind erosion was drought, but the soil had undoubtedly been made more prone to erosion by a decline in organic matter (Fig. 4.5(b)), which had probably been going on for about 100 years. Erosion is an example of an environmental problem that faces people with the difficult question as to how much effort and expense they are prepared to undertake now to safeguard future generations, who will live long after they are dead.

Mineral nutrients

Natural inputs of macronutrients

The elements required in the largest amounts by plants and animals (after C, H and O) are the six ***macronutrients***: *nitrogen, sulphur, phosphorus, potassium, calcium* and *magnesium*. There is also a suite of essential ***micronutrients*** which are required in smaller amounts, including iron, manganese, zinc and molybdenum, but these will not be considered here. Some plants can obtain nitrogen from N_2 gas in the air via symbiotic microbial species, but apart from that, all the supplies of the six macronutrient elements to terrestrial plants have to come from soil. In soil each element can be in various chemical forms, including inorganic chemicals of various solubilities, some perhaps adsorbed to surfaces, also often in some organic compounds. These forms differ in how readily plants can take them up, or how quickly they are converted into forms that plants can take up. On short timescales these processes are crucial for plant survival and growth. However, in this chapter I take a longer-term perspective and concentrate on the nutrient input/output balance of systems.

Nutrient losses

If the fertility of a soil and the productivity of the ecosystem are to be maintained long term, the input of each element must be sufficient to balance losses. Nutrients are lost from natural ecosystems by leaching into rivers, soil erosion, fires, and sometimes by other processes. However, if people are harvesting and removing material, for example cereal grain, meat, milk or timber, there is inevitably an additional loss of mineral nutrients in the products harvested. The main aim of this section is to consider how nutrient losses from such harvested systems can be balanced by inputs.

Nutrient inputs

Box 4.4 summarizes the principal inputs to terrestrial ecosystems of each of the macronutrients. Inorganic fertilizers are not mentioned but will be considered later. The release of soluble ions from the fine rock particles in soil by weathering is regarded here as an input, because it operates over timescales of thousands or tens of thousands of years (Walker & Syers 1976). Box 4.4 is concerned not merely with the immediate sources of input—adsorption of a gas, dry deposition of dust, dissolved in rain etc.—but where that source came from. If it came from plants or soil, then it represents redistribution within the land mass, and much of it may be from a site quite close to where it is deposited. It may therefore be merely balancing an equivalent loss. If, on the other hand, the nutrient came from N_2 in the air, from the oceans, from rock or from fossil fuels, it represents an input in a more fundamental sense, from a pool that is global or formed very long ago.

Inputs of N and S

Among the six elements, two (nitrogen and sulphur) have gaseous compounds at normal outdoor temperatures, and this gives them very different opportunities in transport and cycling from the other four elements. Over much of the world sulphur has not been considered a limiting

Box 4.4. Principal sources of major nutrient inputs to soil–plant systems, in addition to fertilizer.

Nitrogen	**Source**
Biological nitrogen fixation	N_2 in air
Symbiotic (by microbes associated with plant)	
Non-symbiotic (by microbes in soil)	
N-oxide gases	
Dry deposition as gases	
Gases form HNO_3 aerosol	
Gases form NO_3^- in rain	
Sources:	
burning fossil fuels	N_2 in air
lightning	N_2 in air
burning wood	N in plants, N_2 in air
microbial action on organic matter in soil	plant material
Ammonia gas	
Dry deposition as gas	
Forms NH_4^+ in rain	
Sources:	
urine	animals eat plants
decomposition of faeces and plant material	plants
Sulphur	
SO_2 gas	
Dry deposition	
Forms SO_4^- in rain	
Sources:	
burning of	fossil fuels
H_2S, $(CH_3)_2S$ + other reduced-S gases	decomposition of plant materials
sea spray (coastal sites)	oceans
Dust	soil
Weathering	fine rock material
Phosphorus	
In rain	from burning of plant material
Smoke	plant material
Pollen	plants
Dust	soil
Weathering	fine rock material
Potassium	
Weathering	fine rock material
Dissolved in rain	soil dust
(coastal sites)	oceans
Smoke	plant material

[*Continued*]

Calcium	
Weathering	fine rock material
Dissolved in rain	soil dust
Smoke	fossil fuels, plant material
Dust from cement manufacture	limestone
Magnesium	
Weathering	fine rock material
Dissolved in rain	oceans, soil dust
Smoke	plant materials

Further information: Berner & Berner (1996); Butcher *et al.* (1992); Newman (1995)

element for plant growth, and the application of S to crops in fertilizer has not been considered necessary. In many parts of the developed world the largest source of sulphur input is SO_2 from burning of fossil fuels, plus sulphate formed from it. In North America and many European countries SO_2 production has been declining since 1980 (OECD 1997). Figure 4.6 shows the dissolved sulphate-S in rain at Rothamsted rising from 1855 to 1980, then falling steeply. This reduction in 'free fertilizer' could result in increased S deficiencies in crops, but for the moment there seems to be little concern about this.

Nitrate and ammonium dissolved in rain increased through much of the 20th century. Figure 4.6 shows that at Rothamsted N input in these forms rose from 1855 to 1980, but then fell (like sulphate). Dry deposition can contribute more N input than rain. Measurements at Rothamsted in 1996 showed NO_2 gas and nitric acid aerosols to be major components (Table 4.4). Input by biological N fixation will be considered later.

The three major cations, K^+, Ca^{2+} and Mg^{2+}, all occur in seawater (see Box 3.3 for concentrations), and they can get into rain via spray. This is a substantial contribution to Mg input at many sites on land, but for K this is true only close to coasts, and for Ca not even there (Berner & Berner

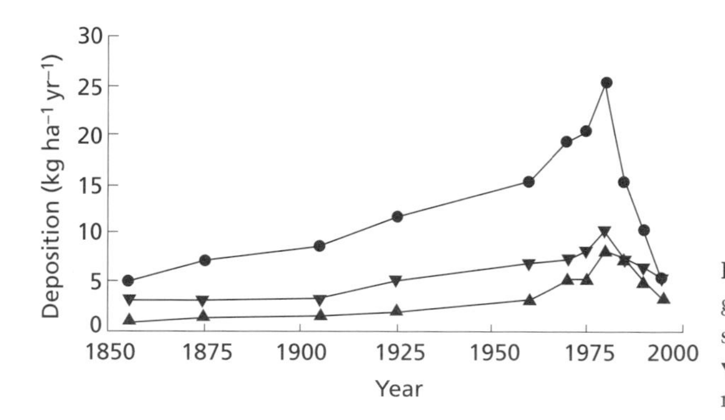

Fig. 4.6 Annual deposition of inorganic nitrogen and sulphur dissolved in rain, at Rothamsted, England, from 1855 to 1995. ▲ nitrate-N, ▼ ammonium-N, ● sulphate-S. Each point is mean for 5 years. From Goulding *et al.* (1998).

Table 4.4 Nitrogen inputs from the atmosphere at Rothamsted, England, in 1996. Biological nitrogen fixation is not included

Source	kg N ha^{-1} yr^{-1}
Dissolved in rain	
nitrate	3.8
ammonium	5.2
total	9.0
Aerosol, nitric acid	16.4
Particulates	2.9
Gases	
nitrogen dioxide	13.6
ammonia	1.6
total	15.1
Total N input	**43.4**

Data from Goulding *et al.* (1998).

Nutrients released by weathering

1996; Asman *et al.* 1981)). At many sites weathering of the mineral particles in the soil is the main input of all three cations to the available pool. Silt and clay consist primarily of aluminosilicates, which have within their molecular structure either one or two of the elements K, Ca and Mg, often comprising 4–12% of the total weight (Cresser *et al.* 1993). Some rocks contain a high proportion of silica (quartz) and little aluminosilicate: these are low in the basic cations and give rise to acid, sandy soils that may be seriously deficient in nutrient cations. But apart from these, most soils contain a mixture of aluminosilicates and hence substantial amounts of K, Ca and Mg. This is in contrast to phosphorus. Weathering of rock is the only genuine input of P—all the other inputs listed in Box 4.4 are recycling within land—but the concentration of P in rocks is rarely above 0.1%, and the rate of release by weathering is therefore slow (Newman 1995); figures for this rate will be given later.

Rates of release of K, Ca, Mg and P into available forms depend on how rapidly the rock weathers, and this varies greatly depending on its chemical nature and also on how long it has been in fine particles subject to weathering. As the mineral material of soil is weathered, K, Ca, Mg and P are either released and lost by leaching or are converted to occluded forms that are very resistant to weathering. It takes thousands or tens of thousands of years for most of this change to occur (Walker & Syers 1976; Crews *et al.* 1995), so soils up to a few tens of thousands of years old count as young, and are likely to weather more rapidly. These include soils formed by water erosion of mountain rocks, e.g. those deposited in northern India (see erosion section) or on the fields along the Nile (see later in this chapter); materials from recent volcanic eruptions; and material ground off rocks by glaciers at the end of the last Ice Age or more recently.

These soils can have high natural fertility. In contrast, large areas, especially in tropical and subtropical regions, have very old, highly weathered soils which release nutrients by weathering only extremely slowly.

The remainder of this section concentrates on two elements, nitrogen and phosphorus. They are often limiting to plant growth, and maintaining their long-term input/output balance is therefore important.

Long-term balance of nitrogen and phosphorus

N and P balance of farms

Table 4.5 gives nitrogen and phosphorus balances of wheat fields, as examples of modern, heavily fertilized arable farming. The figures illustrate several points.

1 Fertilizer was the largest input of N and P; grain harvest was the largest output.

2 The inputs and outputs called 'other' in the table, which are more or less natural, approximately balance each other, so it is not obvious how natural inputs could support N and P removal in harvest. Even if the natural losses could be stopped, the natural inputs are too small to support the N and P removal in grain.

3 Natural inputs and outputs of P are very much smaller than of N. This is true in absolute terms and also in proportion to the amount of the element removed in harvest. There is thus more opportunity for N than for P to improve the balance by increasing non-fertilizer inputs and reducing non-harvest outputs.

Table 4.6 gives information on the amounts of N and P removed in harvesting of products, giving examples of arable food production, animal food production and timber production. The wheat, potato and dairy farms all received fertilizer; the sheep farm and timber plantation did not.

Table 4.5 Nitrogen and phosphorus balance of wheat fields on modern British farms

	$kg\ ha^{-1}\ yr^{-1}$	
	N	P
Inputs		
Fertilizer	146	25
Other*	48	1
Outputs		
Grain	114	19
Straw	26	3
Other†	54	1

* Other inputs. N: dissolved in rain, dry deposition, non-symbiotic N fixation. P: in rain, weathering of rock. † Other outputs. N: leaching, gases. P: leaching, soil erosion. Sources of data. N: figures for long-term wheat field at Rothamsted, from Powlson *et al.* (1986), Dyke *et al.* (1983). P: figures for typical British wheat farm, from Newman (1995).

Table 4.6 Nitrogen and phosphorus concentrations and contents of various plant and animal products. Weights and concentrations are on a dry weight basis except where stated otherwise

Product	Weight harvested ($t\ ha^{-1}\ yr^{-1}$)	Nitrogen concentration ($mg\ g^{-1}$)	Nitrogen amount in harvest ($kg\ ha^{-1}\ yr^{-1}$)	Phosphorus concentration ($mg\ g^{-1}$)	Phosphorus amount in harvest ($kg\ ha^{-1}\ yr^{-1}$)	Notes
Wheat grain	4.5	21.0	94	3.3	15.3	1
Wheat straw	3.8	5.8	21	0.7	2.8	1
Potato tubers	8 (approx.)	14.2	108	2.0	17.8	1
Milk	$14\ m^3\ ha^{-1}\ yr^{-1}$	$5.2\ mg\ ml^{-1}$	74	$1.1\ mg\ ml^{-1}$	15.6	2
Sheep	0.023*	25*	0.58	6*	0.14	3
Tree stems	2.5	1.6	4.0	0.23	0.57	4

* Fresh weight basis. Notes: (1) Field at Rothamsted, England, where crops have been grown with NPKMg fertilizer for more than 100 years. Data from Dyke *et al.* (1983). (2) Typical dairy farm in south-western England; fertilizer used. Data from Haygarth *et al.* (1998), Jarvis (1993). (3) Typical sheep farm in Scottish Highlands. Data from Haygarth *et al.* (1998). N concentration from Crisp (1966). (4) 42-year-old plantation of Douglas fir, Washington state, USA. Figures given per year are total in stems/42. Data from Reichle (1981).

The concentrations of N and P in the milk and sheep meat are expressed on a fresh-weight basis and so are not exactly comparable with the plant materials, which are on a dry-weight basis. Nevertheless, it is clear that among the food materials wheat grain, potatoes and meat are all substantially higher in N and P concentrations than are straw or (especially) tree stems. The tree stems here included bark; if that were stripped off the N and P concentration of the remaining xylem would be even lower (Whittaker *et al.* 1979). The weight of straw produced was almost as much as the grain, but its N and P content was much less. Because of this, whether straw is ploughed back into the field or removed has only a limited influence on the nutrient balance. The data for wheat and potatoes in Table 4.6 come from the same experiment, with the same fertilizer regime, so they are directly comparable. Potatoes had a lower N and P concentration than wheat grain, thereby allowing a higher dry-weight production than wheat while maintaining about the same N and P loss in harvest. (As sold, potatoes have a much higher water content than wheat or flour, so the difference in yield per hectare would appear much larger.) Where the primary food requirement is for energy, this difference between wheat and potatoes could be important. The dairy farm featured was also managed intensively, including the use of fertilizer. Its N and P loss in the product was about the same as for potatoes and wheat. In contrast, the sheep farm's nutrient losses were about two orders of magnitude lower, because the sheep production was so low. This farm was mostly on infertile soils, which received no fertilizer. Finally, the weight of useful product per year from the timber plantation was about half the cereal grain, but because of the low N and P concentrations in the tree trunks the nutrient losses in harvest were less than one-tenth those in the grain, when averaged over the growing period of the trees.

The only nutrient losses shown in Table 4.6 are those in the harvested products. There will be other losses (see Table 4.5). Nevertheless, Table 4.6 raises a number of questions. If high food yields require such large nutrient removals, are inorganic fertilizers essential to maintain those yields? Are there serious disadvantages to the long-term use of inorganic fertilizers? Are there alternative ways of maintaining the soil nutrient status? How did farmers manage in the past, before there were bag fertilizers? Can production systems with no fertilizer input, such as the sheep and timber examples in Table 4.6, be sustainable long term, in spite of repeated removal of the product?

Inorganic fertilizers: advantages and disadvantages

Long-term crop growth using fertilizers

Crop production with inorganic fertilizer can work satisfactorily for more than a century, as shown by experiments at Rothamsted, England (Jenkinson 1991). There, wheat, barley and hay have each been grown continuously since the mid-19th century, with inorganic fertilizers as the only human input of nutrients. Yields have been maintained, and in the case of wheat have substantially increased through the use of improved varieties and of modern herbicides for weed control. Soil organic matter content has also been maintained (Fig. 4.5). This is the longest-running test of growing crops with inorganic fertilizer as the only non-natural nutrient input, but similar experiments have been run successfully for decades in other parts of the world (see Fig. 4.8).

It has been questioned whether continuous cropping with inorganic fertilizer can work long term on old, highly weathered soils in the tropics. Because the mineral material is so weathered and the organic matter content often low, these soils tend to have low cation exchange capacity and strong P-fixing ability. So, added K, Ca, Mg and P quickly tend to cease to be available for uptake by the plants: they are either leached away or converted to non-available forms (Weischet & Caviedes 1993). In an experiment at Yurimaguas, Peru, in the Amazon Basin, crops have been grown continuously for 20 years on a site that was previously secondary rainforest, used for shifting cultivation. The soil was representative of that part of Amazonia: extremely weathered and low in organic matter (Weischet & Caviedes 1993). Figure 4.7 shows yields obtained from a sequence of rice, corn and soybean, grown either with no fertilizer or with a high input of inorganic Ca (as lime, to raise the pH), K, Mg, N, P and S, plus small amounts of four micronutrients. In the unfertilized plot yields declined to zero within 3 years, as often happens in shifting cultivation in the wet tropics. In the fertilized plot the yields have fluctuated, but there is no indication of a declining trend. The yields were mostly between 2 and 4 tonnes ha^{-1}, which (as there were two crops in most years) means an average of about 6 tonnes ha^{-1} year^{-1}, a yield many

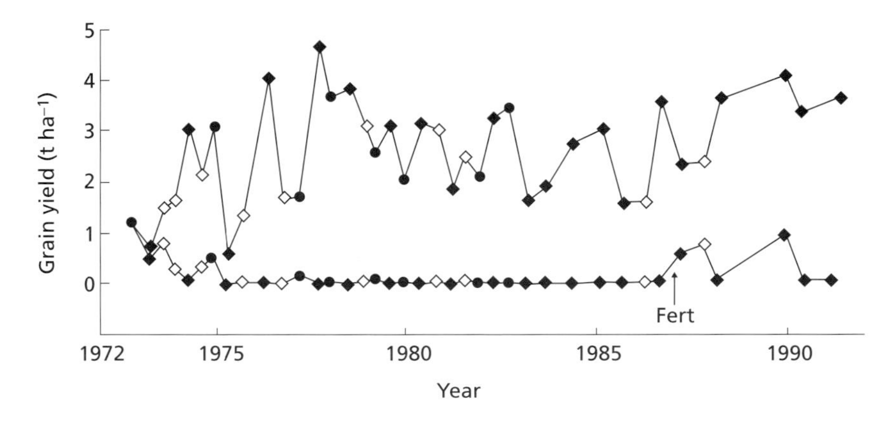

Fig. 4.7 Yields of grain at each harvest in experiment at Yurimaguas, Peru. Upper line: received inorganic fertilizer. Lower line: unfertilized, except when marked Fert. ● rice, ◆ corn, ◇ soybean. From Buol (1995).

farmers in temperate regions would consider very respectable. Fearnside (1987) discussed the results of this experiment, pointing out various difficulties in applying the system in practice in the Amazon Basin. Problems include soil erosion and weeds; but probably the primary reason why no farmers in the region are as yet adopting the system is that the fertilizers have to be brought in hundreds of kilometres over poor roads, and the cost to the farmer on the spot is therefore high.

Fertilizers have disadvantages

Box 4.5 summarizes the main disadvantages of using inorganic N and P fertilizers. Almost all N fertilizer is derived from ammonia, which is made by combining N_2 and H_2 gases at high temperature and pressure. The supply of raw materials is therefore unlimited, but the reaction uses a lot of energy. In a time of concern about the use of fossil fuels (Chapter 2) this is

Box 4.5. Potential disadvantages of using inorganic fertilizers.

Nitrogen fertilizers

1 High nitrate concentrations in food can be a health risk.
2 Leaching losses can give high nitrate concentration in water:
(a) health risk from drinking water,
(b) eutrophication of lakes giving algal blooms and harm to fish.
3 High fuel energy use to make ammonia from N_2 gas.

Phosphorus fertilizers

1 Leaching losses into rivers, lakes and enclosed seas can cause eutrophication.
2 World supply finite.

Both N and P fertilizers

1 If use of organic manure reduced, likely to lead to reduced soil organic matter content, which can have unfavourable effects (see erosion section in this chapter).
2 Fertilizer production localized, so high cost of fertilizer 'at the farm gate' in countries with poor transport systems.

an argument for reducing the production of nitrogen fertilizer. Almost all P fertilizer is made from rock phosphate (apatite), a very insoluble form of calcium phosphate. So the world supply is finite. There are widely varying estimates of how large the world's usable stock of rock phosphate is. At the present rate of use these would predict that supplies will last anything from less than 100 years to more than 1000 years (Williams 1978; Stevenson 1986; Bockman *et al.* 1990). Estimating the world stocks of basic resources is notoriously difficult. There are sources of phosphorus in various parts of the world which are at present not being exploited but which might become exploitable with changed demand and new technology.

Organic manures: advantages and disadvantages

Some people suggest that we should use organic manures instead of inorganic fertilizers. Organic manure can be animal faeces; it can be a mixture of animal excreta and plant material, for example farmyard manure; or plant material alone (*green manure*) can be used. Adding organic manure is likely to increase the organic matter content of the soil (see, for example, Fig. 4.5), which improves soil structure and thereby reduces soil erosion (see erosion section of this chapter). However, the fundamental weakness of organic manure as a source of N and P is that animals do not create these elements: what comes out of the back end of a cow is only a processed version of what it ate. If a plant which has a N-fixing symbiont is used as a green manure, or if animals that eat it provide manure, this is a genuine input of N from the air; but all other N and all P in organic manures is just a way of recycling N and P from soil.

Are nutrients in animal manure readily available to plants?

One reason for using animal rather than green manure is the hope that nutrients may be more rapidly released in plant-available form. A substantial proportion of the N in what a cow or sheep eats ends up as urea in its urine (Church 1988), which is quickly converted by microorganisms in soil to ammonia and ammonium ions, which plants can then take up. However, there is little P in urine: most P leaves the animal in faeces. The release of N and P from decomposing faeces in inorganic form extends over months or even several years. Comparison of the rate of release from sheep dung and from the plant material the sheep ate has shown release to be sometimes faster from dung than from plants, but sometimes slower (Barrow 1961; Bromfield & Jones 1970; Floate 1970a,b).

Experiments comparing the percentage of the N in green manures and in fertilizers that gets into the growing crops have shown no consistent difference between these two types of N source (Peoples *et al.* 1995). Figure 4.8 shows results from a long-term experiment in Oregon in which the fields bore wheat and bare fallow in alternate years for more than 50 years. There was little difference in the effectiveness of animal manure, green manure or inorganic fertilizer as a nitrogen source: grain yield was closely related to the amount of N added, irrespective of the form. However, the only treatment that retained the initial soil organic matter

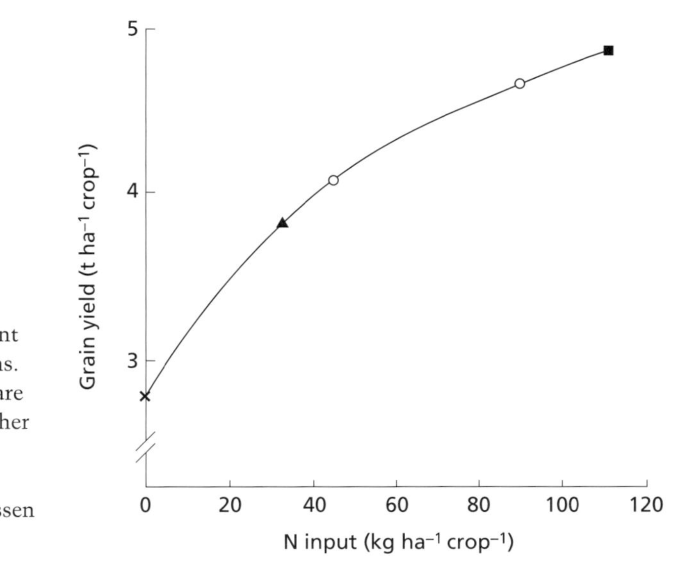

Fig. 4.8 Wheat grain yield in an experiment in Oregon, in relation to nitrogen additions. Experiment started 1931, measurements are mean of 1977–86. One crop per 2 years, other year bare fallow. N input in: ○ NH_4NO_3, ▲ pea (stem, leaf and pod), ■ farmyard manure, × no addition. Data from Rasmussen & Parton (1994).

content was the one receiving farmyard manure; in all the others it decreased. This decrease would occur particularly during the fallow years, when decomposition continued but there was no input from crop residues. Decomposition of the soil organic matter released mineral N, which was probably crucial in maintaining the crop yield in the absence of fertilizer or manure, but the decline in organic matter would be expected to lead to less favourable soil structure and increased erosion.

If some of the N in plant material is lost between its being fed to the animal and their excreta reaching the field, this would be a disadvantage of animal manure compared to green manure. He *et al.* (1994) followed the fate of N when a mixture of vetch, cowpea and rape was used as a green manure for paddy rice, or when it was fed to pigs and their faeces and urine were then applied to the rice paddy. If the plant material was used as green manure, 34% of its N was taken up by the rice plants; but if the plant material was fed to the pigs, only 10% of its N got via their excreta into the rice plants. There were losses as gases and by leaching. Also, 24% of the N in the pig's fodder was retained in their growing bodies—which is not a waste if you want pork or bacon.

Leaching of nitrate

How fertilizers affect nitrate leaching

In Box 4.5 one of the disadvantages listed for inorganic N fertilizer is leaching of nitrate into rivers, lakes and other human water supplies. But if we use organic manures instead, will nitrate leaching be less? Some of the best relevant data come from measurements made from 1878 to 1881 in the Rothamsted long-term winter wheat plots (Table 4.7). The farmyard manure added each year contained about 225 kg of N, substantially more than the inorganic fertilizer, but it maintained a similar grain yield

Table 4.7 Concentration of nitrate in water from drains under winter wheat plots at Rothamsted, England. The figures are mean concentrations of NO_3^- (expressed as mg N l^{-1}) for four periods and for the whole year

Fertilizer or manure added each year	From time of spring fertilizer addition to 31 May	From 1 June to harvest	From harvest to autumn sowing	From autumn sowing to spring fertilizing	Whole year
None	3	0	5	5	4
Inorganic P, K, Mg (but no N)	3	0	5	6	4
Farmyard manure, 35 tonnes ha^{-1} yr^{-1}	4	1	6	10	8
Inorganic N, 96 kg ha^{-1} yr^{-1} (+ PKMg)					
As NO_3^- in spring	50	9	15	8	12
As NH_4^+ in spring	27	1	7	5	7
As NH_4^+ in autumn	7	3	8	28	19

Data from Cooke (1976).

at the time these measurements were made (Johnston 1969). When studying these results we can bear in mind that the World Health Organization's recommended maximum concentration for nitrate in drinking water is 50 mg l^{-1}, although the USA has adopted 45 mg l^{-1}. Expressed as milligrams of N (rather than NO_3^-) per litre, these are 11 and 10 mg l^{-1}. So, any figure in Table 4.7 that is about 10 or higher is cause for concern. The results show several points of interest.

1 All plots showed marked variation through the year. These relate to the time of fertilizer application, and also to the stage of development of the wheat crop: nitrate losses tend to be low when the root system is well developed and active, in the spring and summer (column 2).

2 Plots that had received no nitrogen lost the least nitrate by leaching. Whether it is ever realistically possible to grow crops without any N addition, as inorganic fertilizer or organic manure, will be discussed later.

3 Inorganic N fertilizer resulted in nitrate concentrations well above the recommended safe limit (10 mg l^{-1}) during the period just after fertilizer was applied (and, in the case of nitrate fertilizer, also later on). Ammonium fertilizer resulted in less nitrate leaching than did nitrate fertilizer applied at the same time of year. There was probably some loss of ammonium by leaching (unfortunately that was not measured), but ammonium is less readily leached than nitrate.

4 Farmyard manure resulted in nitrate losses higher than those from unfertilized plots, especially in winter, but in general lower than from inorganic N fertilizer.

These results, and others from more recent experiments, show that nitrogen input, whether as inorganic fertilizer or organic manure, increases nitrate losses in leaching. Organic manure was in this experiment able to maintain as high a yield as inorganic fertilizer, but gave lower seasonal nitrate loss. However, organic manure does still increase nitrate loss, so it cannot be guaranteed as an automatic solution to the problem at all sites. One message from Table 4.7 and other research is that most of the nitrate loss occurs at times when there is no crop present, or when the crop is poorly developed. Control of nitrate loss can therefore be helped by:

1 reducing the length of time between harvesting one crop and sowing the next;

2 promoting the early development of a dense root system;

3 adding the nutrients at several times throughout the season, in amounts related to plant demand and uptake ability.

Method (3) can be achieved with inorganic fertilizers but less easily with most organic manures, which cannot be applied when the crop is well grown. These suggestions are not new: they have already been incorporated into farming practice in the developed world. Earlier sowing of autumn cereals has become widespread. An alternative way to reduce the length of time the soil is bare is to sow a cover crop soon after harvest, a species which develops quickly, takes up any available nutrients, and which can be ploughed in when the next real crop is sown.

Nitrate loss from forest

We should also consider whether leaching of nitrate from forests could play a significant part in their N balance. Useful information on this has come from long-term studies on the Hubbard Brook Forest in New Hampshire, in the northeastern USA (Bormann & Likens 1979; Likens & Bormann 1995). The original mixed deciduous forest was clear-felled in 1910–19 and left to regenerate naturally. The principal tree species are sugar maple (*Acer saccharum*), beech (*Fagus grandifolia*) and birch (*Betula alleghaniensis*). Several adjacent valleys (watersheds) were studied from the 1960s onwards. Because of the impermeable rock, the only way that dissolved substances can leave each watershed is in the outflow

Table 4.8 Nitrogen balance of watersheds in Hubbard Brook Forest, New Hampshire, USA. For Watersheds 5 and 2 the 10-yr period started with clear-felling and continued into the time of regrowth

	kg N ha^{-1} during 10 years		
	Input	Losses	
Management system	In rain + dry deposition	In tree trunks	Leached into stream
Felled trees removed, natural regrowth allowed (Watershed 5)	207	138	67
Felled trees left, regrowth prevented for 3 years (Watershed 2)	207	135*	504
Uncut control (Watershed 6)	207	–	14

* In tree trunks of Watershed 6 at that time.

Data from Pardo *et al.* (1995), Whittaker *et al.* (1979).

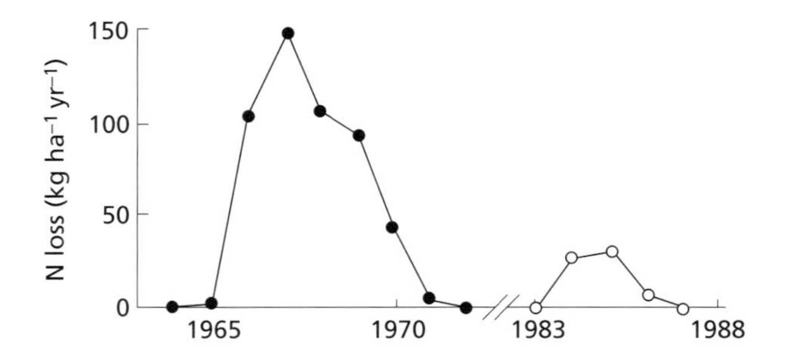

Fig. 4.9 Loss of dissolved inorganic nitrogen (in NO_3^- + NH_4^+) per year in stream water from two watersheds in Hubbard Brook experimental forest, New Hampshire. ● Watershed 2, clear-felled 1965–66, regrowth prevented for 3 years by herbicide, regrowth started 1969. ○, Watershed 5, clear-felled 1983–84, regrowth started immediately. From Pardo *et al.* (1995). Reproduced with kind permission from Kluwer Academic Publishers.

stream. The leaching loss from Watershed 6, which was left uncut after 1910–19, varied between years: during 1964–91 the highest annual loss was 8 kg ha^{-1}, but the loss was usually below 3, and over the 10 years of Table 4.8 averaged 1.4 (Pardo *et al.* 1995). At age 50 years the N content of the tree trunks was 135 kg ha^{-1}, an average uptake rate of 2.7 kg ha^{-1} $year^{-1}$. So the leaching losses were large enough to have an impact on the N balance of the uncut forest. These losses include some ammonium-N, but far more as nitrate.

In two of the watersheds the forest was clear-felled. From one of them (Watershed 5) the felled trees were removed and then it was left for natural regeneration to take place. In the other (Watershed 2) the felled trees were left lying, and herbicide was applied for 3 years to prevent any regrowth of the vegetation; after that regrowth was allowed. Figure 4.9 shows the loss in stream water (mainly as nitrate) during each year. In Watershed 5, where regrowth was allowed at once, the loss in the stream was higher for 2 years and then fell back as the vegetation regrew. In Watershed 2, during the 3 years when regrowth was prevented the losses were much higher, and only after regrowth was permitted did they begin to fall back. Table 4.8 shows N inputs and losses during a 10-year period, starting with the year of felling. When regrowth was prevented for 3 years (Watershed 2) the loss in leaching over 10 years was several times larger than the N content of the tree trunks. If regrowth was allowed at once (Watershed 5) the N loss was still substantially more than the loss from the uncut control watershed, but was less than the loss in tree trunks removed. Input–leaching loss (= 207–67) for this watershed would be just enough to balance the removal in tree trunks. There were probably also losses as gases, but on the other hand the trees had about 70 years to accumulate their N content. The message from this experiment is that rapid re-establishment of growing trees is important. If this is delayed for even a few years the resulting N loss could be serious. Chapter 7 gives more information about tree regeneration after felling.

Nitrogen fixation

N-fixer organisms

The major alternative to fertilizer as a nitrogen input is biological nitrogen fixation, which can be carried out by a range of bacteria, actinomycetes and cyanobacteria. Within the legume family are many plant species that have a symbiotic association with N-fixing *Rhizobium* and *Bradyrhizobium* bacteria. Useful legumes include grain legumes grown for their seeds (e.g. beans), forage legumes grown for animals (e.g. clover), and some trees (e.g. black locust and *Acacia* spp.). There are also some trees and shrubs in several other families that can form N-fixing associations with the actinomycete *Frankia* (e.g. alder and *Casuarina* spp.). Within each of these groups N fixation rates in the field up to about 400 kg ha^{-1} $year^{-1}$ have been reported, though rates within the range 50–200 are commoner (Peoples *et al.* 1995). So, they have the ability to provide enough N to balance normal removals in crop and animal produce (Table 4.6). N fixation is also carried out by non-symbiotic cyanobacteria in water and on soil surfaces, and by heterotrophic bacteria using dead plant material or root exudates as their energy source. Unfortunately there are still no satisfactory measurements of rates of non-symbiotic N fixation under field conditions. Where crops have been grown for decades without any fertilizer or manure, N balance calculations often indicate that there must be an unmeasured N input, and this has been attributed to non-symbiotic fixation (e.g. in rice paddy fields; Kundu & Ladha 1995). However, the N balance calculations do not usually include reliable figures for inputs and losses of other gases (N-oxides, ammonia), so the contribution of non-symbiotic fixation to crops remains uncertain.

Among N-fixer tree species there are a few that have been exploited for timber which grow in tropical forests in Africa (e.g. *Milbraediodron excelsum*) and South America (e.g. *Melanoxylon brauna*) (Sprent 1995). However, no major timber tree of temperate regions is an N-fixer. There may be N inputs to forests from fixation by non-symbionts or by symbionts of non-exploited plant species, but good quantitative information on this is lacking. Removal of N in tree trunks, when averaged over the whole life of the tree, is small compared with removal in many food products (Table 4.6), and may be covered by inputs in rain and indry deposition (Table 4.8). However, many types of plantation forestry have not been going on for long enough for us to be sure that they are sustainable for N and other nutrients on the timescale of centuries.

How N-fixation might be increased

Box 4.6 shows ways in which biological N fixation is currently used to provide N for crops. Almost all of them involve legumes and *Rhizobium*. There is considerable scope for increasing the N-fixing ability of many legume crops (Box 4.6 section 2). The range of effectiveness of different bacterial strains is wide, as shown, for example, by a field experiment in which the same variety of soybean was inoculated with 20 different strains of *Bradyrhizobium japonicum*. N fixed ranged from 60 to 200 kg ha^{-1} (Hardarson 1993). Varieties of the same crop can also differ

Box 4.6. Ways of using nitrogen fixation to provide nitrogen input for crops.

(1) Methods already in use.

1 Grow legumes as food crops (e.g. peas, beans, soybeans). Limitation: seed yields lower than cereals; restriction on our range of foods.

2 Use legume to supply N to other crops. Disadvantage: less space for desired crop.

Mixed cropping: growing a leguminous crop with a non-leguminous one, usually as alternate rows. The legume can be a tree species, used for firewood ('alley cropping').

Rotation: grow a leguminous species every few years and leave the residue.

Green manure: harvest plant material from a sown legume or from grassland containing legumes, and apply it to fields where the crop will be grown.

Animal manure: graze domestic animals on pasture containing legumes, then pen them on the field where the crops will be grown; or feed domestic animals indoors with hay containing legumes, then collect their dung and urine to apply to the fields.

(2) Ways to increase N input

1 Inoculation with *Rhizobium* strains that form more nodules and fix more N per nodule (Hardarson 1993).

2 Breed for legume crop varieties with higher N fixation, in association with *Rhizobium* (Herridge & Danso 1995).

3 Improve transfer of N from microbial fixer species to crop species (Sanginga *et al.* 1995).

(3) Possibilities for the future.

1 Inoculating waste plant material such as straw with a suitable mixture of microbial species can result in substantial N fixation using cellulose as energy source (Lynch & Harper 1985; Roper & Ladha 1995).

2 N-fixing bacterial species have been found in tissues of sugar-cane, maize and rice. Whether they can contribute significantly to the N requirements of the plant is still uncertain (Boddey *et al.* 1991, 1995).

3 It has been suggested that genetic engineering techniques could allow a *Rhizobium*-cereal symbiosis to be produced, or even the insertion of a functional N-fixing gene into the cereal plant itself. Whether this will prove possible is not known.

Further information: Ladha (1995).

substantially in the amount of N fixed. When 20 varieties of common bean (*Phaseolus vulgaris*) were grown at a site in Mexico, all inoculated with the same *Rhizobium*, the N fixed ranged from 7 to 108 kg ha^{-1}. In a similar experiment in Austria the range was 25–165 kg ha^{-1} (Hardarson *et al.* 1993). So there is plenty of scope for selection and breeding.

Among the three 'possibilities for the future' (Box 4.6 section 3), the first can be made to work in experimental conditions but is not yet in practical use. There are very few microbial species that can fix N and also use cellulose as their main energy and C source. So we need a combination of a species that can degrade cellulose to simpler, more soluble compounds (e.g. sugars) and another that can use these products and also fix N. Most cellulose degraders are aerobes, whereas N fixation requires strictly limited oxygen abundance. So N fixation using straw as the energy source requires the right combination of microbial species and the right degree of waterlogging (Roper & Ladha 1995). It has been suggested that inoculation of the crop rhizosphere with N-fixer microbes could provide substantial N to the crop. However, the amount of carbon substrate available in the rhizosphere is probably only sufficient to support fixation of a few kilograms of N per hectare per year (Giller & Day 1985).

The final suggestion in Box 4.6, giving non-legume crops N-fixing ability, of their own or via a symbiont, would be a great prize and has often been discussed. The legume root nodule is a sophisticated structure which, among other things, controls the supply of oxygen to the *Rhizobia* within. Its formation involves complex signalling between plant and microbe, involving a suite of genes in each (de Bruijn *et al.* 1995). So, producing a cereal plant that can form nodules within which N-fixing bacteria can operate efficiently is a major challenge to genetic engineering. Whether that goal can be achieved remains to be seen.

Food production without phosphorus fertilizer

The message from the previous section is that nitrogen input from the atmosphere by fixation can be high, relative to N removals in food, and there are opportunities for increasing present rates of input. For phosphorus there is no corresponding opportunity for large non-fertilizer input. Natural P inputs in rain, in dry deposition and by weathering of rock are much lower than removal in most foods (Tables 4.5 and 4.6). As explained earlier, the P deposited in particles and dissolved in rain comes almost entirely from land sources, some near the site of deposition. To a first approximation we can take these depositions as balancing losses by blowing away of soil dust, plant materials (including pollen) and smoke. So the only real net input is from weathering of rock material.

Rate of P release by weathering

The rate of P input from weathering is difficult to estimate, and figures are available from only a few sites in the world. One source of data is to compare soils in a *chronosequence*, meaning sites where similar fine rock material has been deposited at different times and has then been able to weather and form soil under similar conditions over thousands of years. Examples are freshly ground material deposited by a glacier, and volcanic lava or ash. Data from these and other sources indicate rates of P release by weathering ranging from 0.05 to 1.0 kg ha^{-1} $year^{-1}$ (Newman 1995; Crews *et al.* 1995). Here I take 0.3 kg ha^{-1} $year^{-1}$ as a representative rate. To illustrate the difficulty of supporting sustainable food production by this input alone, suppose that all the people of the world depended

Table 4.9 Phosphorus balance of farming systems of the past which grew cereal crops without using any P fertilizer

Where	When	Grain yield ($t\ ha^{-1}\ yr^{-1}$)	P removed in grain ($kg\ ha^{-1}\ yr^{-1}$)	P input ($kg\ ha^{-1}\ yr^{-1}$)	P source	Notes
Mid-west of USA and Canada	19th–20th century	0.7–1.7	1.1–4.3	3.2	Soil organic matter	1
England	1320–40	0.8	0.9	(0.3)	Weathering	2
China	1929–33	1.3	1.0	(0.3)	Weathering	3
Egypt	2700–323 BC	1.0–2.0	3–6	total 24 dis 1.2 org 2.0	Nile water	4

Yield is per ha under crop that year, but P removed and P input are per ha of whole farm. Data from Newman (1997). Notes: (1) Crops wheat and corn. P input is mean of 8 sites where organic P decline was measured over 30–90 years. (2) One farm in Oxfordshire, growing wheat, oats and barley. Assumes no nutrients in human excreta returned to fields. (3) Mean of many farms, growing wheat, corn, millet and kaoliang, without irrigation. Assumes human excreta were used as manure on fields. (4) Crops mainly wheat and barley. dis = dissolved inorganic P, org = P in suspended organic matter. Remainder in suspended silt.

for their food energy on cereals whose grain contained about 3 mg P g^{-1} (see Table 4.6); 0.3 kg P ha^{-1} $year^{-1}$ would support a harvest of 0.1 tonne ha^{-1} $year^{-1}$. At this productivity, about 3 ha would be needed to feed each person. However, for the present population of the world to be supported by the present arable area requires each 3 ha to grow on average enough food for 12 people (see Chapter 2).

Cereal yields substantially higher than 0.1 tonne ha^{-1} $year^{-1}$ have in fact been produced without the use of P fertilizer: yields of about 1 tonne ha^{-1} $year^{-1}$ have been achieved in various parts of the world (Table 4.9). The key questions are: were these yields sustainable long term, and, if so, how was that possible? The large amounts of P stored in most soils in slowly available forms can support crop production for decades; but if it is not replenished from weathering or other sources, soil fertility and productivity must eventually decline.

P balances: US mid-west, 19th century

I have calculated P balances for four farming systems, all growing cereals without any inorganic P fertilizer, but in other respects very different from each other. Table 4.9 summarizes the results. During the 19th century much of the prairie land of the American mid-West was ploughed and used to grow corn and wheat. As the soil organic matter decomposed and decreased year by year (Fig. 4.5(b)) it released N and P into inorganic form, and this must have been the main source of P to the crops. The amount released during the early decades would have been about enough to balance removal in the crop (Table 4.9). However, the rate of organic matter breakdown decreased over time, so the amount of available P supplied per year would decrease. Figure 4.5(b) suggests that this system could operate for about a century, but clearly not for ever.

England, 14th century

China, early 20th century

The next two examples in Table 4.9 are systems that had continued for a very long time, and one might therefore assume they were sustainable—but my calculations indicate that they were not. For both of them, the figure for P removed in grain is not the total P in the grain harvested, because some of that was fed to animals whose excreta went back on to the fields; also some seed was retained to sow in the following year. In the 14th century English farm some of the area was left fallow each year and some was pasture and hay-meadow. The P from weathering in these areas is assumed to be an input to the farm. In China almost all the farm was under crop every year. Another difference is that in China it has been customary to use human excreta as fertilizer, thereby recycling P from food eaten by the farm family. This may have happened to some extent also in England, but I have not included it in the calculation.

The figures show that in 14th century England and early 20th century China the P removed in the crop was substantially more than the 'typical' amount of P input by weathering, suggesting that there would be a long-term decline of P in the soil. This would lead ultimately to decreasing crop yields—and the yield of wheat on the English farm did indeed decline during the first half of that century (Newman 1997). In both these countries the human population had been increasing for some time, and it seems quite possible that farming systems which had previously been sustainable had become unsustainable because of increased demand for food. In China the problem was later cured by using fertilizer. In England it was 'cured' by the Black Death (bubonic plague), which killed off more than a third of the population of England and western Europe in 1348–49, thereby reducing the demand for food.

Obtaining P from a large area

So, how could farming in the past be sustainable for phosphorus? Basically, by using the P from weathering over a larger area than was being cropped. For example, if this uncropped area is about 10 times the cropped area, and weathering releases 0.3 kg P ha^{-1} $year^{-1}$, it should provide enough P to support a harvest of 1 tonne of cereal per hectare per year from the cropped area (removing 3 kg P in the grains). In shifting cultivation, during the fallow phase (e.g. regenerating forest) P can be released by weathering, and then be taken up by the vegetation or remain in available form in the soil. When the farmer returns to use the site, burning the vegetation converts most of the P in the plants to an available form in ash. However, very old, highly weathered soils, which are common in wet tropical areas, are likely to weather and release P only very slowly. The classic work of Nye and Greenland (1960) on shifting cultivation in West Africa suggested that P availability may be more critical than the total amount of P in the soil: ash raises the pH, which in many tropical soils increases the availability of P (Nye & Greenland 1960; Jordan 1989). One reason for abandoning crop growth on a patch after a few years is that the pH has fallen again; other reasons include weed invasion and increasing risk of soil erosion.

Leaving fields as bare fallow for 1 year between crops has been a widespread practice, for example in the medieval English farm of Table 4.9, and the 20th century American farm of Fig. 4.8. This provides some additional P to the following crop, but a single year of fallow every 2 or 3 years would usually not be enough to make the system sustainable: it was not in the English farm.

Animals as P collectors

Another system, widely used in the past, was to transfer P from a large area to a smaller cropped area in the form of manure. Sometimes vegetation or turfs were collected for use as the manure, but often the large areas involved made this impractical and so animals were used to harvest and transfer the P. Much of the P in what herbivorous mammals eat goes into their faeces, little into the urine. The requirement, then, is that the animals eat in the pastureland but deposit much of their faeces on the arable. The time lapse between material being eaten by a cow or sheep and its appearance in faeces ranges from about 24 h to 5 days (Warner 1981). Cattle and sheep eat mainly during daylight, but defecate fairly evenly throughout the 24 hours (Castle *et al.* 1950; Hardison *et al.* 1956; Church 1988). Therefore, an efficient way to use these animals as P transporters is to have them grazing in the 'outfield' by day, but at night fold them either on the arable land or in a confined area from which their dung can later be collected for use as manure. This daily movement of the animals is time-consuming, especially if the grazing land is far away from the village, but it often had the additional aim of protecting the animals from predators such as wolves. This system has been practised in many countries over many centuries. In Europe it was widespread in the past (Webb 1998), and is still practised locally, e.g. in parts of Italy where wolves still occur (Boitani 1992). It is described in classical Indian literature and still carries on there today (Sopher 1980). The animals and cropland may belong to different people. For example, in parts of Nigeria cattle-owning Fulani people trade dung with Hausa crop-growers, receiving grain and other foods in exchange (Hill 1982). It is usually not known how large the grazed areas were relative to the arable area, since they were often common land, not owned by anyone. In one small Swedish village in the 18th century, of the area available 9.6 ha was carrying a crop each year, 24.2 ha was fallow, 46.6 ha hay-meadow and about 51 ha rough grazing (Olsson 1988). So only about 7% of the total area was bearing a crop at any one time.

A serious disadvantage of all these systems for obtaining P is the large total area required to support the crop production. The 18th century farm in Sweden grew 0.8 tonnes year^{-1} per hectare of cropped area, but only about 0.06 tonnes ha^{-1} year^{-1} if calculated per total used area. The uncropped area may produce other useful things, such as animals or timber, but compared with modern farming the system can support only a low number of people per total area (see Table 2.3).

P from irrigation water

An alternative source of P is irrigation water, which will contain some dissolved P compounds and also suspended organic matter and silt which will contain P. In some areas farming in the past involved complex irrigation systems, e.g. in the Tigris–Euphrates 'fertile crescent' and the Indus Valley; rice paddy irrigation continues today in China and southeast Asia. Other areas received natural flooding: an example is the fields in Egypt beside the River Nile, where food production continued for more than 5000 years without any input except from the annual flooding. This supported a highly organized civilization which left us the pyramids and sophisticated artwork; presumably the farming system was sustainable. Table 4.9 shows a P balance for this farmland in the time of the pharaohs. Each year, until the Aswan High Dam was built in the 1960s (AD), rain falling in Ethiopia caused the level of the Nile in Egypt to rise greatly in late summer and to flood on to the land on each side, carrying dissolved phosphate and suspended silt washed off the mountains of Ethiopia. Of the yields reported by ancient documents (Table 4.9), the lower end of the range could have been supported by the dissolved P plus organic P, most of which would certainly have been available to the crop plants. To support the higher yields would require a small proportion of the P in the silt also to become available, which seems likely. So the civilization of ancient Egypt was made possible by erosion of Ethiopia.

Using human excreta as manure

Where human faeces goes

In most nutrient balances of farms (e.g. Table 4.5) it is assumed that all the nutrients in the harvested grain are lost from the farm, but this is not always the case. Much of the P in what people eat ends up in their faeces, and there are ways of returning this to the fields. In shifting cultivation one function of the fallow may be to receive human faeces, so that it is not lost from the system. In much of the Indian subcontinent the custom has been for villagers to defecate on the fields: health workers have expressed frustration that many are reluctant to give up this habit in favour of using latrines (Pacey 1978). In China, Japan and much of southeast Asia there has been a custom of collecting human excreta to apply to fields or fish ponds. This included exporting it from towns, although that practice has decreased greatly in recent decades (Pacey 1978). McLaughlin (1971) gives an uninhibited account of what happened to human excreta in London, from the Middle Ages onwards. Until the installation of sewers in the 19th century, much of it rotted in the streets; however, some was exported by cart or boat to the countryside and was used as manure. A serious disadvantage of applying raw human faeces to farmland is the spread of disease. The survey of China in 1929–33 which provided data in Table 4.9 also recorded causes of death in the farming population. Among the greatest killers were three faecal-borne diseases, dysentery, typhoid and cholera.

Sewage sludge as a manure?

Modern treatment of waterborne sewage produces a solid material called sludge, which can be used as manure. Box 4.7 summarizes some information about this. The anaerobic digestion process involved in sludge production lasts many days, during which complex chemical reactions are carried out by numerous bacterial species. The temperature depends on the management system, but is commonly within the range

Box 4.7. Sewage sludge and its use as manure.

During sewage treatment, solid material is separated by sedimentation, and more is later produced by microbial growth (see Chapter 9). After anaerobic microbial activity and then drying, this results in sewage sludge. It includes cellulose, organic N compounds, microbial tissue and mineral material (e.g. silt).

Common concentrations of elements (mg g^{-1} dry wt)

N	10–80
P	5–40
K	up to 7
Cd	0.002–1.5
Cu	0.2–8
Pb	0.05–4
Ni	0.02–5
Zn	0.6–20

Advantages in use as manure include:
1 Much of the P (often about half) is in water-soluble forms readily available to plants. However, only about 10–20% of the N is in inorganic, plant-available forms.
2 The organic matter content is high, which has favourable long-term effects on soil structure (see erosion section in this chapter).

Disadvantages in use as manure:
1 Contains parasitic animals, bacteria and viruses that are pathogenic to humans (though far fewer than in raw sewage).
2 Contains heavy metals, in varying amounts (see above), which can accumulate in soil. Only a few plant species can evolve tolerance to high concentrations of heavy metals (see Chapter 11). If plants are able to survive and take up the heavy metals, these will be passed on to animals that eat them and hence to their predators (see Chapter 9).
3 Much human refuse is produced in towns, far from the farms to be fertilized. Because the concentrations of N and P in sludge are much lower than in inorganic fertilizer (about one-tenth as much), the cost of transport per unit of N and P is much greater.

Further information: Berglund, Davis & L'Hermite (1984); Gray (1989); Harrison (1996); Mason (1996)

30–60°C (Harrison 1996; Ford 1993). During this treatment many of the organisms in the sewage that could infect humans are killed, though a few may remain. After drying, the resulting sludge should be safe, from the disease point of view, to apply to crops such as cereals and to hay-meadows, though probably not to green vegetables or to actively used pastureland. The main safety concern is heavy metals: if sludge is applied to soil, especially over many years, heavy metals can reach soil concentrations that are toxic to plants, or concentrations in the plants that are toxic to animals or people that eat them (Gray 1989).

Conclusions

- Rates of erosion vary greatly. Only the slowest are balanced by current soil formation.
- Converting forest or grassland to arable often increases the rate of erosion, though not always.
- Some present-day landscapes have been affected by past erosion caused by human activities.
- Erosion from farmland can often be substantially reduced by altering management methods.
- Soil aggregates make soil less prone to erosion. The particles forming aggregates are held together by plant roots, by fungal hyphae and by organic and inorganic cementing agents.
- Among the major essential elements, N and S have opportunities for wide dispersal in gaseous forms, whereas available K, Ca, Mg and P come mainly from weathering of rock.
- Inorganic fertilizers have disadvantages, including sometimes rapid losses through leaching. However, nitrate leaching can also be rapid even when no fertilizer has been used, e.g. after felling of forest.
- Nitrogen fixation can provide enough N input to support rapid crop growth.
- In the absence of P fertilizer, crop yields in the past have often depended on P removed from other land, e.g. in animal manure or irrigation water.
- Sewage sludge can provide P for crops but presents dangers, including human diseases and toxic heavy metals.

Further reading

Soil, general texts:
White (1997)
Brady & Weil (1999)

Weathering, soil formation:
Bland & Rolls (1998)

Erosion:
Morgan (1995)
Ives & Messerli (1989)

Soil organic matter:
Wild (1988) Chapter 20

Soil structure, aggregates:
Oades (1993)

Soil chemistry, nutrients in soil:
Cresser, Killham & Edwards (1993)

Plant mineral nutrition:
Marschner (1995)

Nutrient cycling:
Berner & Berner (1996)
Butcher *et al.* (1992)

Chapter 5: Fish from the Sea

Questions

- Why does fish production vary so much between different parts of the oceans? Why is production from most of the oceans so low?
- Will removal of fish always cause the fish population to decline?
- Or can ocean fish be exploited sustainably?
- If so, how can we decide how much fish we can catch this year without reducing the stock for the future?
- What is the best sort of rule to ensure that the boats in an area do not catch too much?
- Some major fish stocks have collapsed. Whose fault was it—the scientists', the politicians' or the fishermen's?
- Would we do better to get our fish by fish farming? How can this best be done?

Basic science

- Primary productivity of the oceans and what limits it.
- Food chains leading from algae to fish.
- How to determine the number of fish of a particular species in an area of ocean.
- Fish population dynamics. How population numbers change, and what causes that. Relations between number of adults (stock) and number of new young fish (recruitment). Why this varies from year to year.
- What sorts of diet farmed fish will eat and thrive on.

This chapter is about hunting wild animals. Thousands of years ago, everyone in the world obtained their food by hunting wild animals and collecting from wild plants. Today most food comes from farms and domestic animals, but our main way of getting food from the oceans is still by catching wild fish. This chapter concentrates on fishing for bony fish ('finfish', teleosts) in salt waters—shallow coastal waters as well as the deep open oceans.

Can ocean fishing be sustainable?

The basic questions for this chapter are whether the fish stocks of the seas can be exploited in a sustainable manner, and if so, how that can be done. In other words, we want to catch fish this year, but in a way that allows us to continue to catch fish next year and hopefully for ever. A distinguished 19th-century biologist, Thomas H. Huxley, said in 1883: 'I believe that all the great sea fisheries are inexhaustible' (quoted by King 1995). This was not a ridiculous belief: the world's oceans are very large and our boats seem small and far apart, but his belief has proved to be incorrect. The most dramatic demonstrations of this are the collapses of major fish stocks that have occurred. An example is Newfoundland cod. Fishing for cod off the coasts of Newfoundland has been going on for about 500 years, and it became crucial for the local economy (Hutchings & Myers 1994). Yet in 1992–93 the Canadian government banned cod fishing until further notice, because stocks had almost disappeared. This has been described as 'one of the worst social and economic nightmares in the nation's history' (Walters & Maguire 1996). The causes of this collapse will be considered in more detail later, but there is little doubt that the primary cause was overfishing (Hutchings & Myers 1994). So, this chapter has to consider the possibility that in future ocean fishing will not supply our needs and we shall have to depend more on farmed fish (aquaculture).

Although some fish stocks have collapsed, ocean fishing as a whole has not. Figure 5.1 shows the rise in total catch from the oceans since 1950. There was a plateau during the 1970s and there have been some short-term fluctuations which do not show up on the graph, since its data points are for every fifth year, but the basic trend has been a continued rise. The world catch in 1995 was about five times that in 1950. These totals include species that are not finfish—crustaceans such as lobster and prawn, molluscs such as oyster and squid. In 1995 finfish were 82% of the total catch from the oceans. Figure 5.1 does not include fish caught from fresh waters, which in 1995 was 21 million tonnes. This chapter is mostly about finfish from the oceans. The energy content of fish is about 1.6 GJ per tonne fresh weight (Pimentel & Pimentel 1979), so the present annual catch of about 90 million tonnes has a food energy content of about 1.4×10^{17} J. Comparing this with the total energy content of human food (see Table 2.2) shows that fish caught from the oceans provide only about 1% of total human food energy. However, fish are usually more valued for their protein than as an energy source. If 1.4×10^{17} J is averaged over the total area of the oceans (see Table 1.1), this comes to 0.004 GJ ha^{-1} $year^{-1}$. This is three orders of magnitude lower than modern animal production per hectare on pasture, and even migratory pastoralists in a semiarid region can obtain substantially higher production (see Table 2.3). This average fish yield conceals great variations, between parts of the ocean with much higher production and vast areas with extremely low production. This uneven productivity is, as we shall see, very important for the management and exploitation of the seas. It leads to the questions: why is the fish production of most of the oceans so low? And why are some regions much higher than the average?

Food yield from the oceans is low

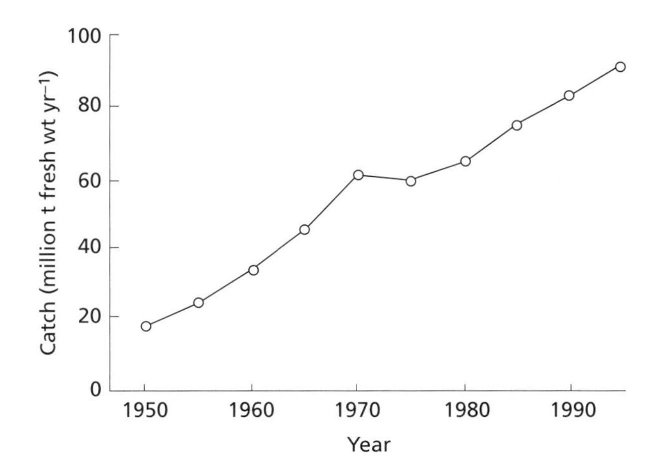

Fig. 5.1 World catch per year from the oceans. Includes crustaceans and molluscs as well as teleost fish. Data from FAO Fishery Statistics Yearbooks.

Productivity and food chains

Animal life in the oceans depends almost entirely on photosynthesis by *phytoplankton*, unicellular cyanobacteria and algae suspended in the water. The *photic* zone, in which light intensity is sufficient to support photosynthesis, is commonly 10–100 m deep. Many animals live within this zone, including commercially important fish species such as herring, salmon and tuna. However, other fish species are *demersal*, i.e. live near the seabed, including cod, sole and haddock. These and the *benthic* animals (which live in the bottom deposit or on its surface) are nevertheless dependent on primary productivity by phytoplankton, which reaches them by downward movement of plankton, larger animals or dead organic matter.

Measuring primary productivity

The primary productivity of phytoplankton has most often been measured by placing a small volume of seawater, with its suspended organisms, in a transparent flask, hanging it at a particular depth in the ocean, and measuring either oxygen production or the incorporation of ^{14}C from dissolved $^{14}CO_2$ into organic matter. Although results from these methods have been very informative, they have the limitation that each measurement is made on one small sample (a few litres) over a short time (usually 24 hours or less). Because the measured rates of production vary greatly, it is difficult to scale them up to mean values for a month or a year and for areas tens or hundreds of kilometres across, which may be important for fish. An alternative is to use data from satellites to estimate the amount of chlorophyll per litre in the surface water (see Box 3.2). This is an important determinant of primary productivity, but not the only relevant factor. Behrenfeld and Falkowski (1997) proposed a model for estimating productivity taking into account, as well as chlorophyll concentration, photosynthetically active radiation reaching the sea surface per day, the temperature of the surface water and the depth of the

Primary productivity: comparing oceans and land

photic zone. By their method, net primary productivity averaged over the whole of the oceans is 2.9 tonnes ha^{-1} $year^{-1}$, which may be compared with some rates of production by vegetation on land given in Table 2.1. It is below the range of production values reported for forests, and near the bottom of the range for savanna and temperate grassland. In energy terms (see Table 2.2) primary production in the oceans is probably about one-third to one-half of the world total, although the oceans occupy nearly three-quarters of the surface area of the globe. So the primary productivity of the oceans does not, on its own, explain why the fish catch from the oceans is so low, compared with animal production on land.

Productivity varies greatly within the oceans

Primary production varies greatly from one part of the oceans to another. Figure 5.2 shows the distribution of primary productivity in the Atlantic Ocean and the southeastern Pacific. It is higher near some coasts (though not all), and in a band across the northern and southern Atlantic. Much of the tropical and subtropical regions of the Atlantic, Pacific and Indian Oceans have productivity in the range 1–3 tonnes ha^{-1} $year^{-1}$, but in the Arctic and Antarctic Oceans it is lower. High primary productivity zones near to coastline have high fish productivity too, and this has important implications for management, as will be explained later.

Mineral nutrients

Mineral nutrients are thought to be the main limiting factors to primary productivity in much of the oceans. Nitrogen and phosphorus are considered the key elements, though there is debate about which of them is the more important (Smith 1984). There is, however, evidence that iron rather than N or P is the limiting nutrient in some areas, including the central equatorial Pacific (see Chapter 2). When phytoplankton die some of their nutrient element content is recycled within the photic zone, e.g. through animals and their excreta, but some is lost downwards in particulate organic matter. If the nutrient status and productivity of the photic zone is to be maintained, these lost nutrients need to be replaced. The higher productivity regions shown in Fig. 5.2 can largely be explained by supply of nutrients to the photic zone. In some shallow coastal waters this happens by input of nutrients from the land, but in most of the oceans it occurs mainly by upwelling (Valiela 1995). This occurs where a deep current meets a coastline and then rises, e.g. off west Africa; and where surface waters diverge and deeper waters rise, as in the equatorial Pacific and the Southern Ocean. The high productivity in the northern Atlantic is due to active turbulent mixing.

If the productivities shown in Fig. 5.2 are compared with values in Table 2.1 we see that much of the ocean has primary productivities towards the low end of the range found on land. However, the low total catch of fish cannot be attributed solely, or even mainly, to low primary productivity. The energy content of fish caught (about 1.4×10^{17} J $year^{-1}$) is about 1/10 000 of the estimated primary production of the oceans. So the main problem is that so little of the energy in the primary production reaches the fish we eat. The main reason for this is long food chains. Because phytoplankton are so small, few fish can eat them directly. The

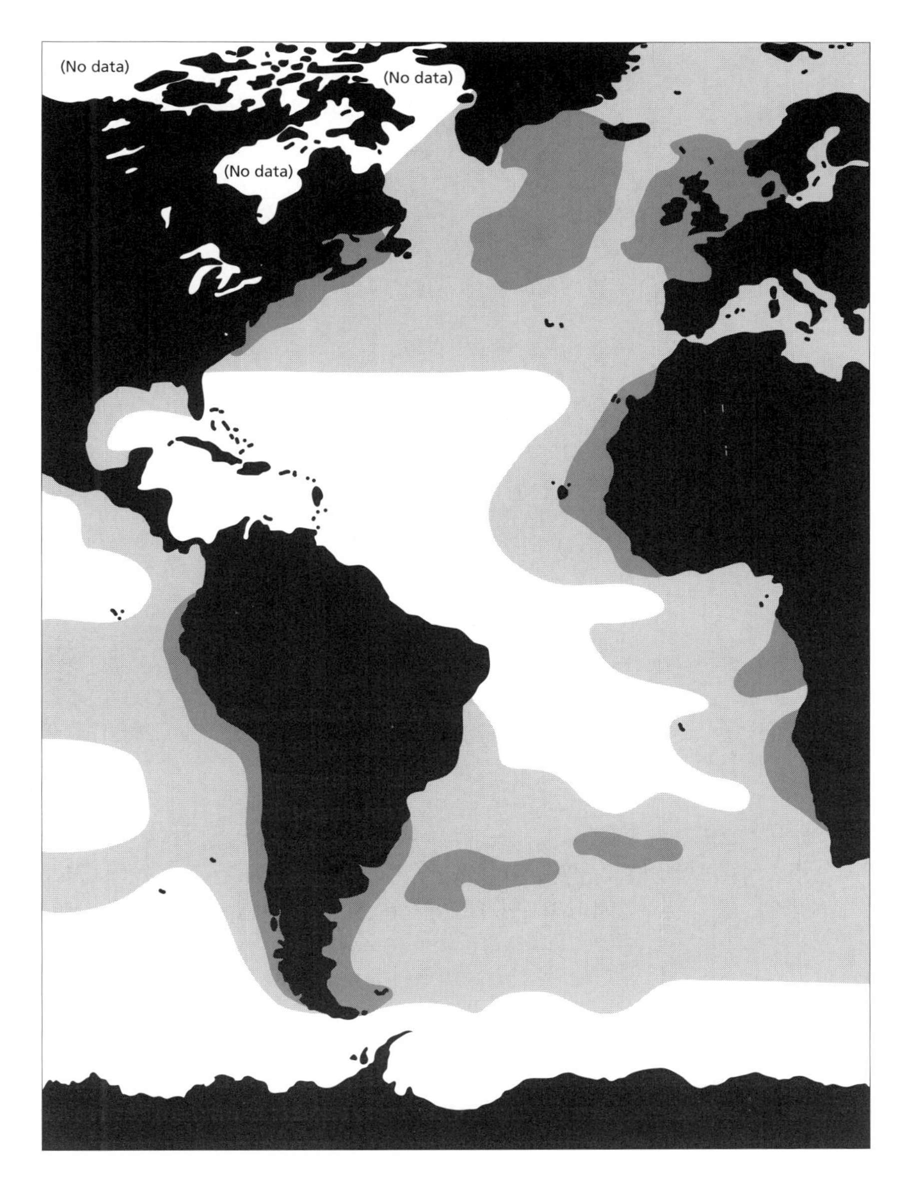

Fig. 5.2 Net primary productivity in the Atlantic and southeast Pacific Oceans, estimated by satellite remote sensing. Productivity in tonnes ha^{-1} $year^{-1}$: dark grey > 6.5; light grey 3.0–6.5; white < 3.0. Simplified from Fig. 9(a) of Behrenfeld & Falkowski (1997).

food we grow on land is either plants or herbivorous animals, but the fish we eat are almost all carnivores.

The sizes of ocean organisms

Box 5.1 gives a simple and widely used classification of ocean species by where they live, whether or not they are photosynthetic and how big they are. Size determines what can eat what, but it has other important effects as well (Fogg 1991; Mann & Lazier 1996). Larger cells sink more rapidly through still water. They also have a reduced ability to acquire

Box 5.1. A simple ecological classification of organisms of the oceans.

Benthic. Living in, or on the surface of, the bottom deposits.
Pelagic. Living in water above the bottom.
Plankton. Small suspended organisms. Although some can swim actively, all are small enough to be much affected by currents.
Nekton. Larger, actively swimming animals. Includes fish.

Classification of plankton by size

Size range (μm)		Main groups involved	
		Photosynthesizers	Heterotrophs
<0.2	Femtoplankton		Viruses
0.2–2	Picoplankton	Cyanobacteria	Bacteria
2–20	Nanoplankton	Flagellates	Flagellates
20–200	Microplankton	Diatoms, desmids	Ciliates
200–2000	Mesoplankton*		Crustaceans
>2000	Macroplankton*		Larvae of fish

* In this size range terminology varies between textbooks, even between the following two.

Further information: Barnes & Mann (1991); Barnes & Hughes (1999).

nutrients, because the unstirred layer immediately around each cell forms a barrier across which the nutrient ions have to diffuse. Some pico- and nanoplankton can partly overcome these problems by being actively motile, or (in some cyanobacteria) by controlling their buoyancy by gas vesicles. But the most favourable places for microphytoplankton (which are the larger phytoplankton) are upwelling regions, where the upward water movement can counteract their sinking and also provide nutrients. In the nutrient-poor regions elsewhere much of the primary productivity is provided by nano- and picophytoplankton.

How many steps in the food chain?

Box 5.2 shows a generalized food web for the oceans. The smaller the primary producers the smaller the animals that eat them, and hence the more links there are in the food chain from them to fish. Phytoplankton release dissolved organic matter, which is probably the main energy source for the very abundant planktonic bacteria. Bacteria themselves, and also the faeces of zooplankton, provide further organic matter which contributes to the *microbial loop*. This adds further steps to the food chain. At each step much of the organic matter taken in by the bacterium or animal is respired to CO_2 and water, only a small proportion being incorporated into its growing tissue. Among a large number of examples from aquatic systems summarized by Pauly and Christensen (1995), the percentage of organic matter transferred from one trophic level to the next was mostly within the range 2–16%, with a mean of 10%. So there are two causes, which reinforce each other, for the great variations in fish production between differ-

Box 5.2. Generalized food web in the oceans.

Only the links leading to fish are shown. In most parts of the oceans all the steps shown here occur to some extent, but their relative importance varies.
For definition of types of plankton see Box 5.1. A major component of each plankton type is given here in brackets, but these are not the only groups involved.
Based on Barnes & Mann (1991), Barnes & Hughes (1999).

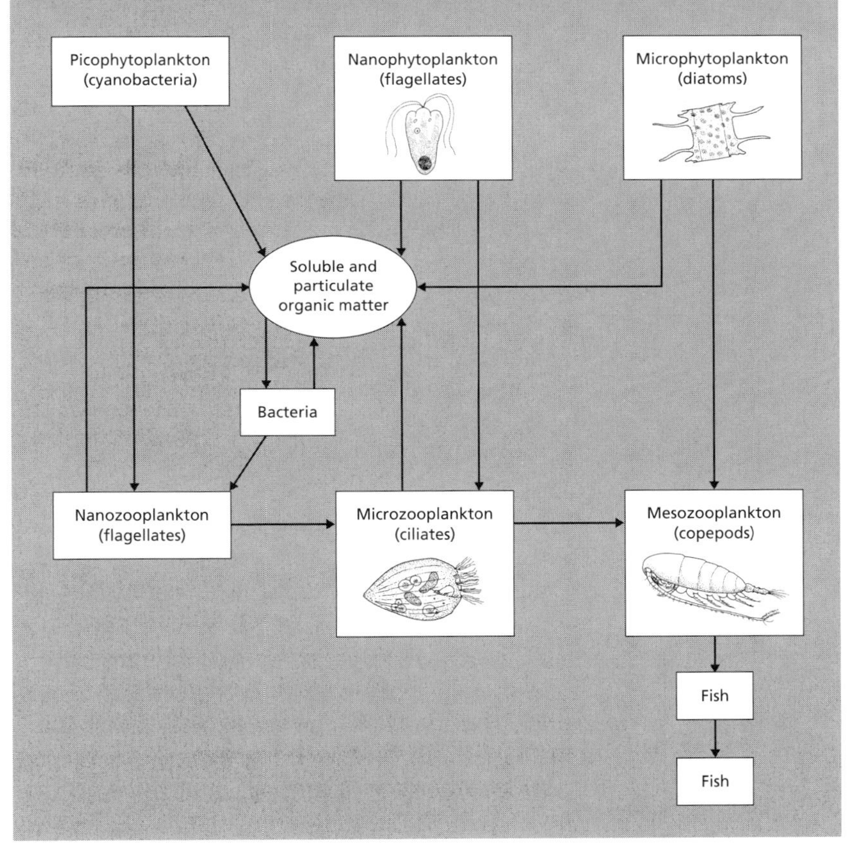

ent parts of the oceans: in regions of nutrient-rich upwelling the primary productivity is higher and it is also provided by larger organisms, so there are fewer steps in the food chain. Although a few fish can use large phytoplankton as food, most are at the third, fourth, fifth or even higher trophic level in the food chain. Their productivity is likely to be several orders of magnitude lower than the primary productivity of the sea where they grow. However, the food chains to fish larvae are shorter than to the adults. Newly hatched larvae are only a few millimetres long and eat smaller food items than the adults. The larvae of some species, e.g. herring and plaice, at first eat phytoplankton, moving on to larger zooplankton as they grow (Grahame 1987). Fish larvae can themselves be eaten by adult fish. The mortality of larvae is high, and this has important implications for the sustainability of fisheries, as we shall see.

Managing fisheries sustainably: predicting how much to catch

Surplus yield

The next question to ask is, can ocean fish ever be exploited sustainably? Will not the removal of fish always reduce the population? This can be put more clearly as an equation. Consider one fish species in one area. The number next year, N_{t+1}, will be

$$N_{t+1} = N_t + R - M - F \tag{5.1}$$

where N_t = the number this year, R = recruitment, i.e. the number of new fish, M = loss by natural mortality, and F = loss by fishing.

Recruitment

The precise meaning of recruitment varies (King 1995): here I take it to mean the number of fish that have newly become large enough to be caught by the fishing fleet. The minimum size of fish that are caught can be controlled by the mesh size of the nets used. Unlike human beings, fish do not stop growing when they reach adulthood, they go on growing throughout life. Figure 5.3 shows an example. Nor do fish have a normal or maximum lifespan. A fish 10 years old has about the same chance of dying of natural causes as a fish 5 years old.

Looking again at Equation 5.1, one might argue that in the absence of fishing the population would presumably be stable from year to year, and therefore R must equal M. If we now cause extra death by fishing, will not the numbers in the stock decline year by year? As we shall see, fish populations are not always stable from year to year. However, putting this aside for the moment, there is a more basic reason why it is possible to remove fish without necessarily causing the population to decrease. Figure 5.4 explains this. Imagine that a few fish arrive in a new area, where the species never occurred before but where food and other requirements are available for it. If no fish are being caught we would expect the population to increase, as in Fig. 5.4(a): at first exponentially, but then, as the fish begin to compete among themselves, following an S-shaped path to a final plateau where the population is limited by food supply or some other requirement. This can alternatively be expressed (part b) as a relationship between population size and rate of growth. If the population is left alone it will increase—i.e. move to the right—until there is no further growth. So there is a range of population sizes where population growth is expected: there is *surplus yield*. This is one of the fundamental ideas of classical fisheries biology, although in fact it is no different in principle from some other biomass/growth relationships (see, for example, Fig. 6.3). It led on to the idea of *maximum sustainable yield*, i.e. that there is a particular stock size at which growth will be fastest and thus the amount we can safely harvest will be largest (King 1995).

Surplus yield

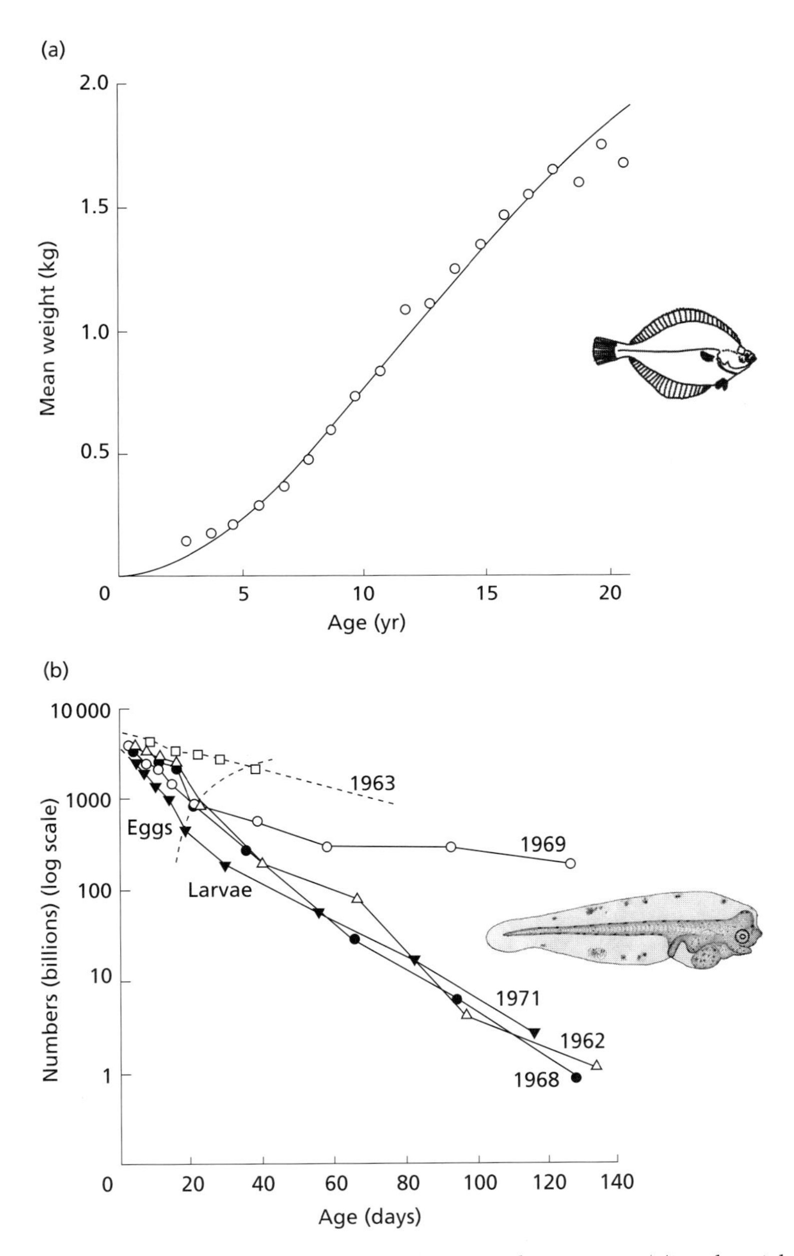

Fig. 5.3 Growth and survival of North Sea plaice, in relation to age. (a) Fresh weight per fish. Mean for 1929–38, from Beverton & Holt (1956). (b) Numbers of eggs and larvae surviving. From Cushing (1981).

Figure 5.4(c) shows two ways in which the surplus yield can be harvested: each year either a set proportion (P) of the stock can be taken, or a set amount or quota (Q). Suppose the stock size is initially *s* on the graph. If fishing is taking the proportion shown by line P, this will remove less than the surplus yield and so the stock size will increase next year. This

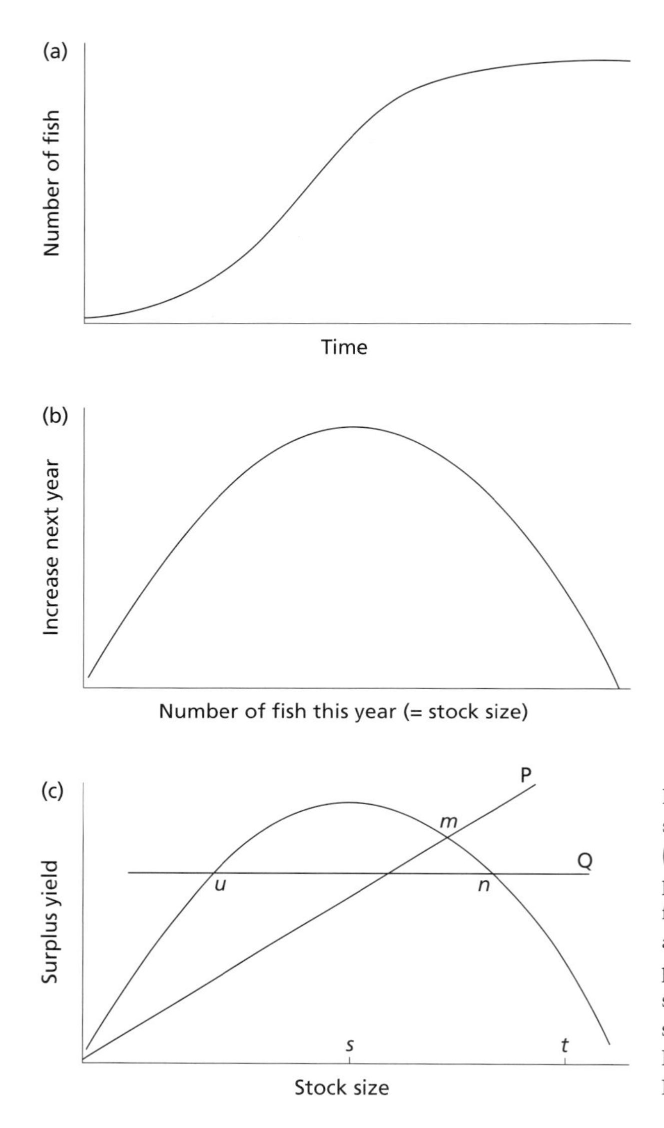

Fig. 5.4 Hypothetical relationships, for a single fish species, between population size (stock size), amount removed by fishing and population growth rate. (a) Population change following arrival of a few fish in a new area, assuming no fishing. (b) Same data as part (a), plotted as growth rate against population size. (c) Curved line taken from part (b); straight lines fish caught, Q = set quota, P = set proportion of stock caught. Points *u*, *m*, *n*, *s*, *t*: see text for explanation.

will continue until we reach point *m*, where removal by fishing just balances surplus yield, so the stock size will remain constant. If the stock size was initially *t*, it will decline to *m* and then stabilize. So *m* is a *stable point*. Suppose, instead, that fishing takes the same set quota each year (line Q). If we start with stock size *s* or *t*, the stock will stabilize at stable point *n*. However, if the stock size ever gets to the left of point *u*, then fishing will each year remove more than the surplus yield, and the stock will decrease each year until it is fished to extinction. So *u* is an *unstable break point*.

Stable and unstable points

Collapse of the Peruvian anchoveta stock

Figure 5.4 is a gross oversimplification of reality, as we shall see. Nevertheless, it presents several ideas that are very important for fisheries management. The danger of taking an approximately set amount of fish

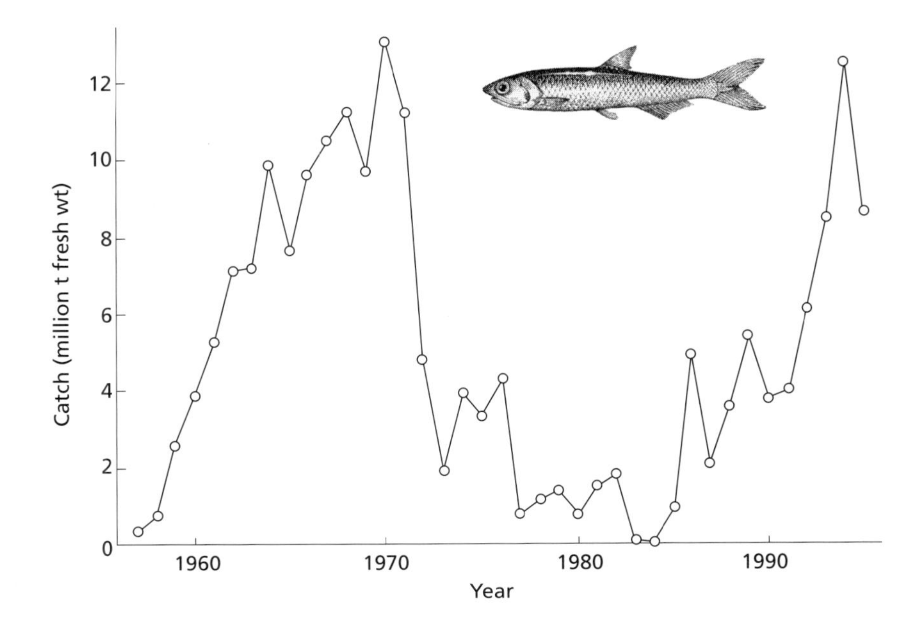

Fig. 5.5 Yearly catches of Peruvian anchoveta from the Pacific Ocean off the coast of Peru and Chile, from 1957 to 1995. Data from FAO Fishery Statistics Yearbooks.

each year can be illustrated by the collapse of the Peruvian anchoveta fishery. The Pacific Ocean off the coast of South America is a region of upwelling with high primary productivity (see Fig. 5.2). Anchoveta eats phytoplankton as well as zooplankton, so its food chain was short and its productivity high (Walsh 1981). Figure 5.5 shows the yearly catches, which increased rapidly after 1957 and by 1962 were contributing more than 10% of the total catch from the world's oceans. Surplus yield models predicted maximum sustainable yield of about 7–10 Mtonne year^{-1}, and the actual catch was near this from 1964 to 1971. This catch might have been sustainable if the weather and ocean currents had remained constant from year to year, but in 1972–73 an El Niño Southern Oscillation Event occurred. This is a complex interaction of atmosphere and oceans that occurs in the Pacific every few years (see Longhurst & Pauly 1987; Mann & Lazier 1996). Off the coast of Peru El Niño causes the upwelling water reaching the surface to be warmer than usual and lower in nutrients. In 1972 this led to extremely poor recruitment of anchoveta. In the absence of fishing the stock might have soon recovered—El Niño had, after all, occurred many times before—but with fishing near or above the maximum sustainable yield the result was a collapse in the stock. In terms of Fig. 5.4(c), the curve moved downwards to below line Q, and by the time the curve moved up again stock size was to the left of point *u*, so the stock could only recover if fishing was greatly reduced. In 1982–83 there was another, more extreme, El Niño event which further reduced the anchoveta stock. The situation was

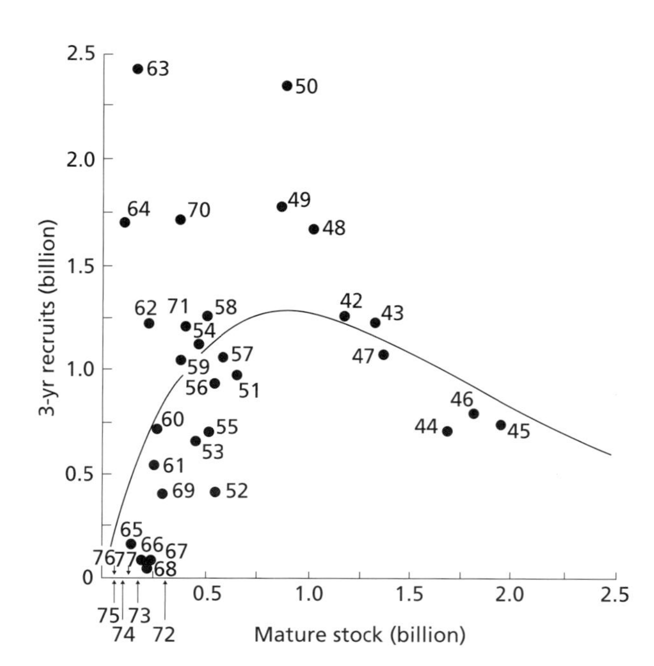

Fig. 5.6 Relationship between number of adult stock and number of new recruits, for cod in the Arcto-Norwegian region of the Atlantic Ocean. Points are for individual years, which are indicated by the numbers. Line is best fit to a model. From Cushing (1981).

. . . but it recovered

complicated by increases in stocks of other species, including sardine. If Fig. 5.5 terminated at 1984, one might perhaps assume that Peruvian anchoveta was essentially extinct but, as the graph shows, within 10 years it recovered, to provide catches as high as those of the 1960s. For further information on the interacting role of fishing and varying environmental factors in the anchoveta collapse, see Longhurst & Pauly (1987) and Mann & Lazier (1996).

Recruitment

Relating recruitment to stock size

If the relationship in Fig. 5.4(b) is to be applied to deciding how much fish of a particular species can be safely caught, we need data for that species relating recruitment each year to stock size. Figure 5.6 shows an example of such data. Much time and effort has been put into collecting such data, for many species in many seas; examples of stock/recruitment graphs are given by Cushing (1981), Pitcher & Hart (1982), Rothschild (1986), Laevastu & Favorite (1988) and Hilborn & Walters (1992). Much time and discussion has also been devoted to trying to agree what shape of curve best fits the data (Hilborn & Walters 1992, Chapter 7). In Fig. 5.6 the fitted curve has a maximum and then bends down to the right, but other models do not have a reduction in recruitment at high stock. A consistent feature of stock/recruitment data is wide scatter. In Fig. 5.6, at the same stock density recruitment can vary more than 10-fold. This is a fair example of published stock/recruitment relationships: some show less scatter about the best-fit line, but some show even more scatter. This has led to serious discussion as to whether it would be more helpful to

Box 5.3. Methods of estimating the size of a stock, i.e. the number of fish of a particular species in a large area of ocean.

1 *Mark and recapture*
Fish are caught, marked with a tag and released alive. When fish are caught later, the proportion tagged can be used to estimate the total number in the population.
2 *Hydroacoustic* (sonar, echo-sounding).
3 *From number of eggs*
Eggs can be counted in water samples, then the number of adult fish estimated if the mean number of eggs laid per female is known.
4 *Catch per unit effort* = (fish caught)/(effort by fishermen)
Usually can give only relative figures, e.g. by what percentage stock has changed since last year.
Effort can be measured by number of boats and how many days they spend at sea. It can also take into account the equipment used, e.g. length of net.
5 *Virtual population analysis*
Requires information on natural mortality rate. Usually only gives information about stock size in the past. For further information see text.

Further information: Barnes & Hughes (1999) Chapter 8; Hilborn & Walters (1992); King (1995)

assume no relationship between stock and recruitment, in other words the points are randomly scattered over the graph (see Gilbert 1997, and replies by Myers 1997 and Hilborn 1997). One method of estimating how much fish can be caught—the yield-per-recruit method (which is explained later)—ignores the relationship between stock and recruitment, thus implicitly assuming that recruitment will always be adequate.

Estimating stock size

Why is there so much scatter? Surely the young fish must come from eggs laid by the older fish, so there must be a relationship between stock and recruitment? One possible reason for the scatter is that the figures for stock and recruitment may be inaccurate. Estimating the total number of fish in a large area of ocean is not easy. Box 5.3 summarizes the principal methods that have been used. Methods 1–3 depend on research vessels (though in mark-and-recapture the recapture can be by commercial fishing). If the area to be surveyed is large and the fish irregularly distributed, in order to achieve adequate sampling the amount of time and effort needed will be great and the costs high. Nevertheless, such research survey methods have been used. For example, annual hydroacoustic surveys have since the mid-1980s been crucial to the management of the anchovy and sardine fishery off the southwest coast of Africa (Cochrane *et al.* 1998).

Virtual population analysis

Methods 4 and 5 of Box 5.3 rely substantially on information from fishing boats. *Virtual population analysis* is a widely used method, described briefly but clearly by King (1995) and in great detail by Hilborn & Walters (1992). It relies upon the fact that the age of a fish can be determined. This is often done by counting annual marks on scales or otoliths (small bony structures in the ears), but can also be estimated from fish size. If we consider a single cohort (age-group), equation 5.1 becomes:

$$N_{t+1} = N_t - M - F_t \tag{5.2}$$

because once the cohort is established there is no recruitment to it, only loss by natural mortality (M) and fishing mortality (F). Virtual population analysis starts with the cohort very old, so there are very few fish left and its actual size does not matter much. The calculation then works backwards in time, estimating the size of the cohort the previous year by:

$$N_t = N_{t+1} + M + F_t \tag{5.3}$$

and so on year by year back to recruitment of the cohort. The mortality due to fishing is determined from catches, allowing for other fish killed but not landed (discussed later). In Equations 5.2 and 5.3 natural mortality rate is represented by M, not M_t, because it is normally assumed to be the same each year. Rates of natural mortality are difficult to determine when fishing is going on. It can be determined from mark-and-recapture, if that is performed in more than one year (Barnes & Hughes 1999). If fishing occurs during only part of each year, natural mortality can be measured in the other period; otherwise, if fishing intensity varies from year to year it may be possible to estimate natural mortality by comparing years (King 1995). Virtual population analysis can be a powerful way of estimating stock size and recruitment, but it does so retrospectively. As we shall see, it has been very useful for working out what went wrong after a fish stock has crashed, but less good at preventing the crash from happening.

Recruitment varies from year to year

Returning to Fig. 5.6, the alternative explanation for the great scatter of points is that recruitment genuinely varies greatly from year to year. There is no doubt that this does often happen. The basic feature of fish biology that allows recruitment to be so variable is that each female fish lays an enormous number of eggs each year (e.g. of the order of a million for the cod in Fig. 5.6), and then there is enormous mortality of eggs and larvae. Figure 5.3(b) shows data for plaice in the North Sea for several years; the numbers surviving are on a log scale. In three of the years, from a batch of several thousand eggs only about one larva would be surviving 120 days later. The graph also shows some marked variations between years. The number of larvae that hatched from eggs was about the same in 1968 and 1969, but the subsequent mortality differed: it was about 7% per day in 1968, but only 1% per day in 1969. That sevenfold difference in mortality produced a difference of more than 100-fold in numbers of larvae by the time they were 120 days old. This high fecundity plus high mortality is very different from mammals and birds, where there are few young per adult per year but low mortality. It is therefore much more straightforward to predict the size of a mammal or bird population next year (see, for example, the relatively modest fluctuations in grizzly bear numbers in Fig. 10.7, p. 302).

Recruitment in relation to environmental conditions

The eggs and larvae of many fish are among the plankton and are exposed to predation, e.g. by jellyfish and by adult fish, including sometimes their own species. Predation is a major cause of the high mortality, but at times food shortage can be equally important (Bailey & Houde 1989). Variations in larval mortality can sometimes be related to environmental conditions. The effect of El Niño on Peruvian anchoveta has already been mentioned, but less extreme changes in sea conditions can also have an effect. One way in which temperature can affect survival is by increasing the length of time in the larval stage and hence the time most prone to predation (Bailey & Houde 1989). Among temperate species the timing of egg laying, hatching and larval development in the spring, and how this matches with the bloom of phytoplankton and then zooplankton, can be crucial in deciding whether the fish larvae have an abundance or a shortage of food (Cushing 1990). Correlations have often been reported between recruitment and temperature, and more rarely with salinity and other environmental factors. Myers (1998) listed cases where such reported correlations had been retested with further data. Sometimes the relationship was confirmed, but quite often it was not. Among the correlations with temperature that were confirmed, most were for populations that were near the northern or southern limits of the species' range.

Thus for most fish stocks the causes of the fluctuations in recruitment from year to year are not well understood. Predictions of recruitment in each year have rarely been used in calculating how much fish should be caught. Because of the difficulty and expense of determining stock sizes, and the weak relationship between stock and recruitment, fisheries scientists have looked for other ways of predicting how much fish can safely be caught.

Catch and effort

Schaefer (1954) proposed a method which is attractive because it requires only data on the *catch* and the *effort* expended by the fishermen. (For definition of effort see Box 5.3.) Figure 5.7(a) shows the relationship predicted by Schaefer. This assumes an equilibrium situation, i.e. effort remains constant from year to year and so does the stock/recruitment relationship. The curve falls to the right because increased fishing has reduced the stock. So it should be possible to predict a level of effort that will maximize the catch. Figure 5.7(b) shows an example of real data. Two alternative curves with different equations (see King 1995) have been fitted to the data. With this amount of scatter, deciding the optimum effort to maximize long-term catch is not straightforward.

If effort remains the same but catch decreases, we should take this as an indication that the stock has declined. As effort rarely does remain the same from year to year, *catch per unit effort* has often been used to indicate whether the stock size is changing. The assumption is that catch per unit effort is proportional to stock size. As we shall see, this is often not

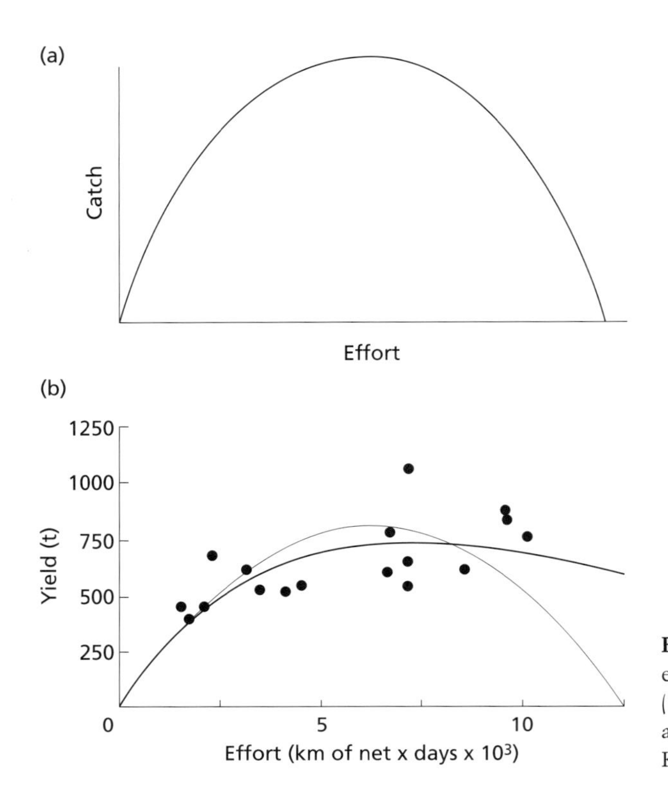

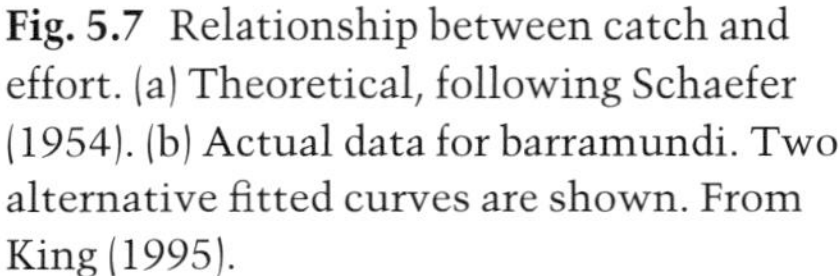

Fig. 5.7 Relationship between catch and effort. (a) Theoretical, following Schaefer (1954). (b) Actual data for barramundi. Two alternative fitted curves are shown. From King (1995).

true. Nevertheless, catch per unit effort data have often been used in fisheries management for monitoring stocks, basically because the figures are available.

Yield per recruit

Another method that has been used to estimate the amount of fish that can safely be caught is based on *yield per recruit* (see King 1995, p. 209). The basic idea is to consider a cohort of fish which has just reached the age and size when it could be caught by the gear available. Our aim is to maximize the total weight of fish we catch from that cohort over its whole lifespan. When should we start taking fish from the cohort, and how much should we take each year? If we catch fewer this year there will be more left to catch next year, when they will be larger. However, some of them will die of natural causes and so never be caught. If we have information on the rate of growth and the natural mortality rate, we can predict the yield per recruit at different intensities of fishing. Figure 5.8 is an example of the relationship. If the aim is to maximize yield per recruit, fishing effort should be F_{max}, corresponding to the highest point on the curve, Y_{max}. Often, however, fisheries managers are more cautious, and during the 1980s it became common to recommend $F_{0.1}$, which corresponds to where the slope of the curve is one-tenth as steep as it is at the origin (at the bottom left corner). Hilborn and Walters (1992) comment: '$F_{0.1}$ is an essentially arbitrary choice of fishing mortality rate, which often appears to be in the right ballpark'. So fisheries management seems

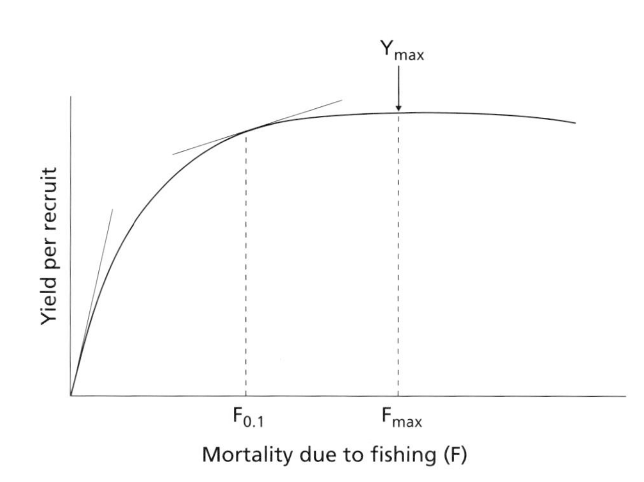

Fig. 5.8 Predicted relationship between fish mortality caused by fishing and yield per recruit. For further explanation see text.

to have moved away from fundamentals and towards arbitrary rules of thumb. The most serious deficiency of basing fishing intensity on yield per recruit is that it pays no attention to how many recruits there are each year, and therefore makes no attempt to guard against recruitment overfishing, i.e. removal of so many fish of spawning age that stock size declines through lack of new recruits.

These methods of predicting how much fish should be caught all have their origins in classical fisheries biology of the 1950s. They are all based on the assumption of equilibrium, i.e. that in the absence of fishing the stock size would be approximately constant from year to year, and that the amount of fish caught is the same each year. Often this is not true. Data available are often from areas where the intensity of fishing has been increasing, and where the stock size may have been decreasing. For example, Fig. 5.6 shows a time trend, with the 1940s in the right-hand half and the 1970s in the bottom left.

Fish species may interact with each other

Another limitation of all these methods is that they apply to one species at a time. Often several species are being caught in the same area, and these species are likely to be interacting with each other: they may be competing for food, or one species may be eating either the larvae or adults of another species. So the amount of one species that is caught this year may well affect the stock size of another species next year. Also there may be interacting effects in the fishing process. Fishing boats with quota for one species may unavoidably catch fish of another species at the same time and tip them back dead. Various methods of multispecies analysis have been proposed (see Hilborn & Walters 1992, Chapter 14). These tend, inevitably, to be more complex than single-species models and to require more quantitative information for input, and so far little use has been made of them for the management of commercial fisheries.

An underlying problem in the management of ocean fish is that we cannot perform properly designed experiments, in which different fishing strategies are applied to fish populations (preferably replicated) whose conditions of environment, food and predation are in other respects similar. For commercially important fish that is not realistically possible.

Management: how to control fishing

Pressures towards overfishing

Imagine a sea on which many boats set out to fish and there are no regulations limiting how much they can catch. Another person buys a boat and fishing gear and joins the fleet. He catches fish, sells them and makes a living (provided he can cover the initial costs of boat and gear). So his joining the fleet is a clear advantage to him. But his catch today reduces the stock tomorrow, and next year too if it is being overfished. So there is less fish for everyone, catch per unit effort is reduced and so is everyone's income. But this loss is shared among all the fishermen, so for each individual it is small compared with the gain by the one new entrant. So another new fisherman joins, and the stock decreases a little further. And so on. This is a classic example of *the tragedy of the commons* (Hardin 1968). If fishing is open to all, and individuals decide whether to fish on the basis of self-interest, there are strong pressures towards the stock being overfished. Eventually the amount of fish to be caught will no longer justify the cost of new equipment, and new entrants to the fishery will be deterred, but probably not until the stock has been severely depleted. Economists predict that under any likely conditions a free market will result in less fish being caught, long-term, than could be obtained by controlled management aiming for maximum long-term catch.

Most governments have accepted for some decades that free market economics does not lead to efficient and sustainable management of ocean fish stocks, and therefore some regulation of fishing is necessary. In the late 1970s many countries that have coastline declared the sea extending 200 miles (320 km) offshore to be their exclusive economic zone, meaning that they declared the right to impose controls on what fishing occurred there by people and boats of any nationality. For reasons explained earlier, many ocean regions near to coasts have high productivity, and these 200-mile zones include many of the world's most productive fisheries. International commissions have been set up to oversee fishing in some areas outside the 200-mile limit, e.g. the Southern Ocean (Everson 1992), but in vast areas fishing is still largely uncontrolled. What follows applies mainly to areas under governmental control.

Limitations: by quota or effort?

Fishing regulations can aim for the amount of fish caught to be a specified proportion of the stock, or can set a quota which is below the maximum sustainable yield but unrelated to the stock size (see Fig. 5.4(c)). In practice, if any quota is set there will need to be continued monitoring of the stock, to avoid getting into the danger area to the left of point u. If a set proportion of the stock is to be taken each year, this can be controlled

Box 5.4. Types of rule which can be used to limit the amount of fish caught.

1 Limits to effort

(a) Limit the number of boats fishing in an area, e.g. by requiring each to have a licence. Boat size can also be restricted.
(b) Limit how many days each year each boat can fish. The dates may be unspecified, or there may be one or more closed seasons.
(c) Place restrictions on what equipment may be used, e.g. the type of net, its length, its mesh size.
(d) Ban fishing in some parts of the area. This could be for only part of the year, e.g. spawning time. It could be on a rotational basis.

2 Quotas

(a) Set maximum *total allowable catch* for whole area for each year.
(b) Quota (maximum catch per year) allotted to each boat. This may be transferable, i.e. it can be sold to another boat.

Further information: King (1995)

either through rules aimed at limiting fishing effort, or by setting a quota that is a certain proportion of the estimated stock size.

Control of effort

Box 5.4 summarizes the sorts of rules that can be used to limit fishing effort. These rules have the advantage of being relatively easy to police. One disadvantage is that they can lead to economic inefficiency, in other words low earnings per boat and per fisher. If boat owners have spent a lot of money on buying and equipping the boat, and maybe have loans to repay, it creates financial problems for them if they are required to keep the boat in port, unused, for many days in the year, or if their catch per working day is reduced by rules that prevent them using modern equipment. Banning fishing from some areas can (if the areas are well chosen) provide an insurance against the stock being made extinct altogether; but if the total fishing effort remains as high and is merely redeployed elsewhere, the size of the total surviving stock may be little different (Horwood *et al.* 1998; Guénette *et al.* 1998).

Figure 5.7 illustrates a further problem with control of effort. If that catch/effort model is being used to decide what effort should be allowed, we need to identify the peak of the curve, and hence the effort that results in maximum sustainable yield. Figure 5.7(b) shows some real data, and two curves fitted through them, which indicate different maximum sustainable yields and associated efforts. Which of them is correct? Actually, neither fits particularly well; an alternative curve would reach a maximum beyond the right-hand edge of the graph. To be sure of where the peak is, and therefore what effort should be allowed, we need data points that go well beyond the peak, declining to the right. If, as often happens, time goes from left to right, with the number of boats and the sophistication of their gear increasing year by year, then the optimum effort will not be decided until there are too many boats. Then, to prevent overfishing, it will be necessary to reduce the number of boats or the number of days they can fish, or to ban some of their gear. This situation is politically unattractive—politicians become unpopular if they reduce people's earnings, or put people out of work altogether, leaving them with boats and gear which nobody else wants to buy.

Control by quota

The alternative system of control involves setting a quota, the maximum amount of fish that can be caught. A *total allowable catch* will be decided for the fishery area for the year. Sometimes the boats have been allowed to fish unrestricted until the quota for the area has been reached, then the fishery is closed for the rest of the year. This leads to a 'race for fish', followed by boats lying in port unused for the rest of the season, which is economically inefficient. Alternatively, each boat may be allotted a quota. Some countries allot *individual transferable quotas*, which can be bought and sold. So, a boat owner can buy more quota if she or he thinks the profit from the extra catch will justify the cost of the quota. This can lead to fewer boats but each allowed to catch more fish, which may be economically more efficient. After the introduction of individual transferable quotas the number of boats did decline in various Canadian fisheries and one Australian fishery, but not off Iceland or New Zealand (Grafton 1996). Nevertheless, the operation of the individual transferable quota system in Iceland and New Zealand waters has been considered a qualified success, at least by some commentators (Arnason 1996, Annala 1996). In any quota system there are difficulties with ensuring that each boat does not exceed its quota (Grafton 1996). Sometimes independent observers travel on the boats, but to police every fishing trip in this way is expensive. More often, policing relies on logbooks of catches kept by the fishermen at sea, and records of fish landed and sold. There is no doubt that substantial amounts of fish are caught and then dumped back into the sea dead. Some of this will be *bycatch*—non-target species which the boat does not have permission to sell, but which were caught along with the target species. Dumping may also be the result of *high-grading*, selecting fish in the size range that will fetch the highest price per kilogram and discarding the rest.

Collapse of the Newfoundland cod stock: a case study

Cod was formerly extremely abundant in the western Atlantic, off the coast of Newfoundland and southern Labrador. It is a demersal fish, confined to waters of the continental shelf and not extending into the deep Atlantic, and so mostly within 200 miles of the Canadian coast (deYoung & Rose 1993). This provided a well established and economically important fishing ground. Yet in 1992 the Canadian government had to close the main cod-fishing regions, lying to the east of Labrador and Newfoundland, until further notice, because the stock had virtually ceased to exist. The other areas, to the south and west of Newfoundland, were closed the following year (Myers *et al.* 1997). Whose fault was the collapse—the scientists', the politicians' or the fishermen's? Or was it caused by some unexpected and unavoidable environmental change? In writing this section I have made substantial use of papers by Hutchings and Myers (1994) and Walters and Maguire (1996).

Cod stock and catches, since 1960

Before 1977 the fishery was open to boats of any nation, and many came from European countries. From the mid-1950s the total catch from the main fishing regions rose greatly, peaking in 1968 (Fig. 5.9). After that it fell steadily until 1977, when Canada declared the 200-mile exclusive economic zone, which included much of the cod fishery and so banned foreign boats from it. A decline in catch, such as occurred from 1968 to 1977, can be due to a decline in either the stock or the fishing effort. Figure 5.10 shows three alternative estimates of how the stock size changed: there is no doubt that it fell greatly between 1968 and 1977. Clearly the fishery was in danger of collapsing. From 1977 onwards foreign boats were excluded and the number of boats dropped suddenly. This provided an opportunity to allow the stock to rebuild, and to plan a fishing regime to exploit the stock in a way that would be sustainable long term. Unfortunately, this was not achieved.

From 1977 onwards a total allowable catch was decided each year, based on the $F_{0.1}$, yield per recruit system explained earlier. The catch rose during the first few years after 1977 (Fig. 5.9), then remained fairly stable, even up to 1990. So the catch figures gave no warning of the disaster to come. The catches were limited by a quota, and the fishermen were trying to fill that quota. As explained earlier, it is easier to estimate stock size retrospectively than at the time. The three alternative estimates in Fig. 5.10 all indicate that numbers rose modestly after 1977 until about 1985, but then started to fall.

So the Canadian government was setting quotas, based on advice from fisheries scientists, but the stock nevertheless collapsed. What went wrong? Basically, the size of the stock was consistently overestimated and so the allowable catch was set too high.

How the quotas were decided

The $F_{0.1}$ system was used to calculate what percentage of the stock could be harvested each year. To set the total allowable catch each year, the size of the stock that year needs to be known. In the estimation of this, catch per unit effort (CPUE) played a crucial part. CPUE rose sharply after 1977 (Fig. 5.11), and this was taken as evidence of a large increase in the stock; we now know that in fact it increased only slowly (Fig. 5.10). After 1985 the true fall in stock size may have been faster than the fall in CPUE. So, in the late 1980s CPUE was indicating that the stock was larger than it had been in the late 1970s, whereas in fact it was smaller and declining. These discrepancies presumably arose because the fishermen were increasing their ability to catch fish in ways not included in the estimates of 'effort'. These could be new equipment, for example new designs of cod trap, sonar, advanced navigation systems; or learning by experiment and experience where were the best places to go to catch the fish. The overall result was that the fisheries scientists consistently and substantially overestimated the size of the stock and therefore the amount that could safely be caught.

The $F_{0.1}$ system, based on yield per recruit, does not guard against inadequate recruitment. The minimum mesh size allowed aimed to prevent fish being caught until they were 3 years old. However, cod do not produce eggs until they are 7 years old, so many female fish were caught before they had been able to reproduce. There has been disagreement about how far low recruitment contributed to the decline and final collapse of the stock, and whether environmental conditions such as low water temperature may have affected recruitment (deYoung & Rose 1993; Hutchings & Myers 1994; Myers *et al.* 1997).

Another uncertainty is whether fishermen contributed to the collapse by catching fish in excess of the quotas. Firm evidence on this is difficult to obtain, but Myers *et al.* (1997) concluded that discrepancies between estimates of stock size by two different methods could best be explained by substantial discarding by fishing boats of 2–4-year-old cod, and misreporting of catches.

Although there have been changes in sea conditions off Newfoundland from year to year and decade to decade which may well have affected the fish, there can be no serious doubt that the principal cause of the collapse of the Newfoundland cod stock was overfishing. The worrying aspect for us as applied ecologists, and for the future management of ocean fisheries, is that the fault lay substantially with the fisheries scientists, who, over many years, persistently gave incorrect estimates of how many fish could be caught without endangering the future of the stock. On the more positive side, cod are not completely extinct off Newfoundland, and a stock reduced to a very low level can still recover, as Fig. 5.5 shows.

Is there a future for ocean fishing?

The story of the Newfoundland cod is an upsetting one for applied ecologists and fisheries managers, as well as for the people of Newfoundland. Could it happen again, elsewhere? Among fisheries scientists some at least (e.g. Walters & Maguire 1996) think the answer is yes, it could.

Fisheries scientists often put part of the blame on politicians, for setting quotas higher than the scientists recommend. South Africa is an example where total allowable catches have been increased above the scientific recommendations for 'socioeconomic reasons': this happened in 5 out of the 9 years from 1988 to 1996 (Cochrane *et al.* 1998).

The European Union has since 1977 controlled fishing in most of the sea off western Europe. The setting of total allowable catches each year has involved bargaining between senior ministers from the different countries of the Union, often ending with the quotas set higher than the fisheries scientists recommended (Corten 1996). The fishermen, in turn,

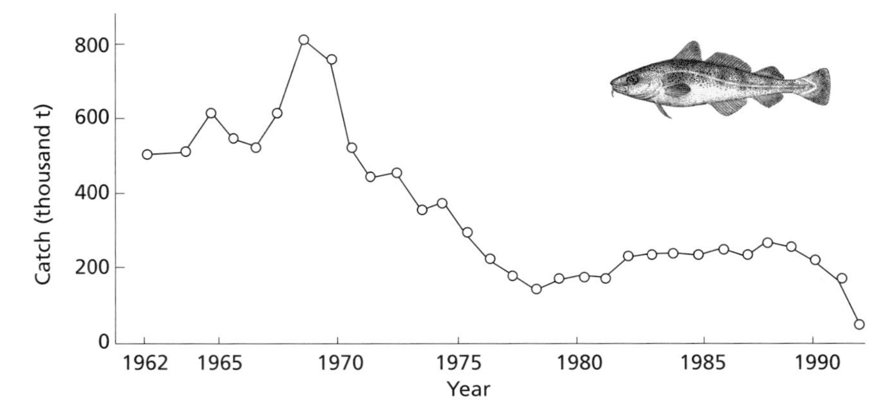

Fig. 5.9 Commercial catch of Newfoundland cod each year from 1962 to 1992. In 1992 fishing was stopped part-way through the year. From Hutchings & Myers (1994).

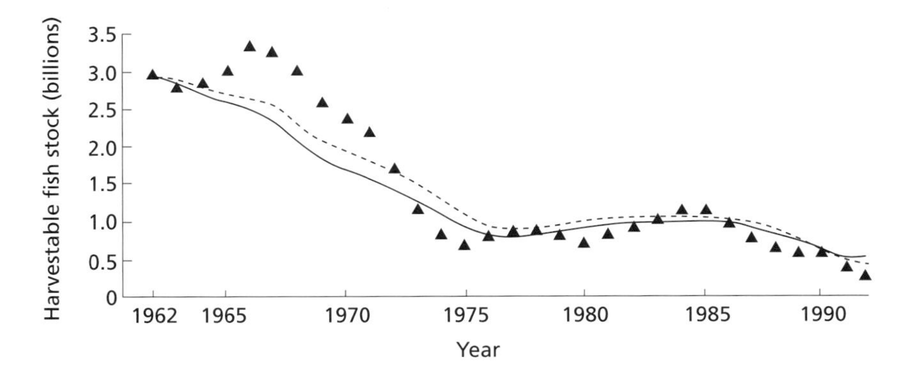

Fig. 5.10 Number of Newfoundland cod of harvestable size, i.e. age 3 years or older, each year from 1962 to 1992, estimated by three alternative methods. From Hutchings & Myers (1994).

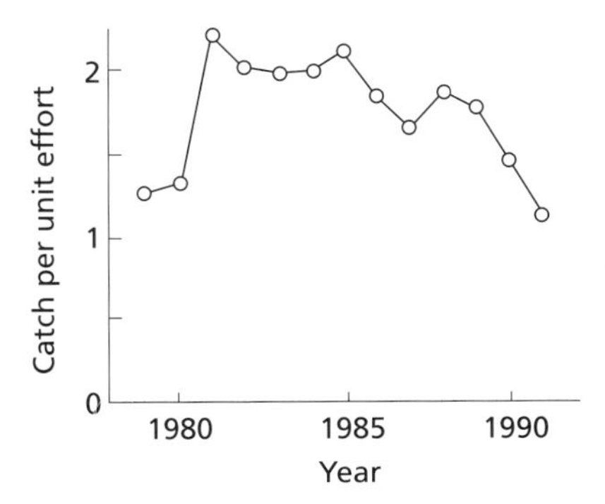

Fig. 5.11 Catch per unit effort for Newfoundland cod, 1979–91, calculated from tonnes of cod caught/total fishing hours. From Hutchings & Myers (1994).

have often exceeded their allotted quotas: estimates for unreported landings of fish from the North Sea during the early 1990s were 60% of quota (i.e. in addition to the quota) for sole and 35% for plaice, but only 10% for cod and herring (Cook 1997). Cook (1997) gave estimates of the North Sea stock and catch of important species from 1982 to 1994, based on independent surveys and so not dependent on reported catches. Total allowable catches were made more restrictive from 1989 onwards. Up to then the stock and catch of haddock and cod had been declining, but they subsequently rose for haddock and levelled off for cod. However, stocks of plaice and herring declined from 1989 to 1994. The future for sustainable fishing in the North Sea remains uncertain.

Questions for the future

In many countries questions have been asked about the future of fisheries management. How should we deal with the uncertainties in the scientists' predictions? How much caution, how much safety margin should be built into the allowable catches? How should we decide between alternative aims, e.g. is it worth accepting a lower mean annual catch if there would be less variation from year to year? What is the best way to bring economics and social factors into the decisions? How should scientists, administrators, politicians and fishermen interact in the decision-making process? These questions have been discussed at length, e.g. by Hilborn and Walters (1992). They are outside the scope of this book.

Another set of questions concerns how fishing affects non-target species and whole marine ecosystems. For example, the gear used for catching demersal fish drags along the surface of rocks and disturbs bottom sediment. This can lead to reduced abundance of species living on the bottom surface, such as sponges, anemones and bryozoans; however, animals living within the sediment often recover quickly (Jennings & Kaiser 1998). Another example is that heavy fishing can lead to a decline in abundance of some sea birds, by reducing their food supply. These and other side-effects of fishing raise the question of whether fisheries management should in future take more account of whole marine ecosystems and all the species in them.

On a world scale, ocean fishing is by no means finished. In spite of substantial declines in catches of important species in some areas, total fish catch from the world's oceans has not, as yet, shown any clear downward trend (Fig. 5.1). Nevertheless, it seems opportune to consider an alternative way to obtain fish for food: fish farming or aquaculture.

Aquaculture (fish farming)

Aquaculture is not a new idea. It has been going on in China for more than 3000 years, and in the mid-1990s China obtained more than half of its total freshwater and sea food this way (Zhong & Power 1997). In 1993–95 17% of the fish (including molluscs and crustaceans) provided as food, worldwide, came from aquaculture (World Resources 1998/9).

Box 5.5. Aquaculture: problems in planning and management.

1 Finding suitable sites
Limitations include requirements for water supply and waste disposal.
2 Water supply
Some farmed species require fresh water (e.g. common carp), some require salt water (e.g. turbot), some can grow in both (e.g. salmon, trout).
3 Maintaining suitable physical and chemical conditions
Includes temperature, concentration of oxygen and ammonia.
4 Food supply
Extensive systems rely on plants and animals that grow in the production ponds; so high primary productivity is important.
Intensive systems require food that has been produced elsewhere.
5 Disposal of waste: fish excreta, uneaten food
6 Diseases of fish
Can spread rapidly because of high fish density. Control by altering conditions, chemical treatments, vaccination.

Further information: Barnabé (1994)

Species that can be farmed successfully include salmon, trout, carp, turbot and mullet. In 1995 the number of Atlantic salmon (*Salmo salar*) in fish farms was more than 10 times as many as in the wild (Gross 1998).

This section, like the rest of this chapter, confines itself to finfish. It considers briefly some applied ecology aspects of aquaculture, especially in relation to food for the fish to eat. Box 5.5 lists problems that need to be addressed.

The facilities can range from artificial ponds to reservoirs to cages in natural lakes and coastal seas. Sites close to rivers, lakes or the ocean are favoured for tanks and lagoons, because they provide a water supply and a place to dump waste. But this can cause problems, if for example the area has special wildlife or scenic value, or the waste causes unacceptable pollution.

Food supply for the farmed fish

Most of the fish species we eat are carnivores, in contrast to all the land animals that provide us with meat, milk, wool and so on, which are herbivores. There are a few exceptions, e.g. grass carp and anchoveta are herbivorous fish that we use; but with the majority, the carnivores, we are dealing with a longer food chain. We need to consider food for the larvae as well as the adults: as described earlier, fish larvae have different food requirements from the adults.

Aquaculture can be either extensive or intensive. *Extensive* means that basically we rely on natural food chains to supply food to the fish. To maintain high primary productivity it is usually necessary to maintain high nutrient status by adding inorganic fertilizer or organic manure. In China it has been traditional to put 'nightsoil' (human excreta) into the fish ponds, as well as animal manure (Zhong & Power 1997). Fish production of 0.4–4 tonnes fresh weight ha^{-1} $year^{-1}$ can be achieved in fertilized pond systems (Barnabé 1994). Taking the energy content of fish as 1.6 GJ $tonne^{-1}$ fresh weight (Pimentel & Pimentel 1979), the food energy production is about 0.6–6 GJ ha^{-1} $year^{-1}$, which is of the same order of magnitude as production by cattle on intensively managed pasture (see Table 2.3). Yields are higher in extensive aquaculture than from natural lakes and oceans, not only because of the fertilizer addition but also because the fish are protected from predators.

Food for the larvae

Intensive aquaculture means that all the food for the fish—larvae and adults—is obtained from elsewhere and fed to them. In their native habitats the larvae of most fish feed on zooplankton. Producing enough zooplankton to feed a large population of growing larvae is a considerable undertaking. The larvae of marine fish are most often fed initially on the rotifer *Brachionus plicatilis*, and later, as they grow larger, on nauplius larvae of *Artemia salina*, a small crustacean. *B. plicatilis* is parthenogenetic, which makes it easier to produce in large numbers. It normally feeds on microalgae, which also have to be cultured in large amounts; however, there has been some success with mixtures of yeast and algae as diet for *B. plicatilis* (Shepherd & Bromage 1988). *A. salina* is not usually cultured: its cysts are collected in large numbers from the banks of lakes, and the nauplii then hatch from them.

The larvae of salmon and trout are easier to feed. Their eggs are larger than those of many other fish, and the larvae at first have a large reserve of yolk. By the time they need food they are able to feed on a diet of dry particles manufactured from components such as yeast and beef liver. There has been research on how to grow the larvae of other fish on manufactured food. One possibility is that they can adjust to manufactured food if given a changeover period when it is mixed with plankton. Rosenlund, Stoss and Talbot (1997) experimented with halibut larvae, which were fed initially on *Artemia* nauplii. Over a 10-day period they gradually increased the supply to the larvae of a microparticulate feed, similar in protein and fat content to *Artemia*, while reducing the amount of *Artemia*. After this changeover period the larvae were fed for another 20 days on particulate food only. Compared with larvae fed on *Artemia* throughout, they had lower mortality and higher final weight. A similar experiment with turbot larvae was less successful. Success may depend on getting the details of the food composition and timing right.

Food for the adults

Adult fish (apart from the few herbivorous species) are usually fed on fishmeal. It may seem that there is little gained by feeding fish on fish, but the fishmeal is made from species that humans do not normally eat, and from parts left over, for example from fish canning factories. Nevertheless, there is a long food chain from the primary producer to the fish we eat, so this is not an efficient way of converting solar energy into food energy. Experiments have been carried out to develop diets for adult fish which are partly of plant origin, and these have been successfully used for some species. Boonyaratpalin (1997) gives a diet that has been used for barramundi which is 70% fishmeal, most of the rest being starch and rice bran. Feeds composed of up to 50% plant materials have been recommended for some carnivorous fish (Shepherd & Bromage 1988). Omnivorous fish can thrive on a diet much higher in plant material, e.g. milkfish on about 90% meal from soybean and grain, and only 10% fishmeal (Boonyaratpalin 1997).

This section on aquaculture has particularly emphasized food supply for the fish, which is one of the major ecological problems affecting future expansion. However, the availability of suitable sites and the disposal of waste may also be limiting factors. It is likely that aquaculture will expand substantially in the future, but it is not yet possible to predict with confidence whether it will in due course overtake ocean fishing as the major source of fish for human food.

Conclusions

- Most fish that we eat are carnivores, and their productivity is reduced by long food chains.
- Relatively high fish productivity occurs in areas of high primary productivity by plankton. Many of these areas are within a few hundred kilometres of coasts, where fishing is controlled by individual countries.
- If fishing is at the correct intensity there can be surplus yield, allowing fishing to continue without destroying the stock. Surplus yield models have predicted the maximum amounts that can be harvested each year.
- The number of young fish joining the stock (recruitment) fluctuates greatly from year to year, and has proved difficult to predict.
- Because of difficulties in determining stock size and recruitment, predictions of how much fish should be caught have often been based on amounts caught in previous years.
- Rules to control fish catches can involve quotas, or can regulate effort (e.g. number of boats, type of equipment).
- Although regulations have helped to maintain many stocks, some stocks have virtually disappeared. Some of these collapses have been due substantially to incorrect information and recommendations from fisheries scientists.
- Aquaculture (fish farming) is now producing a lot of fish. However, there are problems and limitations, including difficulties in producing large quantities of suitable food for the fish.

Further reading

Ecology of the oceans:
Barnes & Hughes (1999)
Valiela (1995)

Fish population dynamics and management:
Hilborn & Walters (1992)
King (1995)

Aquaculture:
Barnabé (1994)

Chapter 6: Management of Grazing Lands

Questions

- Is there a conflict between management of the plant and management of the animal that eats it? How can we manage for high productivity by the plant yet high consumption by the animal?
- Is there an optimum sward height? or structure? How do grazing animals respond to these?
- Does grazing ever increase plant growth?
- Do animals choose to eat only certain plant species? Can this choice ever be modified?
- Does selection by animals alter the species composition of the vegetation? Do they sometimes eliminate the more nutritious species?
- Can heavy grazing lead to irreversible damage to the vegetation?
- Can several mammal species grazing together make more efficient use of the vegetation than a single species?
- How can grazing lands be managed to promote high diversity of wild plants and animals?

Background science

- Relationship between foliage area, photosynthesis and growth of a sward.
- Digestive systems of herbivorous mammals. Differences between ruminants and non-ruminants.
- Diet choice by different herbivore species, wild and domestic.
- Effects of grazing on the diversity of plants and of insects.

Two trophic levels

About a quarter of the world's land surface is grazing land, more than twice the area that is devoted to crop production (see Table 1.1). So growing animals for food involves a considerable commitment of land. Management of grazing animals is a challenge in applied ecology: because of the extra trophic level, compared to cropland, the management inevitably has extra complexity. Most of what the animals eat is green stuff containing photosynthetic machinery, so there is an intrinsic conflict between the requirement of the animals to eat today and their requirement that the plants produce more for them to eat tomorrow or next month. The manager has to consider the needs of both the plants and the animals.

This chapter is about mammals that eat leaves and stems, and about the vegetation they eat. Often I refer to these animals as grazers, although a few are strictly speaking browsers, meaning they feed on material from trees and shrubs. Most are *ungulates*, meaning large, hooved herbivores. Cattle and sheep feature strongly in this chapter, but wild herbivores are also considered.

Inserting a second trophic level in the farm ecosystem inevitably lowers productivity per hectare. The efficiency with which animals convert their food into growth can be expressed as (production/consumption), where 'production' is the energy content of the new animal tissue produced by growth and reproduction, plus other products such as milk and eggs; and 'consumption' is the energy content of the food eaten by the animals. For terrestrial herbivorous mammals this efficiency is commonly about 5% or less, though it can be as high as 17% for intensive dairy farming (Brafield & Llewellyn 1982; Coughenour *et al.* 1985). It follows that animal farming, viewed as a method of converting solar energy to food energy for people, must have an efficiency lower than most systems for the production of plant food. Table 2.3 gives some figures that illustrate this.

Why do we keep animals?

So why do we use animals as food? It is possible for people to live, grow and remain healthy without eating any meat: many people do. Fewer people live without any animal products—milk, milk products, eggs—but it can be done. During the 20 years from 1976 to 1996 the number of cattle in the world increased by 9%, and beef production by a similar percentage. The numbers of sheep and horses scarcely changed during that period, though world mutton production increased by nearly 50% (UN Statistics Yearbooks; Brown *et al.* 1997). So, on a world scale these domestic grazers show no sign of disappearing, though their numbers have not kept pace with the increase in human population (Fig. 1.1). In some countries there are areas where cattle and sheep are grazed on productive pasture on fertile land where crops could be grown, so they can be viewed as competing with arable as a land use. Also, cattle and sheep are sometimes fed on grain, and pigs and poultry often are—an even more direct competition with plant food for people. If food supply becomes short there could be strong arguments for converting some of that land to crop production for human consumption. However, much of the land on which cattle and sheep graze is not suitable for arable cropland, for example because it is too steep, the soil is too difficult to cultivate (e.g. too rocky), or the climate is unsuitable (e.g. low and erratic rainfall). Much of this area has low net primary productivity, and one of the functions of grazing animals is to act as concentrators of energy: the meat in one cow may incorporate energy and nutrients from plants spread over several hectares. In parts of semi-arid Australia commercial cattle production operates with only one animal per 50 hectares (Hodgson & Illius 1996).

A further major advantage of cattle and sheep is that, with the help of their rumen bacteria, they can digest cellulose. Because humans cannot digest cellulose we grow crops for materials such as starch, sugar, fat and protein, and this has led to an emphasis on seed-producing crops, most of which are annuals. Because ungulates can eat and digest leaves and stems, most pastureland can be composed mainly of perennial plants, which tend to be more efficient than annuals at capturing solar radiation on a yearly basis, and also tend to make the soil less prone to erosion (see Chapter 4).

Intensive animal production often involves ploughing, sowing one or more desired plant species, and applying fertilizer when necessary; so the farmer has considerable control over the soil conditions and the species composition. However, in 'rough grazing' or 'rangeland' the unfavourable environment usually makes such practices uneconomic. Often the only management techniques that can be used there involve the animals: deciding which species, how many per hectare, perhaps moving them from one area to another during the year. It is on this aspect of grazing management that I concentrate in this chapter. Management of soils has already been considered at length in Chapter 4.

Interactions of primary production and animal consumption

Figure 6.1 shows an example of the relationship between the photosynthesis per unit area of ground by a grass sward and the amount of green leaf + sheath area per area of ground (here abbreviated to leaf area index, LAI). Photosynthesis is closely related to the amount of incoming radiation absorbed by the canopy. When LAI is low the amount of radiation absorbed, and hence the rate of photosynthesis, is approximately proportional to LAI; but as LAI increases above 1 there is more and more overlap between leaves, so the curve levels off towards a plateau photosynthetic rate.

Effect of grazers on photosynthesis

In Fig. 6.2(a), line P_1 shows the expected influence of the grazers on net primary productivity of pasture. This is part of the curve from Fig. 6.1 reversed left-to-right, on the assumption that the more animals there are per hectare the more of the foliage they will remove. This graph assumes that as the stocking density increases the amount eaten by each individual animal remains the same, so the total amount consumed (C) forms a straight line through the origin. The area between the two continuous lines (P_1 and C) represents primary production not consumed by the farm animals. This graph illustrates the basic conflict in grazing systems: the animals depend on photosynthesis to provide their food, and yet their own feeding reduces photosynthesis. High secondary productivity is associated with rates of primary productivity below the maximum the site could support. Measured data to confirm that Fig. 6.2(a) is correct have not often been produced. It is not easy to measure primary productivity while animals are eating some of the production, nor to measure

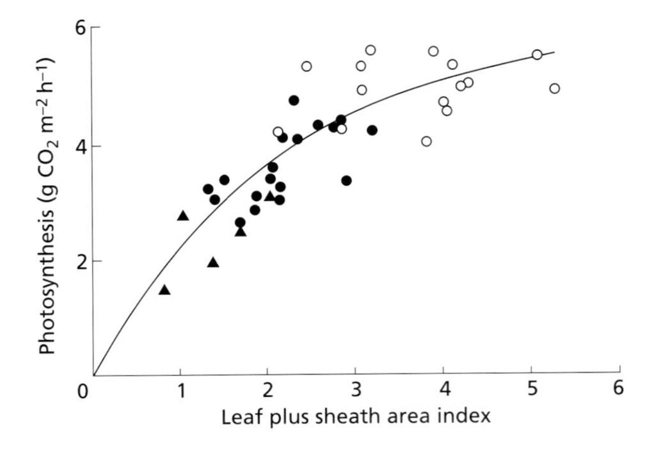

Fig. 6.1 Relationship between rate of photosynthesis (per unit ground area) and amount of foliage in a field of perennial ryegrass in southern England. Each point refers to photosynthesis over a short period; the incoming solar radiation was approximately the same for all. 'Leaf plus sheath area index' = (Area of leaf + sheath)/(Area of ground). The symbols refer to different grazing regimes. From Parsons *et al.* (1983a).

consumption by animals grazing outdoors. Figure 6.2(b) shows results from the same experiment in southern England that produced the data for Fig. 6.1. Sheep were grazed at two densities on perennial ryegrass sward. Photosynthesis was measured by enclosing small areas of the sward in a transparent chamber and measuring CO_2 uptake. Herbage consumption was calculated from measurements of faeces and of the digestibility of the herbage. Measurements were made only during the spring–summer growing season. The leaf area index was usually 2–3 with the lower stocking density and 1–2 at the higher density. Figure 6.2(b) provides confirmation of the key fact that as stocking rate increases the total consumption per hectare goes up, whereas the net primary productivity per hectare declines.

Effect of grazers on sward growth

Figures 6.1 and 6.2 are based on measurements over a few hours or a day, so they cannot give the full story: we need to think what will happen

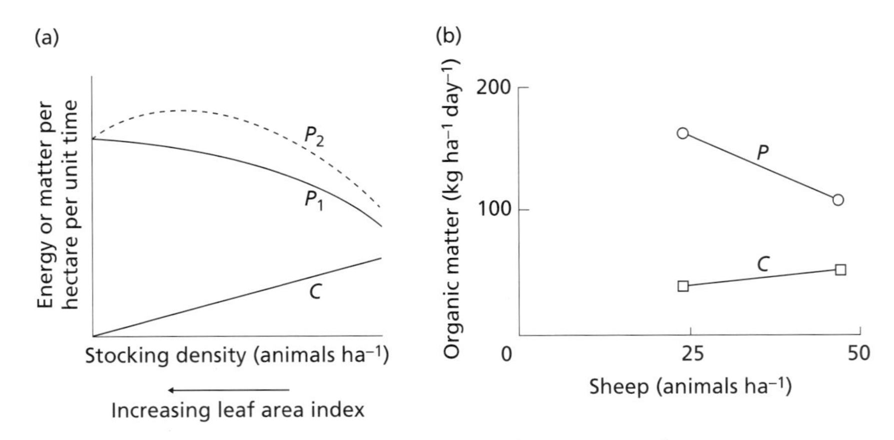

Fig. 6.2 Relationship between stocking density of grazing animals, net primary productivity of sward (P) and amount consumed by animals (C). (a) Theoretical. (b) Experimental results for sheep on perennial ryegrass pasture in England. From Parsons *et al.* (1983b).

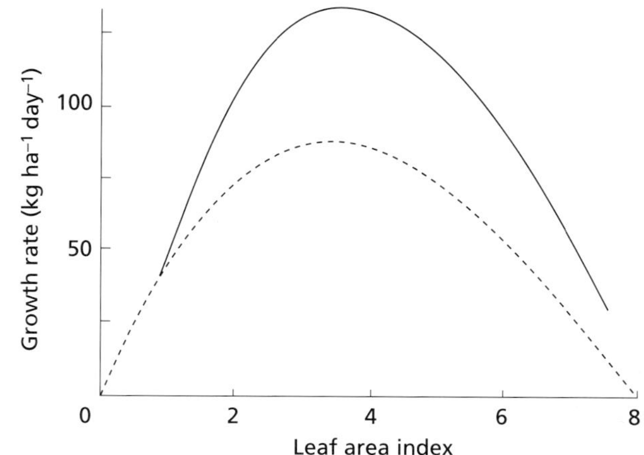

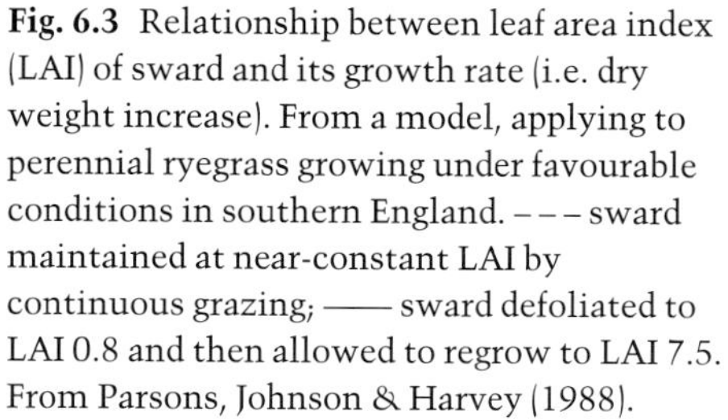
Fig. 6.3 Relationship between leaf area index (LAI) of sward and its growth rate (i.e. dry weight increase). From a model, applying to perennial ryegrass growing under favourable conditions in southern England. - - - sward maintained at near-constant LAI by continuous grazing; —— sward defoliated to LAI 0.8 and then allowed to regrow to LAI 7.5. From Parsons, Johnson & Harvey (1988).

over weeks or a full year. If the animals do not consume all the material produced by photosynthesis, the amount of herbage will not go on increasing indefinitely. Each leaf has a finite length of life, and if a sward is grazed lightly many leaves will die before they are eaten. Figure 6.3 shows that, as a result of this, growth (i.e. dry weight increase per hectare per day) is highest at intermediate LAIs. If the LAI is too low, much of the light will not be intercepted, so photosynthesis will be low. But if LAI is too high that means each patch is grazed infrequently, so there is time for many leaves to die before they are eaten. The dotted line shows growth rate when the sward is grazed evenly and continuously. It indicates that in order to promote high growth rates it would be best (for this type of grassland) to have a density of animals that maintained the LAI between about 2 and 5. It follows that it is better for the grazing to be uniform across the field rather than patchy. If some parts are grazed down to LAI 1, while others are left to grow to LAI 7, the growth will be less than if all are uniformly grazed to LAI 4. In practice, the measurement of LAI is too time-consuming to be carried out frequently and routinely, so it has rarely been used as a basis for management. However, sward height gives an indication of LAI for a particular grass species or combination, and maintaining sward height within a specified range can be used as a basis for management (Hodgson & Illius 1996).

The continuous line in Fig. 6.3 shows how the sward's growth rate changes as it regrows after a severe defoliation. The growth is predicted to be faster than for a continuously grazed sward of the same LAI. This is partly because in the regrowth the leaves are mostly young and little death occurs at first. Also, the young regrowth leaves may have higher photosynthesis per unit leaf area; a possible reason for this is explained later. However, over the whole period of regrowth, from LAI 0.8 to 7.5, the mean growth rate is no higher than the top of the dashed line, for continuous grazing.

If there is an unfavourable season each year

So far, we have assumed that environmental conditions are suitable for plant growth. However, most grazing lands are in regions where the climate is unfavourable for plant growth for part of the year—it is either too dry or too cold. In many intensive systems the animals are given supplementary feed during this season, e.g. hay or silage. In rangeland this is unlikely to be possible, and the number of animals a site can support may well be determined by the forage available during the unfavourable season. In hot, semiarid regions, during the dry season the food ingested may not be enough to support the maintenance requirements of the animal, which therefore can only survive by drawing on reserves within its body, so the bodyweight declines. This is true for both domestic and wild species (see Chapter 3).

Food selection by animals

Response to plant structure and chemical composition

In most grazing land the animals have a choice of more than one plant species. Even sown grassland often contains clover as well as one or more grass species. In pasture consisting of just one species of grass there may be variations between plants in size, structure and chemical composition, and the response of the animals to these could affect how efficiently they use the sward as food.

In Fig. 6.2 there is a disagreement between the prediction of consumption in part (a) and the measurements in part (b). Part (a) assumed that the consumption *per sheep* would be the same at any stocking density, so C is a straight line through the origin. But this was not found to be so in practice: in part (b) line C does not extrapolate to the origin, and in fact the consumption per sheep was 29% lower at the higher stocking density than at the lower. With more sheep grazing, the sward was grazed shorter, and the sheep's eating behaviour presumably responded to this.

How tallness of the sward affects the amount eaten

Allden and Whittaker (1970) prepared plots of annual ryegrass of different heights, by allowing different recovery times after mowing, and then allowed sheep to graze on them. The shorter the sward, the less the sheep obtained per bite (Fig. 6.4). However, over much of the height range this was compensated for by the animals taking more bites per minute. So, over the height range from 8 cm upwards the sward height made no difference to the weight of food the sheep took in per minute. But below 8 cm bite rate failed to compensate for smaller bites, and intake dropped sharply. Black and Kenney (1984) made artificial grass swards by fixing freshly cut grass stems and leaves into holes in boards. In this way they could vary the height and the density (stems per m^2) independently. At high stem density the critical height below which intake rate by sheep started to decline was similar to that in Allden and Whittaker's field experiment, but at wider stem spacings the critical height was greater.

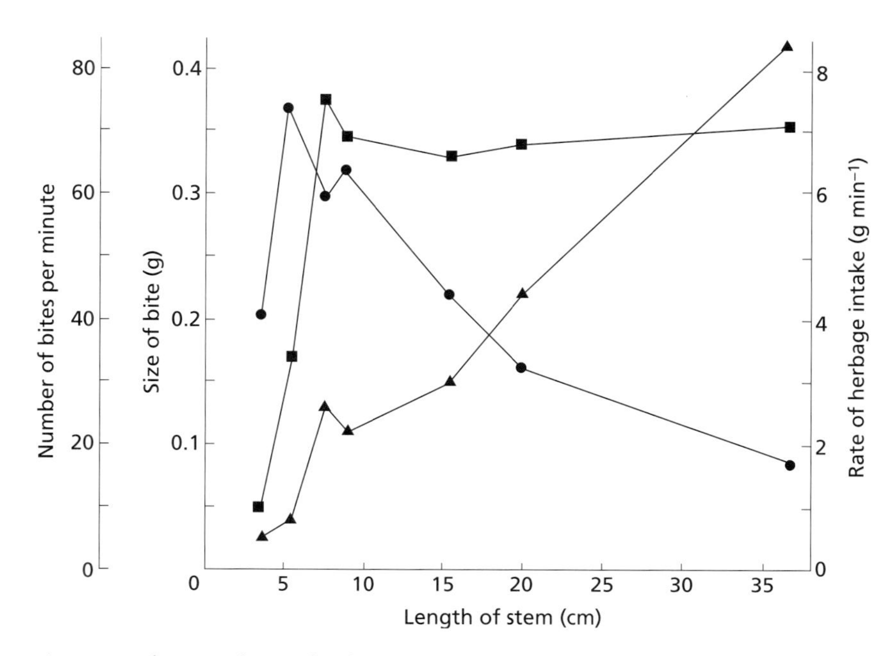

Fig. 6.4 Herbage intake rate by sheep in Australia from annual ryegrass swards of different heights. ▲ amount per bite; ● number of bites per minute; ■ weight of herbage eaten per minute. From Allden & Whittaker (1970).

How low can animals bite?

Grazing animals can be sensitive to the canopy structure of the sward they are grazing. In most pasture grasses the true stem extends little above the ground, but the leaf sheaths enrolled on each other form a 'pseudostem', which is more fibrous than the leaf laminae. In an experiment five species of pasture grasses were grown separately and offered to goats. Figure 6.5 shows that how low the goat bit was apparently not related to the height of the canopy top, but rather to the height of the pseudostem: they stopped about 1 cm above it. The grasses (except for *Agrostis*) held a substantial proportion of their leaf laminae below 3 cm, so this limit to the animals' bite depth would have a strong influence on how much of the foliage they could eat.

Animals' response to chemical composition of plants

Animals can also be sensitive to the chemical constitution of pasture plants. Jaramillo and Detling (1992) applied artificial urine (a solution of urea plus some mineral salts) to semi-arid prairie in Wyoming, in small patches each simulating one 'urination event'. In these patches the nitrogen concentration in above-ground tissue of a dominant grass, *Agropyron smithii*, greatly increased, though it showed little response in growth. The cattle grazed the grass in the urine patches preferentially: a greater percentage of stems was grazed than elsewhere, and they were grazed to nearer the ground. The cattle were presumably responding to the chemical status of the foliage, since there was little difference in sward structure between the urine patches and elsewhere.

Usually the cell wall material in plants is less digestible than the cell

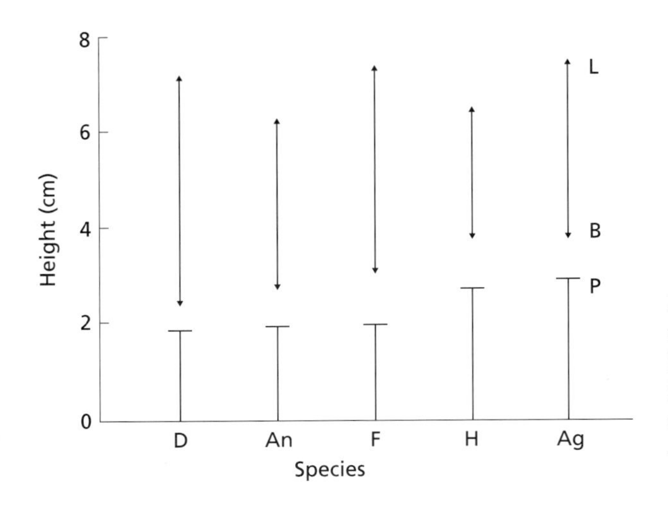

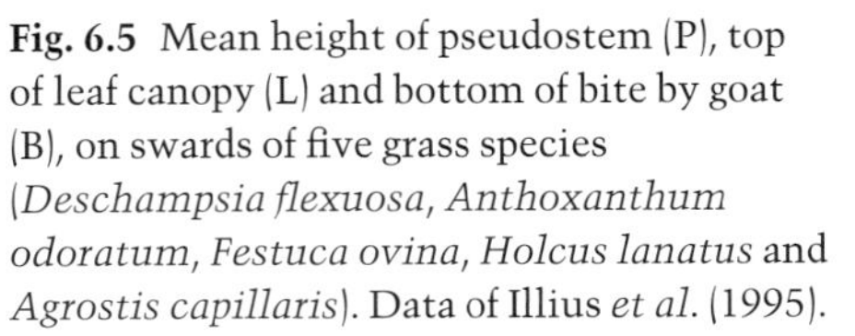

Fig. 6.5 Mean height of pseudostem (P), top of leaf canopy (L) and bottom of bite by goat (B), on swards of five grass species (*Deschampsia flexuosa*, *Anthoxanthum odoratum*, *Festuca ovina*, *Holcus lanatus* and *Agrostis capillaris*). Data of Illius *et al.* (1995).

contents. The cell contents include starch, proteins, lipids and water solubles such as sugars and amino acids. Among the wall components cellulose and hemicellulose can be digested by most ungulates, but more slowly than the cell contents, whereas lignin and cutin are almost indigestible (Duncan 1992). How easily plant material can be digested depends on the relative abundance in it of these major components. Secondary chemicals such as tannins and terpenes also reduce digestibility. It is often suggested that grazing animals choose their diet in line with *optimal foraging theory*. According to this, if their primary requirement is energy they should choose their food so as to maximize their net energy gain (energy intake minus energy costs); energy intake will depend on how much food is ingested and how digestible it is, and the costs will be energy expended in finding the food, eating it and digesting it. Digestibility should therefore be one of the factors influencing food choice, but others could also be involved, such as distance to walk between the most digestible plants, and how much energy is required to bite them. Optimal foraging can also be applied to other requirements, e.g. protein. It is still uncertain how far grazing animals do in fact obey optimal foraging theory. They may not always recognize different plant species or know their nutritive value. They may be constrained by other requirements, such as drinking water (see Chapter 3); and wild animals may need to avoid predators. And in the more productive grazing lands, intake by the animal is determined by how fast it can digest its food, rather than how fast it can find digestible plants to eat. The application of optimal foraging theory to grazing animals is discussed further by Hanley (1997) and Hodgson and Illius (1996).

Optimal foraging theory

Choice of plant species

Choice between two plant species

Grazing animals do not necessarily confine their feeding to one, preferred plant species, even if it is in ample supply. This was shown by Parsons

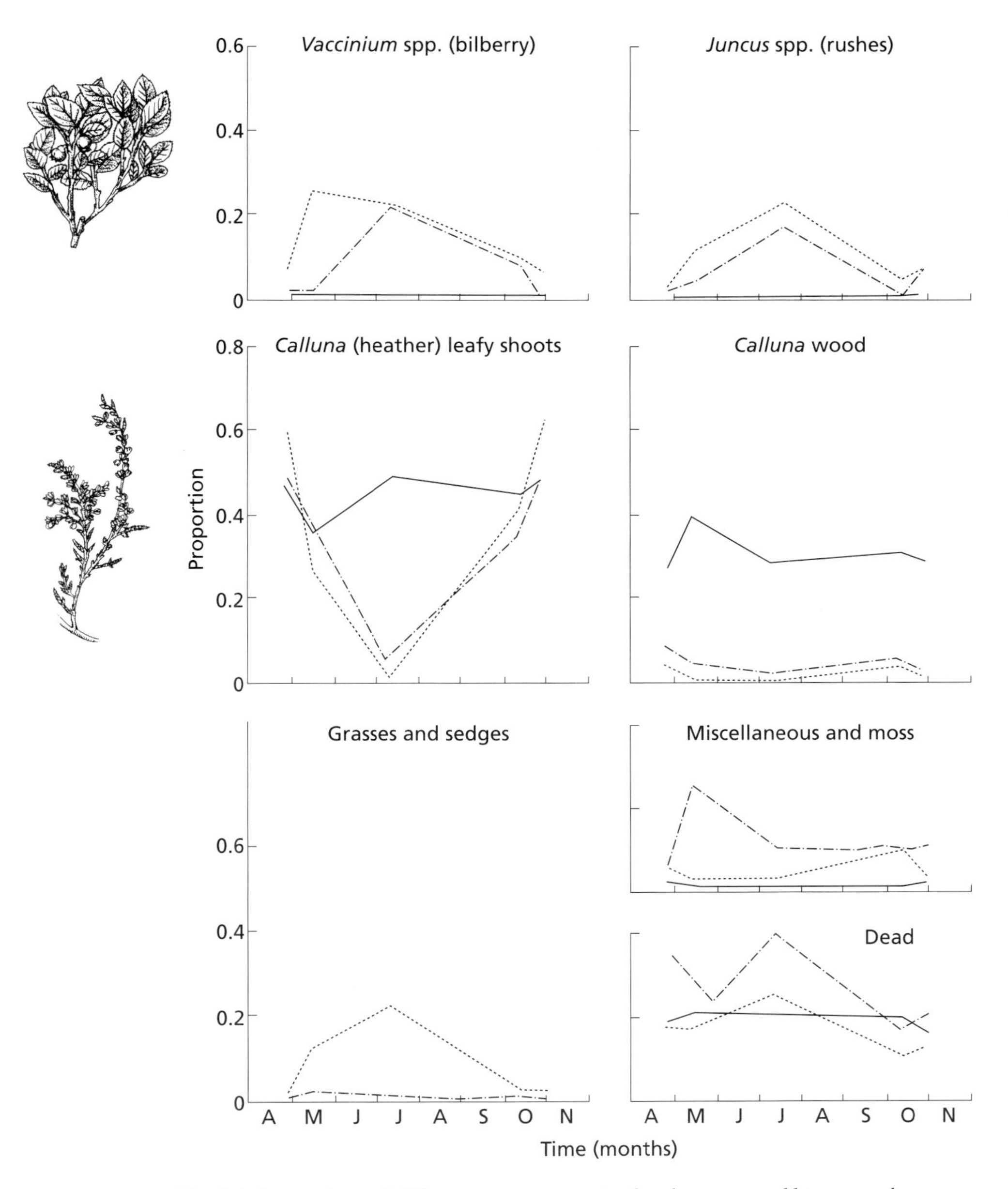

Fig. 6.6 Proportions of different components in the above-ground biomass of vegetation (——), intake by sheep (· · · ·), intake by cattle (– · – · –). From heather moor in upland Scotland. Apart from 'dead' and 'wood', all components included live material only. From Grant *et al.* (1987).

et al. (1994), in an experiment in which sheep could move freely about in a plot containing equal-sized patches of pure ryegrass and pure white clover. The amount of time they spent grazing on each was monitored; time spent doing other things was not included in the calculations. They grazed both species, but spent more time on the clover. For example, on

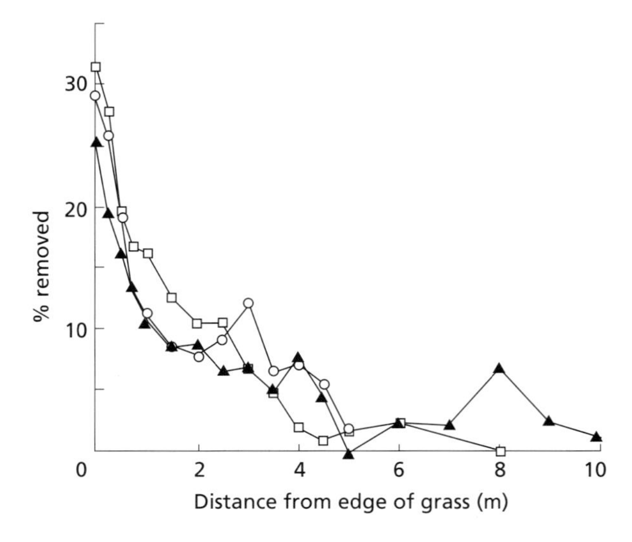

Fig. 6.7 Amount of heather eaten by sheep or deer, measured as a percentage of current year's growth removed, at different distances from edge of nearest grass patch, in moorland in Scotland. ▲ sheep only; □ deer only; ○ deer and sheep together. From Hester & Baillie (1998).

their sixth day in the plot they spent 71% of their grazing time on clover and only 29% on the grass. So they did not walk about and bite at random, they were selective; but they did not confine themselves to eating clover.

Choice among many species

Figure 6.6 shows diet selection by cattle and sheep which were ranging freely over moorland in northern Scotland, from spring through until autumn. Throughout the year, *Calluna* (heather) was much more than half of the plant matter present. However, in summer the animals selected strongly against it, eating more of the soft-leaved bilberry and the grasses, sedges and rushes. The grasses and sedges were so closely grazed that their vegetation mass is not visible on the graph, but they formed a substantial contribution to the diet of the sheep. In spring and autumn the animals were less selective.

What animals eat depends not only on what species are present but how they are arranged. Feeding by sheep and deer was studied in moorland in eastern Scotland. Heather was the most abundant plant species, but there were patches of grassland of various sizes, forming a natural mosaic. The animals spent a lot of time on the grass patches, and when they ate heather it was usually close to a grass patch. Figure 6.7 shows how the amount of heather eaten declined sharply within a few metres of the edge of the grass patch. One might expect from this that sheep and deer will eat more heather if the moorland has many small or narrow strips of grassland, so that all the heather is close to some grass. This was confirmed for sheep by Clarke *et al.* (1995) in an experiment in Scottish moorland, where they created patches of grass of different sizes among the heather. Sheep spent more of their grazing time on heather if there were many small grass patches; however, deer spent about the same amount of time eating heather whether the grass patches were small or large.

This section has summarized only a little of the extensive research that has been carried out on diet selection by grazing animals. However, it is

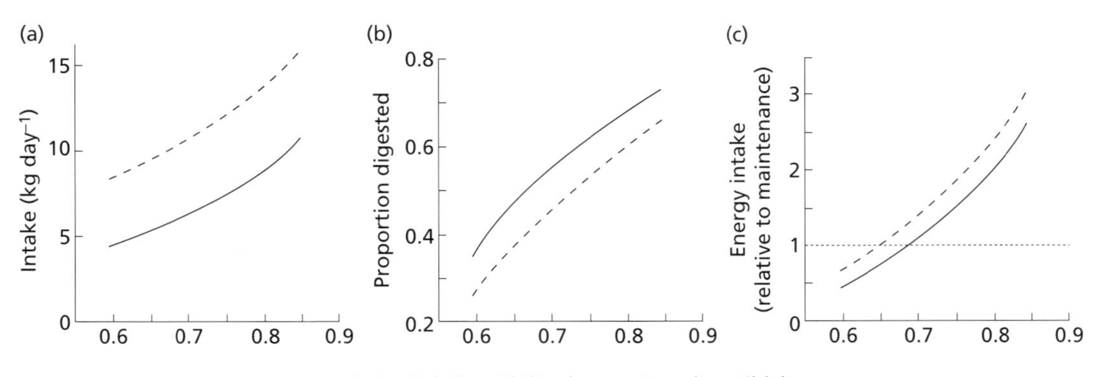

Fig. 6.8 Predictions, for grazing animals of bodyweight 1 tonne, of (a) food ingestion rate; (b) proportion of food digested; (c) energy acquired from food (as a proportion of energy used in maintenance). —— ruminant, – – – hindgut fermenter. From a model of Illius & Gordon (1992).

enough to show that the animals are selective, and that the basis of their selection can be complex. It can include the tallness and canopy structure of the vegetation, its chemical composition and its arrangement within an area; and selection between species can vary throughout the year. Distance from drinking water can also affect what animals eat (see Chapter 3). This complex behaviour by grazing animals provides a challenge in managing both them and the vegetation, so as to maintain vegetation that continues to be productive and has a structure, chemical composition and species mixture that is palatable and beneficial to the animals.

Ruminants and non-ruminants

Up to now, all the animals mentioned have been ruminants. These have special, large chambers in their stomach, the rumen and reticulum, containing bacterial populations that break down cellulose to organic acids, much of which can be absorbed by the animal. Rumen bacteria also contribute to the animal's nitrogen balance. Among farm animals, cattle and sheep are ruminants but horses and pigs are not. Some wild herbivorous

Box 6.1. Examples of large herbivorous mammals which are ruminant or non-ruminant.

Ruminants	Non-ruminants
Cattle	Horse
Sheep	Pig
Goat	
Deer	Zebra
Antelope	Rhinoceros
Bison, buffalo	Hippopotamus
Camel	Elephant
Giraffe	

mammals also are ruminants and some are not (see Box 6.1). One might perhaps suppose that non-ruminants are less able to digest cellulose and therefore favour different diets from ruminants, higher in more easily digestible materials. In fact, many large non-ruminant herbivores are 'hindgut fermenters', which have some ability to obtain energy from cellulose with the help of bacteria in their caecum. This is a less efficient method but they often make up for that by eating more per day than ruminants of the same size, and passing it through their digestive system more rapidly (Duncan 1992; Illius & Gordon 1992). Comparison of the digestion rates and efficiencies of ruminants and non-ruminants is made difficult by the fact that the size of the animal has an important influence. Figure 6.8 shows results from a model that used measured data but made adjustments for the animals' size, to give predictions for a ruminant and a non-ruminant each of 1 tonne body weight (about the size of a rhinoceros). In the graphs the horizontal axis, 'potential digestibility', depends on food quality, i.e. the proportions of more and less digestible material in the food. The vertical axes show the actual amount taken in and assimilated by the animal. Ruminants are able to assimilate more, but this is offset by their lower intake rate, so the hindgut fermenter is predicted to obtain slightly more energy per day whatever the composition of the food. This does, however, assume that enough food is available to satisfy the higher requirement of the hindgut fermenter: if food is in limited supply, then the ruminant could be at an advantage.

Effects of the animals on the vegetation

Effect on plant productivity

In Fig. 6.2(a), line P_1 indicates that grazers reduce primary productivity by removing part of the photosynthetic apparatus. Figure 6.2(b) provides one experimental confirmation of this. Grazing can nevertheless increase the growth rate—i.e. the rate of weight increase—of the sward (Fig. 6.3). This is partly because there is less death of the young leaves regrowing after grazing.

Can grazing increase plant growth?

There has been considerable debate about the circumstances under which grazing on an individual plant can increase its growth, partly because of confusion about how growth should be measured. Is it merely relative growth rate (i.e. relative to the much smaller weight of the remaining plant)? Or does the grazed plant grow fast enough to regain the size of other, ungrazed plants? Or does it actually grow fast enough to exceed them? There is no doubt that the first two do occur. For example, in field experiments in native grassland in Argentina, Semmartin and Oesterheld (1996) showed that following clipping the relative growth rate of the clipped patches was sufficiently increased above that of unclipped patches that after 2 months the green biomass was closely similar in both clipped and unclipped patches.

Table 6.1 Measurements from paddocks at a site in New South Wales in which different numbers of sheep were grazed from 1963 onwards

	Sheep per hectare			
	10	20	30	Standard error
Mean dry weight of above-ground plant material (g m^{-2})*	183	160	71	16
Photosynthesis during 48 weeks (kg CO_2 m^{-2})*	3.86	4.87	3.77	0.42
Sheep biomass (kg ha^{-1})†	528	981	932	36
Wool production (kg ha^{-1} yr^{-1})†	52	102	90	5

* Measured 1969–70
† Measured 1971–74

Data from Vickery (1972), Hutchinson & King (1980).

We should consider whether after grazing photosynthesis can be increased (e.g. as line P_2 of Fig. 6.2(a)), as well as leaf death being reduced. Other questions to consider are: Can grazing cause change in plant species composition? If so, does this alter the productivity of the vegetation, its palatability and digestibility? Does grazing ever cause irreversible changes in the vegetation, i.e. that cannot be reversed by reducing the grazing intensity?

Milchunas and Lauenroth (1993) drew together from the literature 276 cases where grazed and ungrazed areas had been compared. Most were in temperate regions and were grazed by domestic animals. On average the above-ground net primary productivity was 23% lower in the grazed than in the ungrazed areas. However, there was a wide variation among sites, ranging from the grazed areas having less than half the productivity of the ungrazed, to a substantial number where grazed and ungrazed differed by less than 10%, and a few where grazed had higher productivity by 10–30%. At 22 sites root mass was measured: on average grazed areas had higher root mass than ungrazed.

This last analysis of results did not take into account which differences were statistically significant. However, there are some cases where grazing has definitely increased productivity. Table 6.1 shows results from an experiment in New South Wales, where pastures of perennial grass + clover had carried different densities of sheep for 6 years before the first measurements were made. Photosynthesis was measured, by gas exchange, every few weeks over a year. It was the plots with intermediate stocking density—20 sheep per hectare—that gave the highest total photosynthesis over the year, carried the greatest weight of sheep per hectare and produced the most wool. The cause of this higher productivity was not discovered.

There are several possible ways in which grazing could increase photosynthesis. Caldwell *et al.* (1981) and Nowak and Caldwell (1984) investigated how clipping affects two *Agropyron* spp., bunchgrasses that are important in semi-arid grazing lands of the American Intermountain West. Following clipping, photosynthesis by the remaining leaves, per unit leaf area, was increased up to twofold. This was associated with increased soluble protein in the leaves. It has been found for leaves of other species that higher protein concentration and higher rates of photosynthesis tend to go together. This is at least partly because some of the protein is enzymes involved in photosynthesis. Increased protein concentration also raises the nutritive value to the grazing animal. We can ask where the nitrogen for the extra protein came from: from the soil, or from other parts of the plant? If from other parts of the plant, the increased photosynthesis may be at the expense of their growth. So, short-term growth responses do not necessarily show how long-term grazing will influence the plants.

As mentioned earlier, in only a minority of cases does grazing increase primary productivity. If the removal of foliage is sufficiently severe, the sward photosynthesis must inevitably be reduced. In Fig. 6.2(a), line P_2 is shown bulging up, but it still curves down to the right as stocking density increases further.

Effect on species composition

As we have seen, grazing animals can be very selective (e.g. Figure 6.6). This could alter the species composition of the vegetation. If the species favoured by the animals are more nutritious, for example because they have more protein, less fibrous material or less tannin, continual selection could, by reducing the abundance of these species, reduce the overall nutritive value of the pasture. In tropical savanna regions cattle eat grasses but rarely eat the leaves of shrubs and trees. At high stocking densities this can lead to vegetation more strongly dominated by woody plants, with much less grass and hence less valuable for cattle grazing (Walker *et al.* 1981). However, domestic grazers sometimes have the opposite effect, reducing the abundance of some woody species. In South Australia sheep prevent the survival of seedlings of some tree species, and the reduced abundance of some tree species in the 1990s can be related to the abundance of sheep in 1860–1900 (Tiver & Andrew 1997). Deer can also reduce the abundance of some tree species (see Fig. 10.1).

Tree/grass balance in savanna

Species balance in British uplands

In Britain grazing can change the abundance of grass species in upland pasture. *Nardus stricta* is a grass abundant on acid soils, whose leaves are very fibrous. Sheep favour softer-leaved grasses, leading to the danger that over some years grazing will reduce their abundance relative to *Nardus*. Figure 6.9 shows results from long-term experiments in two upland areas which had been grazed by sheep for many years, in which plots were then fenced off to exclude the sheep. After 24 years *Nardus* had almost disappeared from the ungrazed plot, showing that its abundance

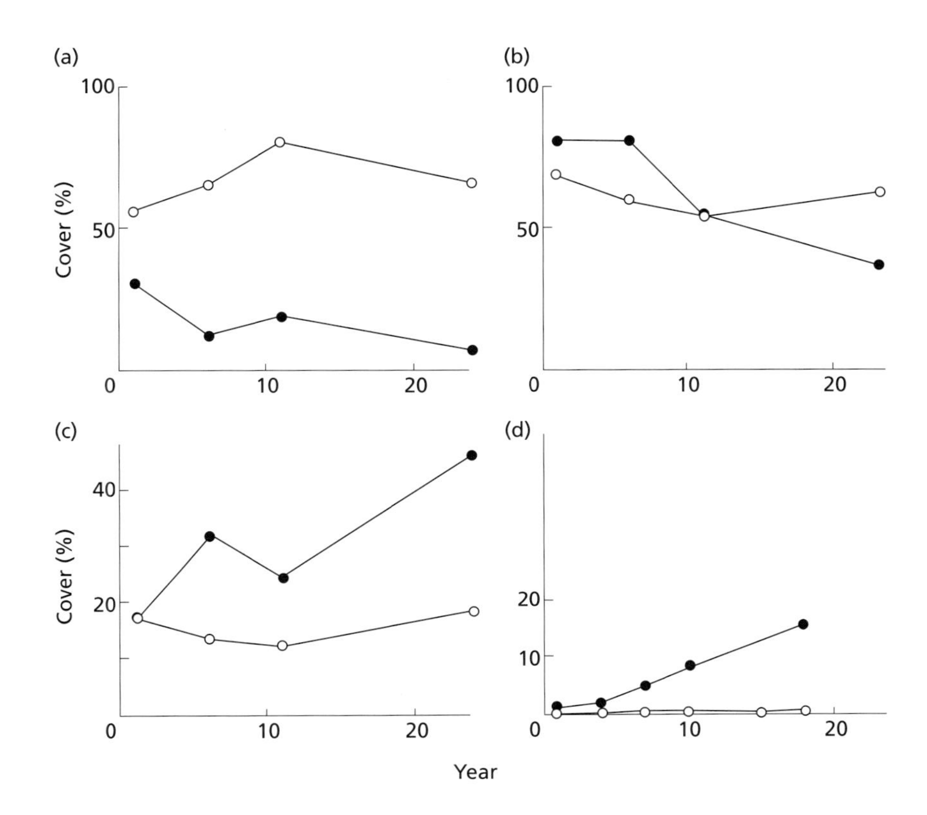

Fig. 6.9 Abundance of species at two sites in upland Britain. ○ grazed by sheep; ● sheep excluded from year 0 onwards. (a)–(c) Three grass species, *Nardus stricta* (a), *Festuca ovina* (b) and *Agrostis vinealis* (c), in Snowdonia, North Wales. From Hill, Evans & Bell (1992). (d) heather (*Calluna vulgaris*) in northern England. From Marrs, Bravington & Rawes (1988). Reproduced with kind permission from Kluwer Academic Publishers. Note difference in scale between upper and lower graphs.

in the heavily grazed pasture outside was dependent on grazing. In contrast, the very palatable *Agrostis vinealis* was sparser in the grazed area and increased when grazing was excluded, to become the most abundant species. However, *Festuca ovina*, which is also favoured by sheep, was abundant in the grazed area and declined somewhat in the exclosure, i.e. it behaved more like *Nardus* than like *Agrostis*. This shows that palatability to the grazers is not the only factor that determines how a plant species will respond to grazing. Two other factors likely to be involved are: (1) the stature of the plant, since species that can grow tall in the ungrazed pasture will be successful competitors for light; and (2) where the meristems are, as plants with apical meristems will be more damaged by grazers than those with basal meristems. Both of these are likely to be important in the way grazing affects the balance between heather and herbaceous species. Figure 6.9(d) gives results from another exclosure experiment, in northern England, where heather was extremely sparse on the grazed hillside but increased steadily once grazers were excluded.

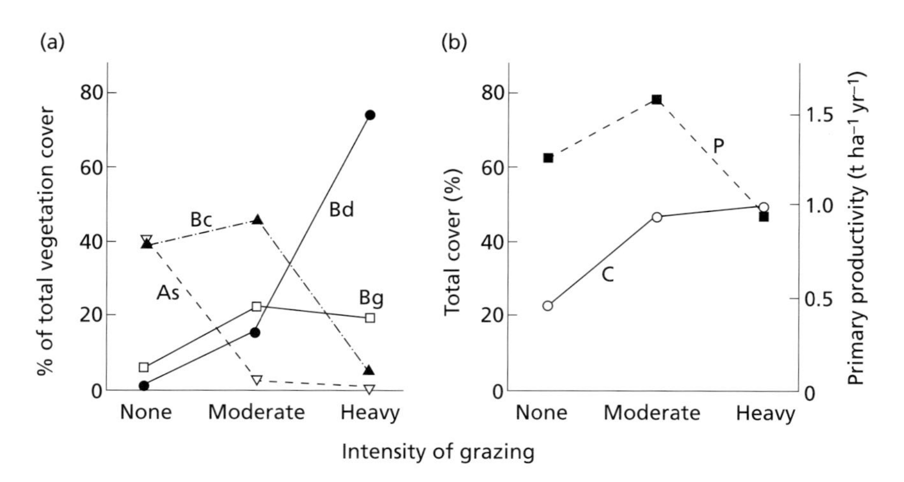

Fig. 6.10 Data on vegetation of three adjacent areas of prairie in western Kansas, USA, subject to no grazing, moderate or heavy grazing by cattle. Data of Tomanek & Albertson (1957). (a) Percentage of total vegetation provided by the four most abundant grass species. As, *Andropogon scoparius* (little bluestem); Bc, *Bouteloua curtipendula* (side-oats grama); Bd, *Buchloe dactyloides* (buffalo grass); Bg, *Bouteloua gracilis* (blue grama). (b) Above-ground dry weight production (P) and total ground cover (C) by all species combined.

This may seem surprising, since sheep strongly prefer grasses and other herbaceous species to heather (Fig. 6.6). In the absence of grazing heather can outcompete most grasses, but when sheep are present even light grazing destroys the apical meristems of heather, thereby reducing its growth and abundance. In contrast, grasses, sedges and rushes have a meristem at the base of each leaf, so growth can continue even if most of a leaf is eaten.

Changes in US prairie

The morphology of the plants was also important in the response of American prairie grasses to cattle grazing. Before the introduction of cattle the prairies had been only lightly grazed by native large herbivores such as bison. This allowed dominance in many areas by bunchgrasses, tall species whose apical meristems are high above the ground and so easily damaged. The effect of grazing on the prairie vegetation was much studied in the first half of the 20th century. The techniques were crude by modern standards, but the principal results were so clear that few people would wish to dispute them. Figure 6.10 shows results comparing different grazing regimes which, as in most of these older studies, were not instigated by the researchers: there is no information on how long they had been happening. The example given is from unreplicated adjacent areas, but the paper also gives similar results from five other sites in western Kansas. Each of the four most abundant grasses responded differently to grazing (Fig. 6.10(a)). The two dominant species in the ungrazed prairie, *Andropogon scoparius* and *Bouteloua curtipendula*, were fairly tall and there was much ground between them that was bare or covered by dead plant material. These two were almost completely replaced in the heavily grazed area by lower-growing, more spreading grasses, and the percentage of the ground covered by vegetation increased substantially under grazing (Fig. 6.10(b)). Perhaps as a result of this increased cover, shoot production was higher on the moderately grazed area than on the ungrazed; however, this was not true at most of the other five experimental sites elsewhere in Kansas.

Can grazing cause irreversible changes?... in USA

This leads on to the question of whether grazing can lead to changes that are irreversible. In Fig. 6.10 heavy grazing led to the virtual disappearance of one of the original dominant grasses, *Andropogon scoparius*, which would make its natural regeneration difficult if grazing were later reduced. Further west in the USA vegetation changes were compounded by the invasion of annual species, mostly natives of Eurasia. Much of the Intermountain West (parts of Utah, Nevada, Idaho, Oregon and Washington) had before the 19th century been dominated by perennial bunchgrasses or by shrubs + bunchgrasses. There were few large herbivores. Between 1880 and 1900 annual species spread rapidly, later taking over as the dominant vegetation in large areas. Cheatgrass (*Bromus tectorum*) became particularly prevalent. Its spread and establishment have been described by Mack (1981, 1986), Stewart and Hull (1949) and Klemmedson and Smith (1964). The question here is how far the conversion from dominance by bunchgrasses to dominance by annuals was due to overgrazing. During the key period, 1880–1930, relevant changes were:

1 increased opportunity for seeds of aliens to be introduced, e.g. due to building of railways, and import of crop seeds probably contaminated with weed seeds;

2 ploughing of land for arable, which was then suitable for invading annuals, growing as weeds;

3 much increased numbers of cattle, which not only damaged the tall bunchgrasses by grazing, but eroded with their hooves the poorly protected soil between the tussocks.

Probably all of these changes contributed to the spread and establishment of annuals. In order to decide how important cattle were, we would need long-term records in areas with and without cattle. These are not available. Daubenmire (1940) described a prairie site in southeastern Washington that had been long protected from cattle by a railway cutting. In it the annual cheatgrass was very abundant, showing that it can survive in competition with bunchgrass in the absence of cattle.

As food for grazers cheatgrass is not a total disaster since it is eaten by cattle, who thrive on it. At a site in Idaho, in some years cheatgrass had greater productivity than the perennial crested wheatgrass (*Agropyron cristatum*), but it fluctuated much more from year to year than did the perennial (Stewart & Hull 1949). Another disadvantage is that cheatgrass provides good grazing only in spring, and dies off by early summer. The dead parts are prone to burn in summer, and summer fires make it difficult for perennials to reinvade. This is an example of how a change in

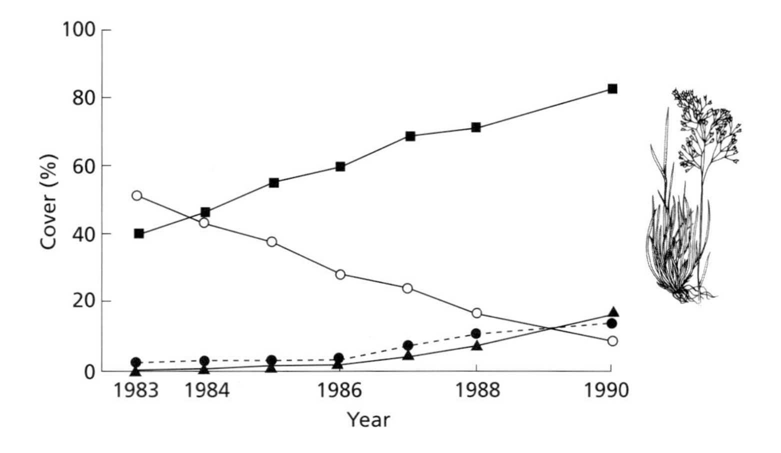

Fig. 6.11 Vegetation in an area of the Peak District, Derbyshire, England, following reduction of sheep density. ■ *Deschampsia flexuosa* (a grass, pictured), ● bilberry (*Vaccinium myrtillus*), ▲ heather (*Calluna vulgaris*), ○ bare ground. Data from Anderson & Radford (1994). Reprinted with permission from Elsevier Science.

species composition can alter the environment in a way that makes the change almost irreversible. Even in the absence of fire it has been found difficult to reintroduce perennial grasses by seeding (Stewart & Hull 1949).

... in England

An example where grazing caused a marked change which had almost reached a point of no return is provided by the Peak District in England. This upland area lies between Manchester and Sheffield and is much used for weekend open-air recreation. In the part described here much of the surface soil is peat, and so prone to erosion. In a vegetation survey in 1913 it was described as heathland dominated by heather and bilberry. By 1982 there was much bare peat, and most of the remaining vegetation was grasses: the heather and bilberry had virtually disappeared (Anderson & Radford 1994). Possible contributory causes to this change were trampling by people, fire, and large numbers of sheep. As described earlier (Fig. 6.9(d)), sheep grazing can cause the virtual disappearance of heather. It was decided to substantially reduce the number of sheep in the area from 1983 onwards. One question was whether heather had become so sparse that it could not re-establish. Figure 6.11 shows changes in vegetation cover that occurred over the next 7 years. Although there was no control plot in which grazing continued, there can be no serious doubt that these large changes were the result of reduced sheep density. By 1990 the amount of bare peat, and hence the risk of erosion, had been greatly reduced. This was due primarily to a great increase in the grass *Deschampsia flexuosa*. However, heather and bilberry were also increasing, and may in due course return to their former dominance. Heather's re-establishment was mainly by seedlings: it has small, wind-dispersed seeds, and seeds may have blown in from areas some distance away, where heather had remained more abundant.

In trying to determine the effects of grazing animals on vegetation, a recurring problem is lack of adequate data. As these examples have shown, vegetation may change slowly, over decades. What is required, then, are sites known to be initially similar in vegetation, soil and other environmental factors, in which contrasting grazing regimes are maintained over decades and in which records of the vegetation are made at the start and at intervals thereafter. Figure 6.9 gives results that come close to this, though it is unfortunate that records were not made until a year after sheep had begun to be excluded. But such long-term records are rare and we have to make do with less satisfactory substitutes. The section on the Sahel in Chapter 3 has already shown how difficult it can be to determine what changes in vegetation have occurred, and whether the cause is grazing, other human activities or something else such as climate change. Another problem when interpreting vegetation change is lack of information on how this changes the digestibility and nutrient value to the animals.

In spite of these limitations in the available evidence, this section has shown that grazing at a moderate intensity can sometimes lead to increased primary productivity by the vegetation and to changes in the species composition that are beneficial to animals. However, moderate grazing has in other places had harmful effects on the vegetation, and if grazing intensity continues to increase it is certain to become harmful sooner or later.

Several ungulate species together

So far we have considered one animal species at a time. Would it be possible to increase total meat production by having a mixture of ungulates? This could be sheep plus cattle; or one of those plus native species; or would it be better to have game ranching with a mixture of native species?

Arguments in favour of mixtures mostly centre on differences in diet. If each animal species eats different plant species, we might expect them together to make more efficient use of the vegetation than would any one of them on its own. There has been much research on the diet of coexisting ungulate species, especially in African savanna where many species coexist. Lamprey (1963) made an early and influential study in Tanzania. In the Tarangire Game Reserve 14 species of large herbivore coexisted. Table 6.2 summarizes the diets of four species, all of which spent much of their time in open woodland. They each have a diet with a different composition, though with substantial overlap. The diet is not related to whether the species is a ruminant: buffalo and giraffe are ruminants, warthog and elephant are not. As explained earlier, there is no consistent difference in diet between ruminants and non-ruminants.

Diet of African large mammals

Table 6.2 Diet of four large herbivore species in Tarangire Game Reserve, Tanzania

	% of diet			
Species	Grasses	Other herbaceous species	Shrubs	Trees
Buffalo	94	1	5	0
Warthog*	82	15	1	1
Elephant†	12	1	30	56
Giraffe	1	1	12	86

* also ate roots and bulbs
† also ate fruit and bark

Data from Lamprey (1963).

Although there were differences between all the four species in Table 6.2, it is sometimes useful to classify large herbivores as grazers (which eat mainly grasses and other herbaceous plants) and browsers (which eat mainly leaves and other parts of woody plants). Buffalo and warthog are grazers; elephant and giraffe are browsers. The physiological and anatomical differences between grazers and browsers relating to this difference in diet are still being investigated (Gordon & Illius 1994). Perhaps the most obvious expectation is that browsers would be larger, to enable them to reach up to taller trees. In fact the reverse is true among African species: many browsers have body mass below 50 kg, e.g. bushbuck, springbok, whereas many grazers are larger than 50 kg, e.g. wildebeest, buffalo (Gordon & Illius 1994). However, giraffe and elephant are clear exceptions, being large browsers. There are consistent trends among ungulates relating body size to diet and digestion (Hodgson & Illius 1996; Hanley 1997). The smaller the animal, the more energy it requires per unit body mass. This is one reason why smaller species tend to be browsers, because browse tends to have a higher percentage of the easily digestible cell solubles (though also of lignin), whereas the grasses are higher in cellulose. Smaller animals, because they have smaller mouths, are better able to select their food on a smaller scale and may therefore be able to choose a more nutritious diet.

Domestic cattle and sheep sometimes have markedly different diets from the most abundant wild species. In Zimbabwe cattle were grazed in savanna woodland where impala and kudu were the most abundant wild ungulates. The cattle were eating almost entirely grasses and herbaceous species, whereas the impala and kudu were mostly browsing on the woody vegetation (Fritz *et al.* 1996). To take a temperate example, a study was made in prairie in northeastern Colorado of the diet of cattle, sheep and two native ungulates, bison and pronghorn antelope. The vegetation was classified as grasses, forbs or shrubs, and the grasses and forbs were further divided into warm season (which grow mainly in summer) and

Four grazers in prairie

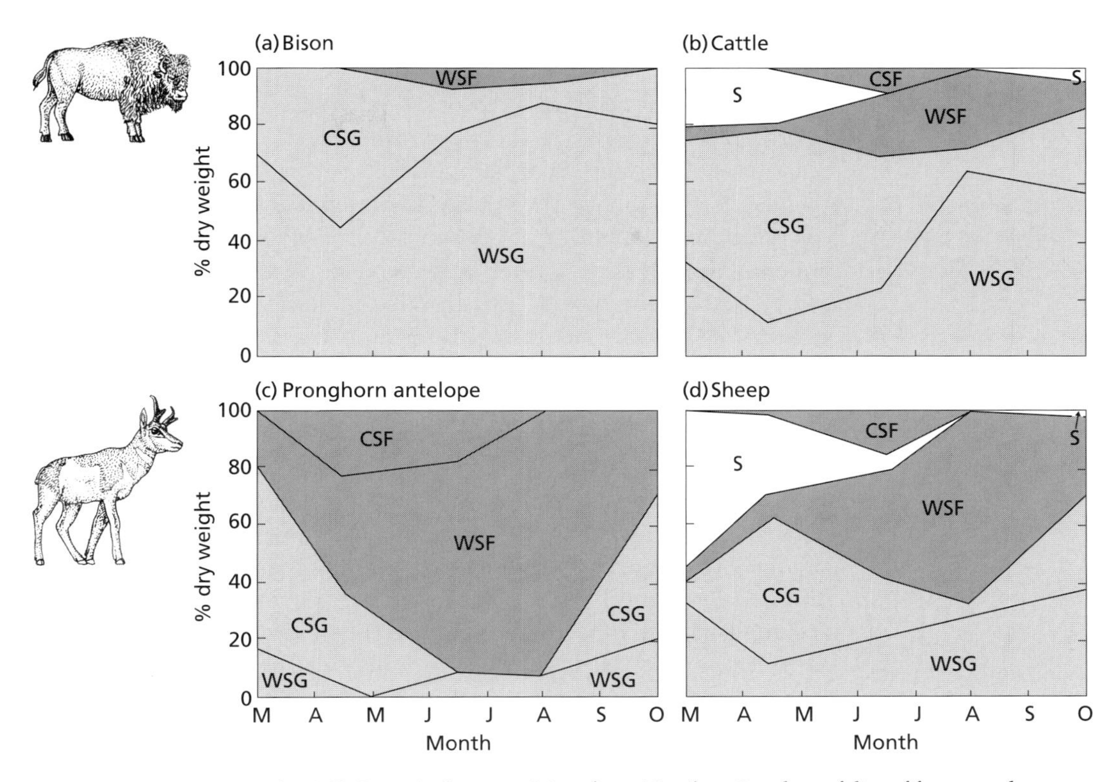

Fig. 6.12 Botanical composition, from March to October, of diet of four ungulate species in prairie in Colorado. CSG = cool-season grasses, WSG = warm-season grasses, CSF = cool-season forbs, WSF = warm-season forbs, S = shrubs. From Schwartz & Ellis (1981).

cool season (which grow substantially in spring and autumn). Figure 6.12 shows that bison ate predominantly grasses and scarcely any forbs. In contrast, during the summer pronghorn ate predominantly forbs, and never ate much of the warm-season grasses. So there was a marked difference in diet between these two species. Cattle and sheep differed little from each other in food choice. Unlike the two native species they ate some shrubs in spring, but in other respects their food choice was intermediate between the two natives.

This example suggests that there would be little advantage in grazing sheep and cattle together. However, more detailed studies have shown differences between them in what they eat. At the site featured in Fig. 6.6 sheep were able to eat far more of the fine-leaved grasses than cattle, but cattle ate more moss and dead plant material. This relates to the size and shape of their mouths: sheep are able to bite closer to the ground and obtain food from a short turf, whereas the larger mouth of a cow takes in a more mixed sample, which can include less nutritious dead material. Figure 6.9 shows how sheep, by selective grazing, can lead to an increase of the wiry grass *Nardus* at the expense of the softer-leaved *Agrostis*. Research at another site (Grant *et al.* 1985) showed that sheep selected strongly against *Nardus* and in favour of softer-leaved grasses, but cattle ate each group approximately in proportion to its abundance in the sward. *Nardus* forms small tussocks, and the small-mouthed sheep can select the other grasses in between, whereas the larger-mouthed cattle tend to take a mixture in each mouthful.

Do mixtures of animals have higher productivity?

There is thus plenty of evidence that ungulate species living together can have diets that differ, at least in the proportion of the different components, though there is often considerable overlap. However, this does not prove that the overall growth rate of animals is higher if there is a mixture of species. Evidence on this is more limited. The Turkana people of northern Kenya provide an example of the use of several animal species (Coughenour *et al.* 1985; Coppock *et al.* 1986). They live in an area of low seasonal rainfall; the vegetation is open savanna woodland. They are almost entirely dependent on their animals for food, so milk is an essential source of daily energy. The grasses die down in the dry season, leaving little fodder for grazers. Traditionally part of the Turkana's solution to this was to migrate with their herds, following the movement of the rains. Such migration is practised by fewer people in Africa nowadays, although in Australia there is an increase in the practice of trucking cattle from one part to another to follow the rains (Hodgson & Illius 1996). The other part of the Turkana's solution has been to keep a mixture of animals. Their cattle eat almost entirely grasses plus some forbs; the camels eat mainly woody plants; and the sheep, goats and donkeys have a more varied diet, eating a mixture of woody and herbaceous species. The cows give a lot of milk in the rainy season, but when grass growth stops in the dry season they stop lactating. The camels, however, continue to give milk, which is a crucial food source for the people at that time. The combination of strongly seasonal climate and very varied vegetation makes this a situation where the advantage of having more than one animal species is clear.

African pastoralists who keep five species

Domestic or wild species?

Fritz and Duncan (1994) drew together data from numerous sites in eastern and southern Africa and used an analysis of covariance to find out whether the biomass of ungulates per hectare was related to rainfall, soil fertility, number of species, and to whether the animals were predominantly wild or domestic species. They excluded high-input animal farming, e.g. involving irrigation or supplementary feed. There was no significant difference in biomass domestic versus wild animals, and much of the variation was accounted for by rainfall and soil fertility. In addition, there was a smaller but significant relation of higher biomass to the presence of more ungulate species. This was not just a concealed comparison of wild species versus (fewer) domestic species, since the species-richness relation still showed up when wild species were analysed on their own. These results could suggest that having more animal species leads to more animal biomass; on the other hand, more biomass could allow more species. The results do not indicate that, overall, more meat can be obtained in Africa by game ranching than by grazing cattle.

In a similar comparison in grasslands of southern South America, herbivore biomass, at a particular rainfall level, was about 10 times higher if there were cattle or sheep than if there were only wild animals (Oesterheld *et al.* 1992). This relates to South America's having a low abundance of native ungulates. At the end of the Ice Age many large mammal species became extinct there (Stuart 1991), and so there are fewer species of large ungulate in South America than in Africa. What restricts the abundance of the remaining South American species is uncertain.

These studies in Africa and South America used animal biomass data, not growth rate or production. Animal biomass per hectare is an indicator of production, but species do differ in their growth rates.

Sheep and cattle together

In temperate regions meat production by mixtures of sheep and cattle has sometimes been compared with that by sheep or cattle alone. Connolly and Nolan (1976) reported results from an experiment in which cattle and sheep grazed separately or together, in various densities and proportions, on grass–clover pasture in Ireland. The weight gain by the animals was measured, and from the results it was calculated that in 1974 the greatest total weight gain per hectare during the season—302 kg ha^{-1}—would be achieved by having 16.6 sheep per hectare with 4.2 cattle per hectare. For cattle alone the highest gain was 275 kg ha^{-1}, and for sheep alone 235 kg ha^{-1}. However, in 1973 sheep performed poorly, and cattle alone gave a higher weight gain than any sheep + cattle combination. In a later, similar experiment on well-fertilized ryegrass pasture, averaged over 4 years, the greatest weight gains were by mixtures of sheep + cattle (Nolan & Connolly 1989). The authors suggested this was because the sheep graze the taller vegetation close around cow dung, which cattle avoid. In contrast to these results, Abaye *et al.* (1994) in Virginia found a greater weight increase by sheep than by cattle or by the two together. These experiments showed that a sheep + cattle mixture can sometimes give a higher meat yield than either species separately; but this is not always true, and reasons for the differences in results are not clear.

Grazing management and biodiversity

This chapter has so far assumed that the sole aim of grazing management is the production of food for people. However, grazing lands are also home to wild species. The previous section considered wild ungulates and their differing food requirements. Here I discuss how grazing can affect the diversity of plant species and of insects.

Grazing can increase plant species diversity

Table 6.3 shows results from experiments in grassland in various parts of the world, in which fencing allowed neighbouring areas to be with and without grazing for some years. At most sites the grazed grassland had higher plant diversity, whether assessed by number of species or by the Shannon Index. One likely reason for this is that, by keeping the sward short, grazing allows low-growing species to survive, whereas in tall,

Table 6.3 Plant species diversity in field experiments in which grassland areas with and without grazing have been compared

				Plant species diversity			
				Number of species		Shannon Index†	
Location	Features	Grazers	Duration (yrs)*	Ungrazed	Grazed	Ungrazed	Grazed
(1) N. England	One hay cut per year	Cattle + sheep	4	15.6	20.9		
(2) N. England	Upland	Sheep					
	pH 5.4		24	24	31		
	pH 4.1		23	5	7		
	pH 4.0		23	10	10		
(3) Netherlands	On chalk	Sheep	11	14.5	41.6	0.71	3.10
(4) East-central Argentina	Wet pampas	Cattle	7	15.2	24.5	0.82	0.98
(5) Serengeti, Tanzania	Savanna	Native spp.	5				
	short-grass					1.04	1.97
	mid-grass					0.79	1.43

* How long the contrast of grazed and ungrazed was maintained.
† See Box 10.6, p. 311.

Sources: (1) Smith & Rushton (1994); (2) Rawes (1981); (3) During & Willems (1984); (4) Rusch & Oesterheld (1997); (5) Belsky (1992).

ungrazed grassland they are shaded out. As an example of this, Table 6.4 gives more information from the most species-rich of the three upland English sites (pH 5.4) of Table 6.3. After grazing was excluded most of the bryophytes and lichens decreased; among the angiosperms that decreased many were of low stature. Species that increased were predominantly tall or straggling grasses.

Grazing animals can have other effects besides keeping the vegetation short: they deposit urine and faeces in patches; they also create gaps in

Table 6.4 Effect of preventing grazing in an area of grassland in the Pennine Hills, northern England: number of plant species that changed significantly in cover percentage during 22 years inside a fenced area from which sheep were excluded. Based on records from 1000 point quadrants

Effect of no grazing	Grasses	Other angiosperms	Bryophytes	Lichens	Total
Significant increase in cover	5	3	0	0	8
Significant decrease, but still present at end	2	6	12	2	22
Decreased to zero	0	4	7	3	14

Data of Rawes (1981).

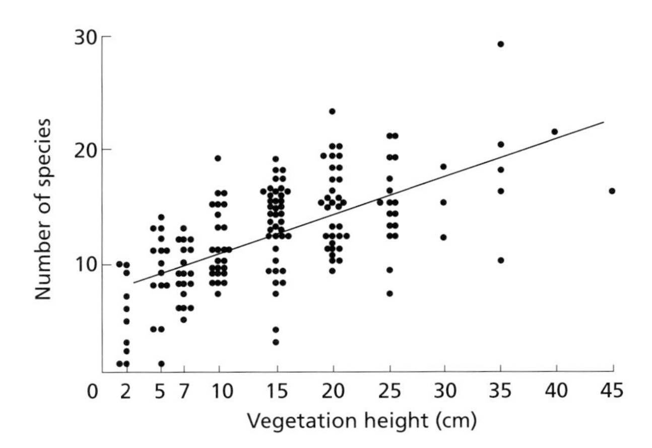

Fig. 6.13 Diversity of leaf-hoppers (Auchenorhyncha) at English grassland sites, in relation to tallness of vegetation. The line is the best fit linear regression. From Morris (1971a).

the sward by trampling, in which seedlings establish, and this can alter the plant species composition (Williams & Ashton 1987; Silvertown & Smith 1988; Williams 1992; Bullock *et al.* 1995).

Tall grassland is not always species poor: old hay-meadows cut only once or twice a year can be rich in plant species (see Fig. 10.13(b)). That graph also shows that there can be a conflict between productivity and diversity. In that long-lasting experiment the addition of fertilizers increased productivity but reduced the number of plant species.

If vegetation is managed to promote increased plant species diversity, will this also achieve high animal diversity? The answer for insects is often yes, because many species of herbivorous insects eat only one or a few species of plant (see Chapter 10). However, it may be less true of grassland than other vegetation types, because grasses contain few of the secondary chemicals that are responsible for much of the specificity in insect-plant relations (Harborne 1993). Species diversity in animals is often also promoted by structural diversity in vegetation. Figure 7.6 gives an example of this for birds in forests. If this applies also to invertebrates in grassland, one might expect that taller grassland, even if less rich in plant species, would contain more species of invertebrates that live above ground. This is certainly sometimes true, and Fig. 6.13 shows an example. Among more than 100 English grassland sites, whose vegetation tallness ranged from 2 cm to about 40 cm, there was a clear and statistically significant tendency to have more species of leaf-hopper in taller vegetation. This was true also for two other groups, Heteroptera (plant bugs) and weevils. Experiments comparing grazed and ungrazed plots confirmed that for these invertebrate groups grazing is unfavourable to diversity (Morris 1969, 1971b). However, this is not true for all invertebrates in all grassland. In northern Colorado, large prairie plots have since 1939 received heavy, moderate or light grazing from cattle, or no grazing at all. In the early 1970s the diversity of macroarthropods (above- and below-ground) was highest in the moderate-

But grazing may reduce insect diversity

Table 6.5 Abundance of three bird species that nest in prairie, in plots in Colorado that had received different intensities of cattle grazing for about 30 years. Figures are breeding pairs per hectare

	Grazing intensity		
Species	Light	Medium	Heavy
Horned lark	0.3	0.6	0.9
Mountain plover	0	0	0.2
Lark bunting	0.7	0.4	0

Data from Milchunas *et al.* (1998)

grazing plots, lower if there was heavy grazing or none (Milchunas *et al.* 1998).

Different species are favoured by different grazing regimes

Grazed and ungrazed grassland provide different habitats: it is therefore unlikely that any management will be best for all species. Grazed areas provide habitat for dung feeders, ungrazed are good for decomposers of dead plant material. Thomas (1991) described the requirements of six butterfly species that occur in grassland in southern England. Two of them, both small and brown, illustrate different responses to grazing. The larvae of the silver-spotted skipper (*Hesperia comma*) will eat only one plant species, the grass *Festuca ovina*, and only if it is 1–5 cm tall and preferably next to bare ground. That is where the adults lay their eggs. The Lulworth skipper (*Thymelicus acteon*) larvae eat a different grass, *Brachypodium pinnatum*, which must be at least 10 cm tall and preferably 30 cm. Chapter 10 gives more information about how the food requirements of the silver-spotted skipper influence its distribution.

Grazing management also affects birds. In the long-term grazing plots in Colorado, some species of birds that nest in prairie were favoured by light grazing, some by heavy grazing (Table 6.5).

When managing grassland to promote wildlife we need to define our aims: do we want maximum α-diversity, maximum β-diversity, or to promote a few particular species? If we want maximum species richness in every square metre we may have to decide whether plant or invertebrate diversity is more important. But if we favour diversity at the landscape scale, we can provide habitats for more species by a mixture of managements—some areas heavily grazed, some lightly grazed, some as hay-meadows.

Conclusions

- Grazing animals can sometimes promote plant growth, but most often they reduce it by removing foliage. However, in fertile grasslands there are stocking densities of animals that allow productive swards as well as high consumption by the animals.

- How much the animals eat is influenced by the tallness, structure and chemical composition of the vegetation.
- Animals are selective among plant species. Grazing can affect the species composition of the vegetation, sometimes in ways that reduce its future grazing value, but not always.
- When several grazing mammal species inhabit the same area they often have different diets. However, there is only limited evidence that a mixture of animals can achieve higher total productivity than a single species.
- Shorter, grazed grassland often has a higher diversity of plants than ungrazed, but may have a lower diversity of insects. However, species differ in their response to grazing, and conservation of all wild species of grazing lands requires contrasting management in different areas.

Further reading

Grazing mammals: physiology, behaviour, diet selection:
Church (1988)
Duncan (1992)

Vegetation and grazing mammals: ecology and management:
Hodgson & Illius (1996)
Hudson, Drew & Baskin (1989)

Management in relation to conservation of wild species:
Hillier, Walton & Wells (1990)
Curry (1994, Chapter 5)
Sutherland & Hill (1995, Chapter 8)

Chapter 7: Management of Forests

Questions

- Can the world's present forests supply our timber needs on a sustainable basis? Or is loss of forests inevitable?
- Is forest area decreasing? Why? How fast?
- Can we obtain our timber needs from forests without undue harm to wildlife? Or should we aim to obtain most of our timber from plantations, and conserve forests elsewhere for wildlife?
- Before human interference, were all forests mixed-age, regenerating by small gaps? Or were large areas even-aged?
- Is it satisfactory to clear-fell large areas? Or is it better to remove individual trees to leave small gaps? How do wild species respond to these two felling methods?
- Can natural regeneration re-establish forests satisfactorily after felling? Or do we always need to replant?

Background science

- Amounts of wood used per year worldwide.
- Timber yields that can be obtained from plantations of different species. The best age to harvest them.
- Population dynamics of trees. How this affects forest structure.
- Influence of gaps on tree regeneration and on animals.
- Forest fires in the past: where they occurred, how frequently, their effects on forest structure and on animals.

Forests have many functions; Box 7.1 lists nine. The list is not intended to be in order of importance: different people will have different views on that. For people who live in forests, carrying out shifting cultivation and hunting, functions 1–4 may be essential. However, many readers of this book will no doubt give high priority to function 5, preservation of the high diversity of wild species in forests. As Box 7.1 notes, some of the functions of forests are dealt with in other chapters. This chapter considers only two functions, timber production and wildlife. The key questions for this chapter are: can an adequate amount of timber be obtained in a sustainable manner? And if so, how? Is this sustainable timber production compatible with wildlife conservation?

Box 7.1. Major function of forests.

1 They provide wood, which can be used for
(a) making things, or
(b) fuel (see Chapter 2).
2 They can provide other useful products, including fruits, tanning materials, latex.
3 Domestic animals can graze in them.
4 Some mammals and birds living in forests can be hunted for food.
5 Forests are a habitat for many wild species of plant and animal, whose continued existence may be important in its own right.
6 They influence the environment, e.g. conversion of forest to another type of vegetation can increase soil erosion (see Chapter 4) and alter water balance (see Chapter 3).
7 They are involved in carbon cycling and storage. Clearance of forests can alter the world's carbon balance (see Chapter 2).
8 They provide areas for recreation.
9 They contribute to the appearance of the landscape.

The meaning of 'sustainable' can range from narrow to broad interpretations (see Chapter 1). The most limited meaning would be that, at least when viewed over a large area, wood can continue to be extracted at the same rate year after year, always leaving enough growing trees to produce the wood for the future. This is a definition in terms of yield alone. Box 7.2 sets out more extensive requirements for sustainable production. A still more extensive definition would take into account maintenance of all the functions in Box 7.1: people and wild species living in or dependent on the forest would therefore be considered, as well as wider environmental effects such as continental climate and global carbon balance. Like all the other chapters in this book this chapter is selective: it concentrates on requirements 1 and 2 of Box 7.2—timber yield and tree species composition—and on whether these can be compatible with preserving wildlife. Requirements 3–5 of Box 7.2 are covered in other chapters. In Box 7.2 *maintenance* of production is the expressed aim; but sometimes we may consider whether improvements are possible, e.g. increasing the abundance of desired species.

Box 7.2. Requirements for sustainable timber production.

1 Adequate growth rate of harvestable timber, providing a steady supply.
2 The quality of timber is maintained. In mixed-species forests this involves maintaining a suitable complement of tree species.
3 Maintaining soil fertility (see Chapter 4).
4 Maintaining satisfactory water balance (see Chapter 3).
5 Damage by pests and diseases maintained at an acceptable level (see Chapter 8).

Has timber production in the past been sustainable?

Continuing timber production requires that after trees are harvested new trees replace them; and by having trees of different ages growing up there will always be some approaching a suitable age for felling. This is not a new idea: the Romans had forest management systems that probably succeeded in achieving sustainable production. There have been continuous traditions of sustainable forestry in parts of Europe since the Middle Ages, e.g. in parts of Germany (Heske 1938), but in other parts of the world at other times standing forests have been treated as a non-renewable resource: people or organizations have extracted timber and then moved on. Williams (1989) provides maps that show how timber felling moved across the United States. During the early part of the 19th century it was concentrated in the northeast, but as those forests became worked out felling spread to the Lake States, the southeast and, by the early 20th century, to the Pacific Northwest. Clear-felling as a one-off activity happened in other countries too, for example in Australia. Some people assume that once the trees have been cut down that is the end of forest on that site for ever. But new trees can establish and grow. We should not assume that because the timber extractors took no interest in what happened to the cleared-felled areas after they left, those areas were never forests again. Forest regeneration is considered in detail later in this chapter.

Management systems

When we consider the whole world, a wide variety of management systems for forests has been used; others have been tried out experimentally.

Box 7.3. Systems of forest felling and regeneration.

Felling

1 *Selection systems.* Fell and remove individual trees, selected by species, large size or both.
2 *Shelterwood systems.* Fell strips or areas, leaving other strips or areas in between uncut.
3 *Clear-felling systems.* Fell large areas.

Regeneration

1 *Vegetative regrowth* ('sprouting'), from remaining stumps ('coppicing') or roots.
2 *Advance regeneration.* Seedlings and young trees that had already established under the large trees are left to grow on after the large trees are felled.
3 *From seed*:
(a) already in the soil ('seed bank');
(b) 'seed rain' from living trees nearby; these can be in patches, strips or as individual 'mother' trees;
(c) or sown.
4 *Young trees* grown in nurseries and planted out.

Particularly important features are how trees are selected for felling and how the establishment of new trees (regeneration) is brought about. Box 7.3 lists the main alternatives. The system that most closely emulates farm crops involves clear-felling and then replanting with seeds or seedlings. This gives considerable control, for example over the genetic complement of the new trees, and over their spacing. But even in selection systems there are things the foresters can do to influence how the forest develops in future. Desired species may be encouraged by, for example, leaving gaps of particular size (discussed more later) or removing undesired species by cutting or poisoning. Such aspects of management are components of *silviculture*. This chapter discusses the systems listed in Box 7.3, and considers which of them is the most appropriate for forests of different types in different parts of the world.

Plantations

In a plantation many young trees are planted in the same year, so that an even-aged stand grows up, usually all of one species. After some years they are clear-felled, i.e. a large area is felled in one year. The area is then replanted. If this system is to provide a continuous supply of timber there will need to be a *rotation*, in which different parts of the whole forest are planted in different years and so become ready for felling in different years. The *rotation length*—the time required for growth between planting and felling—is clearly crucial in the planning of such rotations. How is this rotation length decided?

Annual growth in plantations

After a plantation is established, during the early years growth is slow, primarily because much of the solar radiation falling on each hectare is not intercepted by the foliage of the small and widely spaced young trees. The rate of wood production per hectare per year—called the *current annual increment* (CAI)—at first increases year by year (Fig. 7.1). However, in due course the plantation reaches an age when the current annual increment peaks and then starts to fall. You might think that this is the best moment to harvest the trees, but not so. If the aim is to grow a sequence of tree crops on the site, and to maximize the timber yield over several rotations, then the *mean annual increment* (MAI) is the crucial thing to measure.

Mean annual increment = volume of timber in stand/age of stand

This is what determines the timber yield per 1000 years if there is a continuous series of rotations on a site. As Fig. 7.1 shows, when CAI peaks and starts to fall it is still above MAI, and current growth is therefore still pulling MAI upwards. Where the two lines cross is the maximum MAI, and that is the time to harvest if the aim is to maximize long-term timber yield.

When to harvest

Table 7.1 gives examples of the ages at which plantations of various species, in various parts of the world, have been found to reach the

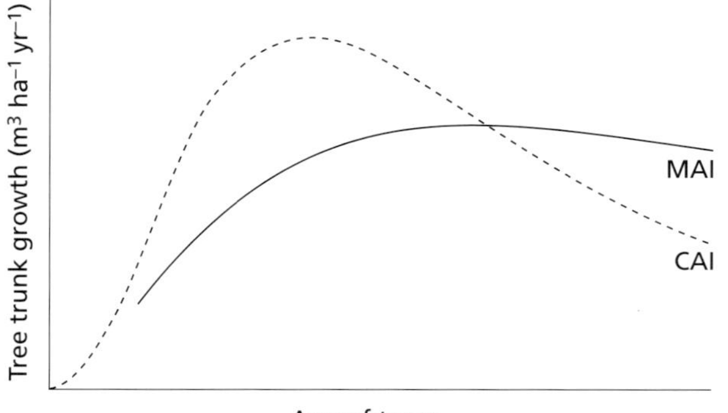

Fig. 7.1 Relation of tree trunk growth to plantation age. CAI, current annual increment; MAI, mean annual increment.

maximum mean annual increment, and what that MAI is. For each species a range is given, because even within the same geographical area the MAI of a species can vary greatly, depending how favourable the individual site is. These figures exclude very low yields caused by major catastrophes such as fire or pest epidemic. The figures in Table 7.1 show that there are big differences between species in the time to harvest and the MAI achieved. In general, species with lower MAI require longer until harvest; slow growers take a long time to reach the peaks of the curves in Fig. 7.1.

How delay affects the economics of forestry

In Table 7.1 the time from planting to harvest ranges from 4 to 130 years. The attitude of a landowner to a crop will be markedly influenced

Table 7.1 Timber yields of some tree species in plantations in various parts of the world

Species	Location	Mean annual increment $m^3\ ha^{-1}\ yr^{-1}$	Mean annual increment $t\ ha^{-1}\ yr^{-1}$	Age at harvest (yrs)*
Fagus sylvatica (beech)	Britain	4–10	(3–7)	120–75
Picea sitchensis (sitka spruce)	Britain	6–24	(2–10)	70–45
Pinus sylvestris (Scots pine)	Central Spain	5–10	(2–4)	100–80
Pseudotsuga menziesii (Douglas fir)	Northwest USA	2–18	(1–7)	130–60
Pinus radiata (Monterey pine)	Chile	(18–38)	(7–15	30–25
Eucalyptus globulus	Portugal	5–40	(3–26)	20–12
Eucalyptus camaldulensis	Morocco	3–11	(2–9)	1 6–9
Eucalyptus grandis	Transvaal	13–46	(7–25)	12–9
	Uganda	24–52	(14–29)	7–4
Gmelina arborea	Tropics	19–39	10–20	17–7
Albizia falcata	South-east Asia	20–60	8–23	?
Tectona grandis (teak)	Tropics	6–18	3–12	61–14
Coppice:				
Populus trichocarpa (poplar)	Britain	15	7	5

* Age to give maximum mean annual increment (except for coppice, where MAI was still increasing at final harvest).

Figures in parentheses are approximate. Sources of data: Hamilton & Christie (1971), FAO (1979), Cannell (1980), Curtis *et al.* (1982), Golley (1983), Whitmore (1984), Stage *et al.* (1988).

by such differences: it requires a special attitude of mind to take the trouble to plant trees which will not provide timber until after the person planting them is dead. Time delays are common in life: there is a delay in many activities between starting to invest effort and finally reaping the reward, whether it is growing a wheat crop, training for a job or building a dam for hydroelectricity. In plantation forestry delay is serious, because much of the investment is near the start (obtaining the land, fencing it off, making roads, preparing the ground, growing the seedlings and planting them), followed by a long period when there is little or no reward until the final harvest. Economists have given much attention to time delays. They generally assume that money has to be borrowed to finance an activity and that interest has to be paid while the money is on loan. One way of applying this to forest plantations is *discounting*. This says, for example: if that forest will be worth £1000 in 10 years' time, how much is it worth now (what is its *net present value*)? To calculate the net present value an interest rate is assumed. For example, if the interest rate is 5% and the value in 10 years' time will be £1000, the net present value is £614. In other words, if instead of setting up the plantation you invested the £614 in some other way at 5% compound interest, after 10 years you would have £1000. Discounting is one way to measure the 'cost' of rotation length. Sticking with the example of a final value of £1000 and interest at 5%, the net present value is: for harvest in 5 years' time, £784; 10 years, £614; 40 years, £142; and 100 years, £7.60. The greater financial rewards of fast-growing trees arise not only from their greater timber production per year, but also from their being ready for harvest sooner. For example, in Table 7.1 compare beech with sitka spruce, or *Gmelina* with *Tectona*.

The economics of long delays has a strong influence on what is considered worth doing to a plantation, and therefore on how it is managed. We might ask, for example, is it worth applying fertilizer to a young plantation? If the rotation length is 100 years and the interest rate 5%, the figure given above shows that spending £7.60 on buying and applying fertilizer when the trees are young will only be worthwhile financially if it increases the final value of the harvested trees by £1000. Other possible sources of yield increase, such as genetic improvement, are discouraged for the same reason.

Table 7.1 gives one example of wood production from *coppice*, in other words allowing stumps to resprout and harvesting the regrowth shoots when they are still quite young and narrow. This is not a new system: a prehistoric trackway several thousand years old, found preserved in peat in Somerset, England, was evidently made from poles from coppice growth (Clapham & Godwin 1948). The example of coppice in Table 7.1 has an MAI no higher than can be achieved from spruce in the same geographical area, but the rotation length is much shorter. Economics suggests that where narrow stems are acceptable, e.g. for making paper, coppice may have a future.

Can worldwide tree growth supply world timber needs?

If timber is being produced sustainably, trees are cut down this year but next year the total area of forest, and the total volume of timber standing on it, must remain just as high. Therefore, a basic requirement for sustainable timber production is that the amount of new wood produced each year by growth of tree trunks in the world's forests must be at least as great as the amount of wood harvested. Is this so?

The amount of wood we use

Table 7.2 shows that in 1995 the amount of wood harvested throughout the world totalled more than 3 km^3, an imposing volume if it had been a single cube. More than half of this was used for fuel; for many people, especially in large parts of the tropics, fossil fuels have not replaced wood as their primary fuel (see Table 2.5). The remainder of the wood harvested is classed as 'industrial roundwood': some of it is made into paper, some into chipboard and other composites, and some is used as intact wood for making furniture, wood-framed houses and many other things. Some people have suggested that the use of wood will in future decrease, as it is replaced by other materials such as concrete, metals or plastics, but so far there is no sign of that happening. Table 7.2 shows that worldwide use of wood increased by about 1.5% per year between 1975 and 1995. You may think that the use of paper will decrease, because the printed word will be made obsolete by electronics—telephones, compact discs, the Internet and so on, but again so far that has not happened: world paper production increased by nearly 50% over the 10 years from 1983–85 to 1993–95 (World Resources 1998/9). The fact that you are reading this book at this moment shows that little black marks on pieces of paper are not yet entirely outmoded as a way of conveying information.

To decide whether annual wood growth in the world's forests exceeds annual harvest, we need to know the total area of forest and its productivity. The area of forest and woodland given in Table 1.1 is not quite the figure we need, because it includes substantial areas of open woodland which are not useful as timber sources. What we want to know is the area of closed-canopy forest; Table 7.3 gives figures for that.

Could we get all our timber from plantations?

Table 7.2 suggests that world wood requirements will quite soon reach 4 billion m^3 $year^{-1}$. Supposing we planned to get all that from plantations. Table 7.1 suggests that it should be realistically possible to grow and

Table 7.2 Wood harvested in 1975 and 1995, in the whole world. In billion m^3

Use	1975	1995	% increase 1975–95
Fuel wood and charcoal	1.28	1.92	50
Industrial (i.e. for making things)	1.30	1.54	18
Total	2.58	3.46	34

Data from Brown, Renner & Flavin (1997).

Table 7.3 Total area of forest in 1995

Location	million km^2
Tropics	17.3
Non-tropics	17.2
World total	34.5

Data from World Resources 1998/9.

harvest an average of 10 m^3 ha^{-1} $year^{-1}$ on a continuing rotational basis. To produce 4 billion m^3 each year would then require a total of 4 million km^2 of plantation; that is, about 12% of the world's present forested area, some of which is already plantations. So it is not a ridiculous suggestion that we should plan to grow all our timber needs in plantations, and reserve the remaining 88% or so of forests for other things, including wildlife.

... or from existing unplanted forests?

Up to now much of our wood harvest has been from existing, natural or seminatural forests, not plantations. If we rely entirely on them, can they provide all our requirements sustainably? If the whole world's forested area of about 35 million km^2 could be used, the average wood growth required to produce 4×10^9 m^3 $year^{-1}$ would be 1.2 m^3 ha^{-1} $year^{-1}$. We do not have sufficient information to estimate reliably what is the average wood growth rate of the unplanted forests of the world. Bowen and Nambiar (1984) reported rates from 0.5 to 17 m^3 ha^{-1} $year^{-1}$ from various sites in the tropics. Among the temperate deciduous forests of the Great Smoky Mountains (North Carolina and Tennessee, USA) wood growth rate ranges from 1.2 to 14 m^3 ha^{-1} $year^{-1}$ (Whittaker 1966). These and other data suggest that there is sufficient growth in the world's unplanted forests to supply our present timber needs. Another way to look at it is that there are large areas of forest that are not being exploited at all for timber at the moment, for example in remote parts of Siberia and Congo, where all the trees die naturally and the wood rots; so there is much wood production that is not being used by people. All this does not *prove* that it is possible to exploit natural and seminatural forests for timber in a sustainable way—this raises other problems that we have not yet discussed—but it does at least suggest that it may be a realistic possibility.

We should now consider the question, if forests can be exploited sustainably, why is deforestation occurring?

Deforestation

Among the figures published for rates of deforestation there are some wide disagreements, especially for the tropics. One reason for this is different meanings attached to the word 'deforestation'. It is sometimes assumed that if forest is cut down, that is the end of forest on that site. However, if the site is left alone after the trees have been cut and removed, often new forest will regenerate and grow there. The eastern USA provides clear examples of this: much of the extensive present-day forest is regeneration after clear-felling during the last 200 years. This includes parts of the Shenandoah and Great Smoky Mountains National Parks (see Chapter 11). Whether all the wild species survive in regrowth forests is an important question to be considered later. For the moment I concentrate on the presence of forest, and define deforestation as meaning that the vegetation cover of an area changes from forest to something else. So, changing from one sort of forest to another is not considered here to be deforestation: if, after felling, trees of any sort are planted or allowed to re-establish naturally, that is not deforestation.

Does collection of fuel wood destroy forests?

In many tropical countries wood is the main source of domestic fuel for cooking, and there is concern that as human populations increase tree growth will not keep up with demand; this will cause the disappearance of forest and open woodland (Leach & Mearns 1988). In some areas the collection of fuel wood is regulated by local rules and customs, which have so far resulted in the supply of wood being sustainable; Shackleton (1993) gives an example for an area in Transvaal, South Africa. There is concern that if people return more frequently to cut wood, this will prevent tree regeneration and so kill the trees. Whether or not the trees are killed depends on whether they regenerate vegetatively or by seed, and how they respond to cutting. Trees that regenerate well from the base can give a high yield if cut every 5 years (Table 7.1). This was the basis of long-term fuel wood production in Europe in the past: in medieval England coppice rotations of 4–8 years provided a supply of fuel year after year (Rackham 1990). So it is not clear how far the disappearance of fuel wood sources in parts of the tropics are due to increased collection, and how far to clearance of the forests for farming and to increased grazing preventing tree regrowth.

How much forest have we lost?

Table 7.4 part (a) provides information, for four areas, of how much deforestation has occurred since people first began cutting down trees. It is, however, expressed as area of forest remaining, not area removed. England today has little forest cover, and of that only about one quarter is classed as ancient semi-natural; most of the rest is plantations. The great Domesday Survey carried out across England in AD 1086, on the orders of King William I (William the Conqueror), gives us a basis for calculating how much of the country was forest-covered then. Rackham (1986) has calculated that only 15% of the land area was forest + woodland, and some of that was fairly open. So most of the original forest had already gone. In the other three example areas deforestation started more recently, and more of the forest remains. The figures for the two tropical areas are already substantially out of date because of continued deforestation. They serve to make the point that the percentage of forest still remaining varies greatly between different parts of the tropics. A major influence on this is access. In spite of the widespread publicity about deforestation in the Amazon Basin, most of the forest is totally inaccessible and still remains.

Table 7.4 Estimates of forested area remaining, and current rates of deforestation, in several areas of the world

(a) Present forested area, expressed as percentage of original area of forest. 'Present forest' based on assessments during 1980s.

Location	Present forest area (%)	Main period of deforestation	Notes
England	8 (2)	Before 1086	1
USA	60	Last few centuries	2
Peninsular Malaysia	52	20th century	3
Brazilian Amazonia	94	Last few decades	4

(b) Recent rates of change in forest area (percentage of area per year).

Location	% per year	Period	Notes
UK	+ 0.5	1990–95	2
USA	+ 0.3	1990–95	2
Peninsular Malaysia	−1.4	1980–85	3
Brazilian Amazonia	−0.4	1978–88	4

Notes: (1) Figure in parentheses is ancient seminatural forest only. Assumes England originally 90% forested. Present forest area from Spencer & Kirby (1992). (2) From World Resources 1998/9. (3) From Collins, Sayer & Whitmore (1991). (4) From Skole & Tucker (1993).

Present rates of deforestation

There has been great concern about the rates at which deforestation is occurring at the moment, and various figures have been published. There are difficulties in making such estimates, which are discussed by Grainger (1993). If we are comparing forests only a few years apart, when changes are likely to be relatively small, there is the opportunity for errors in measurement to have a large influence on the calculated rate of loss. Remote sensing from satellites has been a valuable source of information on changes in forest area but can be subject to error, for example if there have been changes in technique (e.g. in the resolution), or if different parts of the area were obscured by cloud on different occasions. Table 7.4 part (b) shows rates of change in four regions that featured in part (a). These are expressed as a percentage of the remaining forest, not of the original forest. In both the temperate countries the forested area increased, whereas in the two tropical regions it decreased. In recent decades loss of forest has been widespread in the tropics. Between 1980 and 1990 the loss over the whole tropics averaged 0.8% per year, with losses in tropical Asia, Africa and America all contributing (Whitmore 1997). If such rates continue for decades into the future, they represent substantial losses.

In temperate regions changes in forested area have been more variable. Between 1970 and 1995 forested area decreased in the USA, Canada and Japan, but increased in Australia, New Zealand and much of Europe (OECD 1997, Table 5.1B). That publication does not give data for Russia or other countries of the former USSR, but almost all other European countries had a larger area of forest in 1995 than in 1970. This includes the four major timber producers of western Europe—France, West Germany, Sweden and Finland—where there is a long tradition of sustainable timber production. So it is certainly possible to harvest substantial amounts of timber without reducing the amount of forest remaining long term.

Why is forest being lost?

So why is the area of forest decreasing in some countries? The primary reason is because the land is wanted for other uses. In each of the four areas in Table 7.4(a) much of the cleared forest has been converted to food production—arable land or grazing. However, even if after felling the area is not used for farming, forest regeneration sometimes fails to occur. There are two major causes of this.

1 Browsing by domestic or wild animals may kill young seedling trees or vegetative regrowth. This was a common cause of forest changing to heathland or rough grassland in medieval England (Rackham 1990). Browsing by deer has affected forest regeneration in parts of the USA (see Chapter 10).

2 Soil erosion is likely to be greater after felling (see Chapter 4), and soil loss can sometimes be severe enough to prevent any new forest establishing.

Tree population dynamics and forest structure

The previous section made clear that the ability of new trees to establish after old ones are felled is crucial for sustainable forestry. Different patterns of felling and subsequent regeneration will lead to different structures of forest—for example, are all the trees about the same age and size, or very varied in age and size? This will have important implications for future timber production and for wildlife, so we now need to consider it in more detail.

Even-aged forests can become mixed-age

Imagine a large area of forest dominated by a single species of tree, then suddenly all the trees are killed by some major event, for example a hurricane, or clear-felling. Suppose there are many viable seeds in the soil, which germinate and soon give rise to a dense stand of seedlings of the dominant tree species. What will happen during the succeeding years? You might perhaps think that the trees will all grow up together as an even-aged stand until they all reach the end of their normal lifespan; then they all die approximately simultaneously and are replaced by another even-aged stand. In reality this is not what happens. One reason for this is that very often the number of seedlings that establish after a major disturbance is far greater than the number of mature trees that can survive in the area. As the young trees grow larger, competition between them intensifies and some of them die. However, even after that the remaining trees do not have a 'normal lifespan'. We take the idea of a normal lifespan for granted, because in a loose sense humans, along with

Trees do not have a 'normal lifespan'

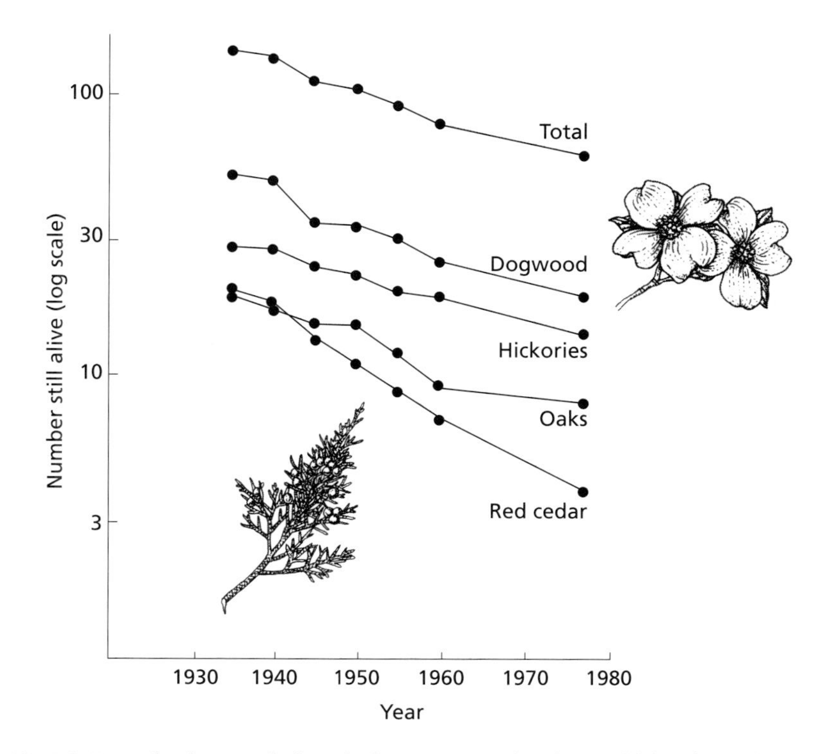

Fig. 7.2 Records of survival of marked trees in a 0.1-ha plot in old deciduous forest in North Carolina, USA. Individuals were marked in 1935; the graph shows the number still alive over a subsequent 42-year period. Dogwood (shrub, upper picture) = *Cornus florida*; hickories (tree) = *Carya* spp.; oaks (tree) = *Quercus* spp.; red cedar (shrub, lower picture) = juniper = *Juniperus virginiana*. From Peet & Christensen (1980). Reproduced with kind permission from Kluwer Academic Publishers.

many other mammal species, have one. In reality we do not all drop dead on our 75th birthday; but our mortality risk is related to our age. If you are aged about 20 and you have a grandmother who is still alive, you know that you have a better chance than she does of still being alive in 10 years' time. Your granny knows it too. But in tree species that have been investigated, after the seedling stage mortality risk is not much related to their age. Figure 7.2 shows an example of this: survivorship data for trees and shrubs in a mixed-age, mixed-species temperate deciduous forest. The lines show some wiggles, as one would expect with real data, but the slopes do not become consistently steeper or shallower with time. Because the vertical axis is on a log scale, the steepness of a line shows the proportion of the remaining trees that die per year. For example, the total number of trees in 1935 was 144; by 1945 32 (22%) of them had died. In the next 10 years a further 20 died, which is 18% of the number present in 1945 (so not much different from 22%). So although there is no normal age of death, each species has a *half-life*. Of the red cedars present in 1935, about half had died by 1955: it had a half-life of about 20 years. For dogwood and oak the half-lives were about 30 years, and for hickory about

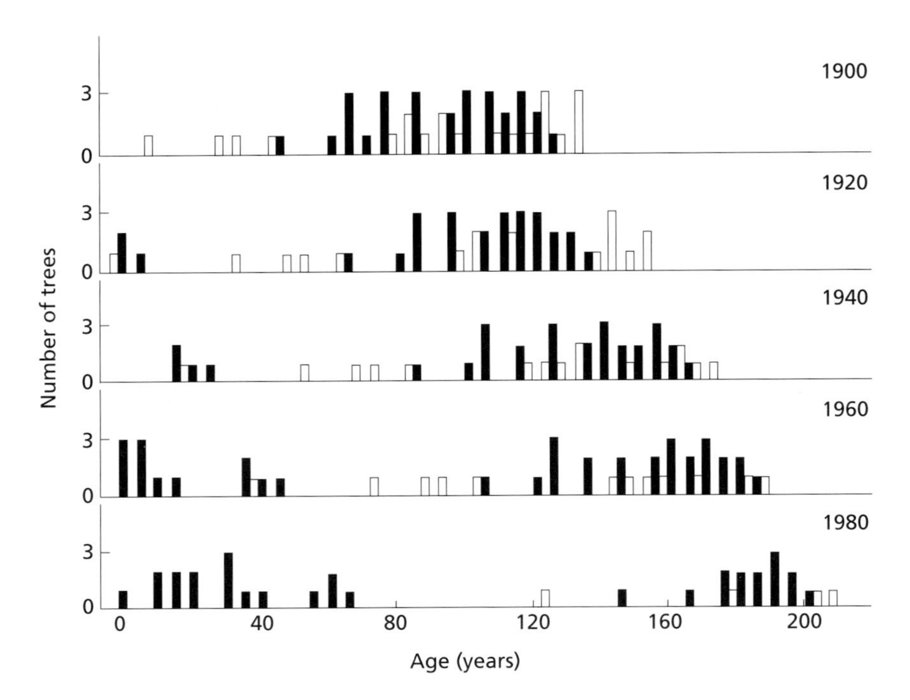

Fig. 7.3 Ages of trees alive in particular years in a conifer forest in the Rocky Mountains in Canada, where the previous forest had been killed by a fire in 1760. Black bars: Engelmann's spruce (= *Picea engelmannii*); open bars: lodgepole pine (= *Pinus contorta*). From Johnson *et al.* (1994).

40 years. Studies of tree mortality in tropical rainforest have found mortalities within the range 1–3% per year, corresponding to half-lives in the range 25–70 years (Lieberman *et al.* 1985; Manokaran & LaFrankie 1990; Korning & Balshev 1994; Phillips 1997).

This lack of relation of mortality to age results in an even-aged forest developing, with time, into a mixed-age forest. Because the trees die at different times, gaps are formed, by the death of a single tree or of several neighbours that die at about the same time. If a single small tree dies the neighbours can grow to fill the gap, but larger gaps will require new individuals to colonize them. Because these gaps do not all appear at once, the new stand that develops will be mixed-age. Seeds are not the only means by which trees can regenerate. If most of the above-ground parts have died but the stump or roots remain alive, new vegetative growth (coppice growth) may occur from them. Many angiosperm trees have this ability, but few conifers do.

How long it takes a forest to become mixed-age

Figure 7.3 shows a forest changing from approximately even-aged to mixed-age. This is in the Canadian Rocky Mountains, in an area where all the trees were killed by a large fire in about 1760, and a new forest dominated by spruce and pine then developed naturally. The ages of trees that were alive in 1900 and afterwards were determined by examining the rings of those still alive, standing dead and fallen dead. Figure 7.3 shows

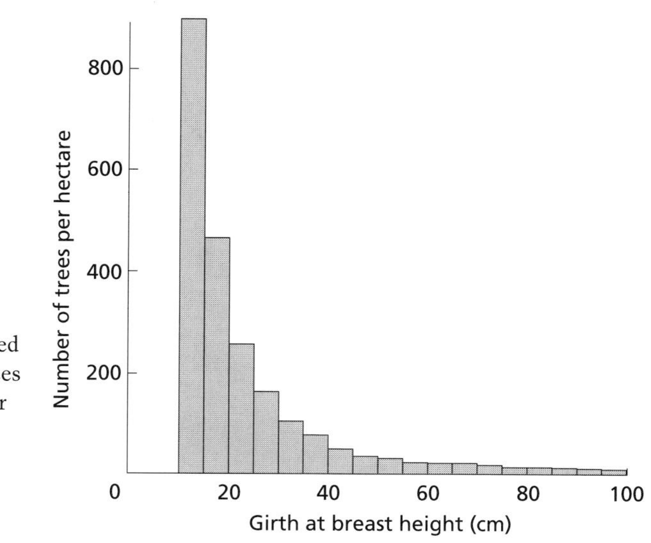

Fig. 7.4 Abundance of trees in 5-cm girth classes, in tropical rainforest in Sabah, Malaysia. The forest had been little disturbed by people. Based on measurements of all trees of girth at breast height (gbh) 10 cm or larger in two 4-ha plots. There were 63 trees ha^{-1} with gbh >100 cm. Data of Newbery *et al.* (1992), replotted from figures supplied by D. M. Newbery.

that in 1900 most of the trees were aged 70–140, so they established during the 70 years after the fire. There were very few younger than that. Some of this cohort continued through to 1980, but many of them died during the period, and younger trees were establishing. So by 1980, 220 years after the fire, the forest had become substantially mixed-age.

To call the forest of Fig. 7.3 in 1900 'even-aged' may seem to stretch the meaning of the term a little. However, we can contrast that age structure with the forest shown in Fig. 7.4, a tropical forest that has probably never been subject to any large-scale disturbance. The age of the trees cannot be determined, but girth is an indicator. The concave shape of the curve is similar to what the near-straight lines in Fig. 7.2 would look like if they were plotted on an arithmetic (not log) scale: it is what we expect in a long-undisturbed forest if mortality risk is unrelated to age. It is clearly very different from the age distributions of Fig. 7.3.

From studies such as that of Fig. 7.3, and knowledge of natural tree mortality rates (mentioned earlier), we can conclude that the development of a forest from even-aged to mixed-age will take of the order of one to several centuries. Therefore, if large-scale disturbance, killing all the trees, occurs more frequently than that, there will be a series of even-aged stands. If, on the other hand, large-scale disturbances are much less frequent than that, mixed-age stands will be the norm. This raises important questions for management. Are mixed-age stands the natural state of all forests? If so, would it be best, for wildlife and for sustainable timber production, to extract timber in a way that maintains mixed-age stands? Is it in fact practicable to extract timber in this way sustainably? On the other hand, if even-aged stands are natural in some types of forest, is this an argument for clear-felling there?

On a world scale the primary cause of death of large areas of forest—apart from felling by people—is fire. The next section provides information on the frequency of fire in different types of forest and its effects on the forests.

Fire in forests

Forest fires can be started naturally, by lightning. But in most parts of the world people have used fire for tens of thousands of years, and in some parts for over 100 000 (Russell 1967). In the recent past, and probably for much longer, hunters have used fire to manage vegetation, to keep it more open in order to make hunting easier. It is therefore very difficult to tell whether or not past fires were caused by people. For our purposes it may not matter: the main point is that fire has been influencing vegetation so long that it has not only affected the distribution of species but given time for some of them to evolve fire-tolerance characteristics.

Recent major forest fires

Some large fires have occurred in recent decades in North American conifer forest. A dramatic example was the Yellowstone fires of 1988. Yellowstone National Park is an area of forests (mostly coniferous), rivers and lakes, about 80 × 100 km in size, in the northwest corner of Wyoming. It was the world's first national park. In 1988 much of the USA had an unusually dry summer. Several fires that started in or near Yellowstone burnt with such ferocity and spread so rapidly that they proved impossible to control, and by the time the first snows of autumn finally quenched them, about 40% of the national park area had been burnt. Large areas were left entirely black, thousands of charred dead trees still standing above the blackened ground. The fires and their immediate effects were described in more detail by Romme and Despain (1989). Turner *et al.* (1997) described the beginnings of vegetation recovery.

The Yellowstone fires were not unique in modern times. There were, for example, severe fires in Boise National Forest, Idaho, in 1994 (Macilwain 1994). These modern fires raise the question, how often have fires occurred in the past? Evidence on past fires can sometimes be obtained from written records. Two other sources of evidence are *fire scars* on trees and *charcoal* in lake mud. If a fire is severe enough to damage the bark on one side of the trunk but not to kill the tree, the tree will continue to grow but the fire scar will remain in the trunk and can be dated by counting annual rings in the xylem. Deposition of mud at the bottom of lakes often varies through the year, so that annual layers can be recognized. If there is a fire nearby, charcoal will be incorporated into that year's layer of mud, so it can be dated.

Fires during the last few centuries

One of the most extensive studies of past fires was carried out by Heinselman (1973) in the Boundary Waters Canoe Area, 4170 km^2 of forests and lakes in northeastern Minnesota. The forests are mostly coniferous (pine, spruce, fir and larch), and about half the area has never been logged. Fire has occurred there for at least 10 000 years, as shown by charcoal in the mud of one small lake (Swain 1973). Heinselman (1973) determined the dates of fires since 1610 over the whole unlogged area. His paper pro-

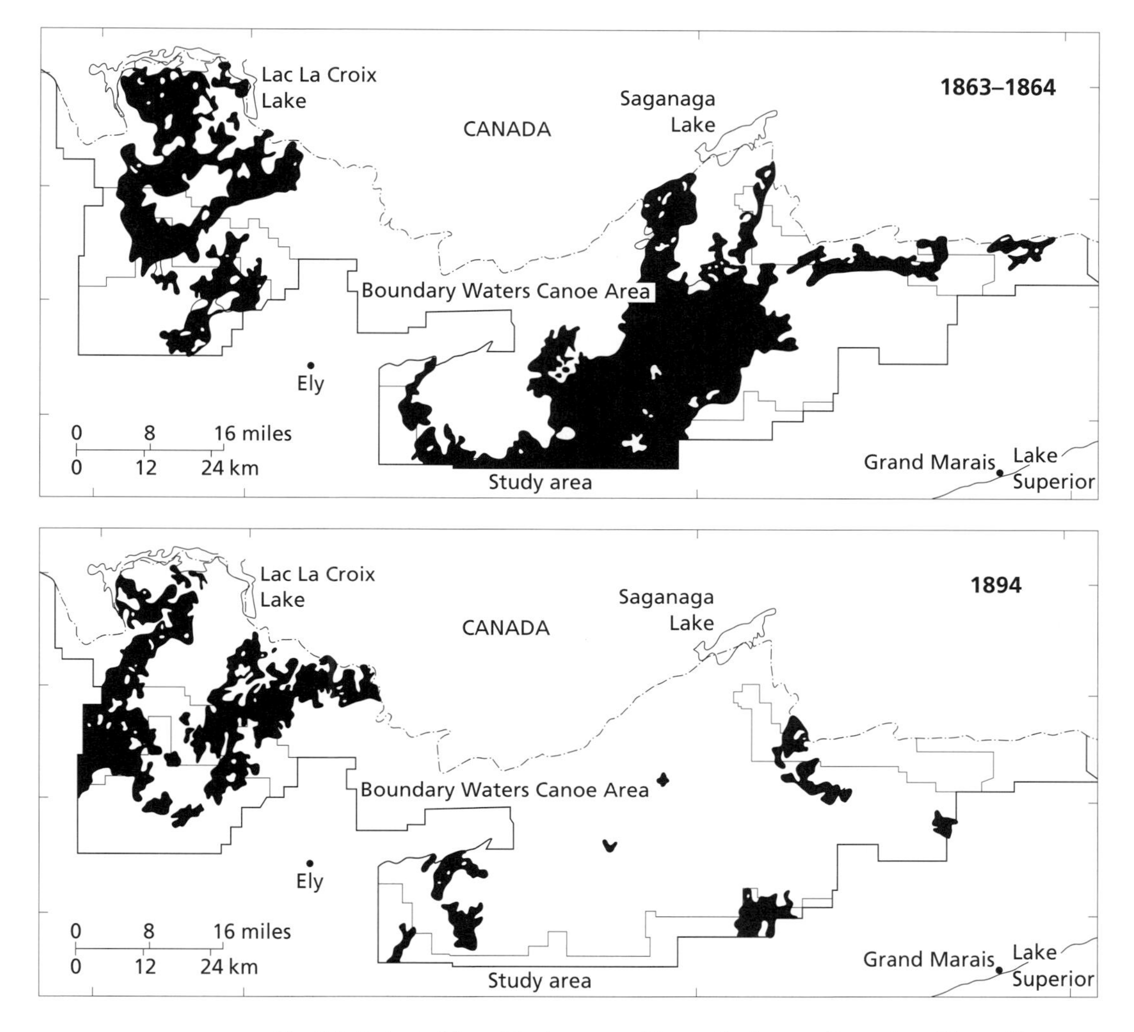

Fig. 7.5 Extent of fires (black on maps) in conifer forest of Boundary Waters Canoe Area, Minnesota in two major fire years, 1863–64 (upper) and 1894 (lower). From Heinselman (1973).

vided maps of the major fires from 1610 onwards; Fig. 7.5 shows two of them. It is evident that in each of these years fires must have started at more than one place. Of the unlogged forest standing in 1973, 65% was even-aged stands that had established after fires in 1863–64, 1875 or 1894; and most of the rest of the forest was made up of smaller even-aged stands attributable to fires in other years. After 1910 the US Forest Service operated a fire control programme and there were no very large fires.

Were these fires started by people or by lightning? There were people in the area during this time, first Chippewa and Sioux, and later fur trappers. However, fires can start naturally in the Boundary Waters Canoe Area: during the 15 years 1956–70, 113 fires caused by lightning were reported within the area.

Table 7.5 Mean time between recurrences of forest fires

Location	Predominant tree species	Mean interval between fires (yr)	Methods	Source
Alberta	Pine	40	S, W	1
	Spruce	80–100		
	Aspen	40		
Alberta	Spruce	70	C	2
Canada and Alaska	Pine	30–160	?	1
	Spruce	100–170		
	Aspen	60		
Sweden	Pine + spruce	80	S	3
Michigan	Pine	80–260*	W	4
	Pine + oak	170–340*		
	Hemlock + pine + beech + maple	1400–2800*		
New York State	Spruce or spruce + pine	90	C	5
	Hardwoods	no major fires		

* Estimated fire interval within this range.

Methods: C, charcoal in lake mud; S, fire scars on trees; W, written historical records.
Sources: (1) Larsen (1997); (2) Larsen & MacDonald (1998); (3) Zackrisson (1977); (4) Whitney (1986); (5) Clark *et al.* (1996).

It is likely that in the past many of the fires in conifer forests were much smaller in extent than those shown in Fig. 7.5. Clark (1990) made a detailed study of 1 × 1 km of unlogged forest, mainly pines, in another part of northern Minnesota, and was able to map the extent of fire in each year from 1650 onwards. Most fires burned only part of the square, their distribution was patchy, and some fires were less than 100 m across.

How frequent were fires?

Table 7.5 summarizes available data on fire return times, obtained by different methods. For spruce, pine and aspen the return times are usually too soon for a mixed-age forest to have developed, and so most of the forest would be even-aged stands of various ages that regenerated after the last fire. Information on fire return time in temperate deciduous forests is sparser. Whitney (1986) used records made by survey teams in the mid-19th century, in an area of Michigan where some of the forest was dominated by pines but in other parts there was a strong admixture of deciduous species ('hardwoods'). Where the hardwood species were mixed with pine, the fire return times were longer than in pure pine forest; and where the commonest hardwoods were beech and maple the return times were clearly long enough to allow mixed-age stands to predominate (Table 7.5). Clark *et al.* (1996) used charcoal layers in mud in a lake in northern New York to determine fire frequency over the last 10 000 years. During that time the vegetation changed from conifer to hardwoods; their method detected nearby fires during the conifer phase but none in the subsequent hardwood forest. In contrast, a 13 000-year record from mud in two lakes in southern Switzerland showed charcoal input remaining as high or even increasing when oak and alder replaced fir as the predominant trees (Tinner *et al.* 1999). Most of these Swiss fires were probably started by people. On balance, it is likely that in the past most temperate deciduous forests suffered large-scale disturbance infrequently enough to be mixed-age. This conclusion is based more on studies of the age structure of present forests than on information about past fire frequencies, which is inadequate.

Which past forests were even-aged?

The message is that fire was formerly frequent enough in most northern conifer forests to maintain even-aged stands, but in temperate deciduous forests it was normally infrequent enough to allow mixed-age stands to predominate.

What about tropical forests? It is not possible to generalize about them. They range from rainforests in regions where rainfall is normally substantial in every month of the year, to others where there is a dry season several months long and which may border on savanna where fire is frequent. Fire risk in tropical forests today has been increased by human activities: selective felling leaves gaps of various sizes, some much larger than most natural gaps, where unwanted tree remains may be left to dry after felling. Patches of pasture created within forest areas pose further fire risks. But there is evidence that fires occurred in tropical rainforest in the past. For example, the tropical rainforest of the Gogol Valley, Papua New Guinea, was extensively burnt in 1890 and again in 1930 (Saulei & Swaine 1988). At those times there were no roads into the area: the only inhabitants were living in small, widely separated villages and practising shifting cultivation (Lamb 1990). Sanford *et al.* (1985) collected soil cores from 96 locations near San Carlos, Venezuela, in tropical rainforest, and examined them for charcoal. They found charcoal in 63 (two-thirds) of them. The oldest known evidence of people living in the area is pottery remains dated 3750 BP. Of 10 charcoal samples that were radiocarbon dated, nine had dates ranging from 3080 to 250 BP, and so could be the result of slash-and-burn farming. The tenth, however, dated from 6260 BP. Overall, the evidence points to most fire in tropical forests in the past being associated with shifting agriculture, probably often localized, but perhaps in occasional dry years spreading substantially. However, there is no doubt that in spite of these fires, and also the destruction of other areas of forest by hurricanes from time to time (see Nelson *et al.* 1994), most of the area of unexploited tropical forest has been and remains mixed-age and mixed-species.

Implications for forest management

The previous section presented evidence that the natural state of much northern conifer forest is large even-aged stands, but that most temperate deciduous and tropical forest is naturally mixed-age. Is this a basis for arguing that most tropical and temperate deciduous forest should be exploited by selection felling, leaving gaps the size of one or a few trees, but that in conifer forest it is better to clear-fell large areas? Would these practices be best for sustainable timber production? Would they be best for wildlife?

Selection felling

In this system the trees are normally of mixed age and size, and only trees above a certain size are selected for felling. If there is a mixture of species, often only those with more desirable timber are chosen. There are practical problems (described later); but selection systems have been used over long periods in single-species and mixed-species forests in Europe, removing the largest trees at intervals. The system is still practised in conifer forests in some steep, mountainous areas in France and Switzerland, where clear-felling is not favoured because of concerns about soil erosion (Peterken 1993). An example of selection harvesting over more than a century is provided by the jarrah (*Eucalyptus marginata*) forest of Western Australia (Abbott & Loneragan 1986). Features that have favoured the success of the system there include:

1 Most of the forest is dominated by that one species.

2 The open forest structure makes it easier to remove large trees without damaging smaller ones.

3 Because of the stony, infertile soil there has been little pressure to clear forest for other uses.

Selection felling in tropical forests

Most tropical forests comprise a large number of tree species, whose wood is likely to differ in features such as density, strength, colour and amount of resin, which all affect the value of the wood. This is one major reason why extraction has been selective. Another is size: large-diameter tree trunks are much more valuable than thinner ones. As we have already seen, a forest that is left undisturbed for a long time is likely to develop a 'reverse-J' size distribution, as seen in Fig. 7.4. If the forester requires boards or posts at least 20 cm across, this means trunks of about 70 cm girth or more. In the forest of Fig. 7.4, and most other tropical forests, the vast majority of the trees are narrower than that. So, timber extractors have been looking for a few valuable species, and only the largest specimens, and so have taken few trees per hectare.

Damage caused by selection felling

This system sounds as though it should cause only limited damage to the forest and therefore be sustainable. However, there are potential problems. One is that by removing the most valuable trees we may in the long term make that species less frequent. The hope is that by leaving smaller trees of the species there will be enough individuals to grow up and to provide adequate seed source; but this does not always happen. In a study by Verissimo *et al.* (1995) of mahogany extraction in Brazil, although only 0.3–2.1 mahogany trees per hectare were extracted, the number of smaller ones left at each site was less than the number that had been extracted. Allowing for subsequent natural mortality, this does not sound promising for maintaining the abundance of the species in the future. Mahogany evidently has special requirements for regeneration: its population structure did not conform to the reverse-J of Fig. 7.4.

Another problem is that the damage to the forest is usually much more than just a gap where the valuable tree was: many other trees are also killed or injured. Johns (1992) gave figures from five sites, in three continents, at which the percentage of trees extracted for timber ranged from 0.6 to 10%; the other trees killed during logging ranged from 25 to 60%. This damage is caused by making roads and access tracks, manoeuvring heavy machinery, the fall of the cut tree on to others, and then its being dragged through the forest. As well as trees killed at the time, there is the danger of longer-term harm through soil compaction, erosion and impeded drainage. It has been shown in several areas that this damage can be reduced, given the will (Johns 1992; Whitmore 1998). Procedures that help include careful planning of access tracks, felling trees in a direction that minimizes damage to others, moving logs to the road by cable rather than tractor, and extracting uphill rather than downhill to reduce soil erosion.

After the timber extractors have finished, they leave roads into the forest. These provide access for hunters of large mammals and birds, and also for people to move in to clear the forest for farming. Chatelain, Gautier and Spichinger (1996) used satellite photographs to show how this led to deforestation in southwestern Côte d'Ivoire. The area was formerly tropical forest, sparsely inhabited by people, who were carrying out shifting cultivation, and was virtually inaccessible to outsiders until a road was built through it in 1968. However, it was not until two sawmills were built in 1977–80 and extensive timber felling started that there was a major influx of people. They used the side roads left by the timber extractors, and much clearance of the forest for farming—clearly visible on the satellite photographs—took place.

There is widespread concern about whether selective extraction from tropical forests can be made sustainable, and, if so, how to do it. Gomez-Pompa *et al.* (1991) describe some management systems that have been tried. So far, none has been operated long enough to show clearly that it would be sustainable.

Selection felling and wildlife

Is selection felling favourable for wildlife? If selection felling causes so much damage in tropical forest, does it simulate natural tree fall adequately, or is it very harmful to wildlife? An early indication of the importance of maintaining a complex forest structure was provided by MacArthur and MacArthur (1961), who compared 13 forest sites, including northern coniferous, temperate deciduous and tropical. 'Foliage height diversity' was calculated from the way the foliage was distributed vertically: the lowest diversity occurs when all the vegetation forms one layer, the highest when about equal amounts of foliage occur in low vegetation, shrubs and in the tree canopy. Figure 7.6 shows that there was a strong correlation between bird diversity and the structural diversity of the forest.

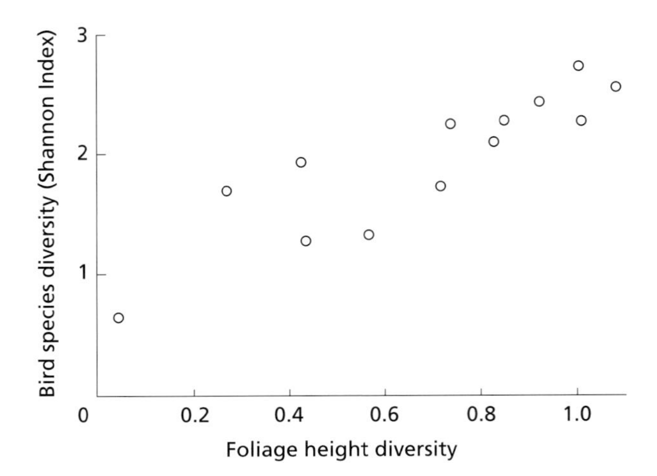

Fig. 7.6 Relationship between bird species diversity and forest structural complexity at sites in eastern USA and Panama. Shannon Index: see Box 10.6. Foliage height diversity: see text. From MacArthur & MacArthur (1961).

Taking into account plant species diversity did not help to account for the remaining scatter on the graph. This and other, more recent papers (e.g. Julliens & Thiollay 1996) have shown the need for concern, that if selection felling alters the structure of forests this may influence wild species.

Table 7.6 summarizes results from two studies of the effects of selective logging in tropical rainforest on bird species diversity. In the Guianan

Table 7.6 Bird species in tropical forest: comparison of unlogged areas with areas selectively logged

(a) French Guiana	
(i) Total number of bird species	
Unlogged	241
Logged 1–2 years before	166
Logged 8–12 years before	173
(ii) Number of species showing substantial difference in abundance	
Logged > unlogged	41
Logged < unlogged	131
Data from Thiollay (1992); based on 336 sample plots each site, 20 min recording per plot.	
(b) Sabah, Malaysia	
(i) Total number of bird species	
Unlogged	103
Logged 6 years before	133
Logged 12 years before	132
Logged twice, 14 and 9 years before	129
(ii) Number of species showing substantial difference in abundance	
Logged > unlogged	46
Logged < unlogged	15
Data from Johns (1996); from transect surveys, mean number of species per 75 h recording	

Table 7.7 Mean number of primate individuals (per km^2) in selectively logged and unlogged forest in Uganda

Primate	Unlogged	Logged	Statistical significance
Blue monkey	15.6	58.2	*
Redtail monkey	8.3	46.4	*
Colobus monkey	27.0	44.2	*
Baboon	14.0	11.0	ns
Chimpanzee	3.2	2.8	ns

* Difference between unlogged and logged significant ($P < 0.001$); ns, difference not significant.

From Plumptre & Reynolds (1994).

study logging led to a reduction in bird species diversity, whereas in Sabah diversity increased. At neither site was there any clear sign of recovery within 12 years. The reason for the different results between the two regions is not known, but an important factor is likely to be the effects of felling on forest structure, and the nature of the resulting gaps. Bird species diversity tends to be higher in natural gaps than in neighbouring forest (e.g. in Panama; Schemske & Brokaw 1981). At both sites some species decreased in abundance or disappeared following logging, whereas other species increased or appeared for the first time (Table 7.6). So, counting the number of species present is not on its own an adequate assessment of the effects of logging: some species characteristic of undisturbed forest are sensitive to logging, even though other species may be favoured.

Plumptre and Reynolds (1994) studied primates in a forest in Uganda where different parts had been selectively logged at various times since 1935, but some had never been logged. The abundance of the five main species was assessed by observations along transects. Table 7.7 shows that the three monkey species were all more abundant in the logged areas, whereas chimpanzees and baboons showed no significant difference in abundance between logged and unlogged areas.

These examples show that selective logging is not always harmful to all wild species, and can in fact be beneficial to some. This may be a response to felling providing increased habitat diversity, with plant species that colonize gaps perhaps providing favourable food for some animals. However, some species are harmed by selective logging, so if we want to ensure the survival of all forest species it will probably be necessary to preserve some unlogged forest.

Influence of gap size

If different tree species have different optimum gap sizes for regeneration this could give an opportunity in selection felling, by choosing the sizes of the gaps formed, to influence subsequent tree species composition,

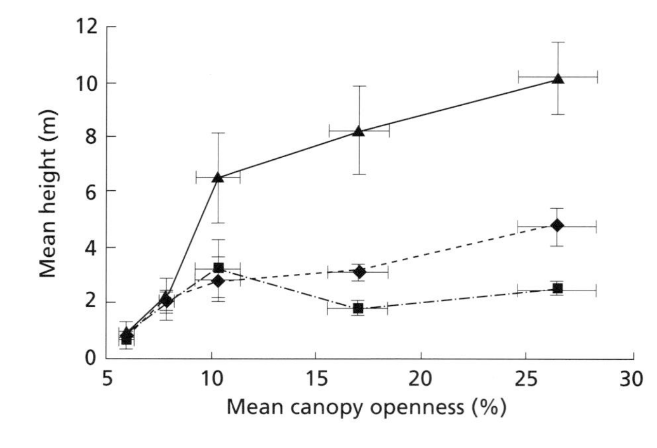

Fig. 7.7 Mean height of young trees of three species under closed canopy (left-hand points) and in gaps of different sizes, in tropical rainforest in Sabah, 6 years after gaps were created. Bars are 95% confidence intervals. ◆ *Hopea nervosa* (a shade-tolerant species), ▲ *Shorea johorensis* (light demander), ■ *Parashorea malaanonan* (light demander). From Whitmore & Brown (1996).

either to favour diversity or to select for desired species. It has been traditional in forest ecology to distinguish two groups of species: (1) *pioneers* or *secondary forest species* which can establish only in large cleared areas; and (2) *non-pioneers* or *primary forest species,* which can establish in small gaps formed, for example, by the death of a single tree. The seedlings of many primary forest species can survive for a long time under the canopy of undisturbed forest, though with little growth; this constitutes the *advance regeneration*—young trees that are already present when a gap forms and thus have an advantage over pioneers, which could only then germinate. It has further been proposed that there are species which establish best in gaps of intermediate size, e.g. in eastern North America (Bormann & Likens 1979).

Response of young trees to gap size

There has been much research on the responses of young trees to gap size (see Coates & Burton 1997). However, the results tend to be difficult to interpret because most of the research has involved observations on naturally occurring gaps; these are likely to differ in age as well as size, and perhaps also in topographical position and causal mechanism. Repeated recording in a 50-ha forest plot on Barro Colorado Island, Panama, allowed Dalling, Hubbell and Silvera (1998) to identify gaps of varying size and light regimes, but all formed within a 2-year period. They recorded the abundance and height growth of seedlings of 24 pioneer species and one shade-tolerant species. The shade-tolerant species occurred more often in small gaps with low light, but among the pioneers there was no significant difference in the way their abundance or growth responded to light in the gaps.

An alternative approach is to create gaps of different sizes experimentally. Gaps of different sizes were cut in tropical rainforest at a site in Sabah, and the subsequent establishment and growth of tree seedlings was studied over 6 years (Kennedy & Swaine 1992; Brown & Whitmore 1992; Whitmore & Brown 1996). There was no consistent difference between gap sizes in germination or subsequent mortality. All the three main species grew faster in gaps than in closed forest (Fig. 7.7).

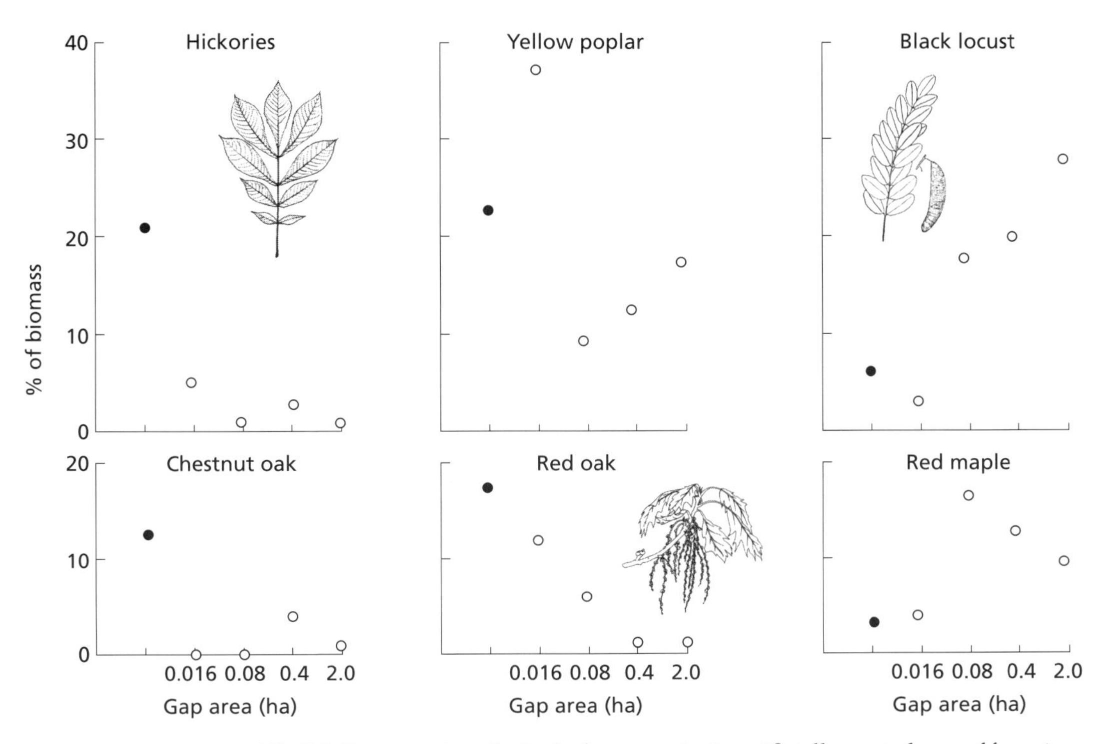

Fig. 7.8 Regeneration of principal tree species in artificially created gaps of four sizes in mixed-species deciduous forest in North Carolina, USA. The abundance of each species is expressed as a percentage of the total biomass. ● Mean value before cutting; ○ second year after cutting. Data of Phillips & Shure (1990).

One species grew much faster in large gaps than small; the other two showed less difference between gap sizes. These responses did not differ in the way one would expect from the species' categorization by foresters as 'shade tolerant' or 'light demanding' (see legend to Fig. 7.7). In this experiment no species showed a clear 'preference' for smaller gaps, whether assessed by germination, survival or height growth.

In mixed-species deciduous forest in the southern Appalachians in North Carolina, USA, Phillips and Shure (1990) felled areas of four different sizes, from 160 m^2 to 2 ha, and measured the regrowth over the first 2 years after the cutting. In all gap sizes most of the regeneration was by sprouts from stumps and roots that had survived from the previous trees; only a small proportion came from advance regeneration and newly germinated seeds. Figure 7.8 shows results for the six species that formed most of the tree biomass before cutting or after 2 years of regrowth. The original forest was dominated by two oak species (*Quercus* spp.), hickories (*Carya* spp.) and yellow poplar (*Liriodendron tulipifera*). Of these, chestnut oak and hickory regrew poorly in all gaps; red oak regrew only in smaller gaps, but yellow poplar regrew fairly well in all sizes of gap. Among species rare in the previous forest, black locust and red maple grew well in large gaps. Thus the gap size influenced the composition of the forest that developed: in the smallest gaps red oak and yellow poplar predominated, but in large gaps it was black locust, maple and yellow poplar. Not all of these early colonizers would be expected to dominate indefinitely: black locust in particular is a successional species which would in due course be replaced by others (see Fig. 11.1).

Control of future tree species composition by the size of gaps created during selection felling is an attractive idea, but so far there is inadequate information on the response of individual species.

Felling strips

Compared with single-tree felling, felling of a patch or strip containing many trees has the advantage of greatly reducing damage to unharvested trees. The subsequent forest will have a patchy structure very different from that of a mixed-age unexploited forest, but the surrounding forest may still provide some protection against soil erosion, and against wind damage to the new young trees. It may also act as a seed source for the establishment of new trees, and a place for animals to survive which can later reinvade the felled area. Shelterwood systems have been operated for centuries in parts of Europe (Peterken 1993). These 'high forests' rely on natural regeneration from seed but aim to produce even-aged stands. In parts of France and Germany there are forests of beech and oak still being managed in this way, some involving rotations as long as 220 years. In some, strips or wedges are felled and then left to natural regeneration. In other areas the system involves an initial thinning of the forest, allowing seedlings to establish under a more open canopy, which provides seeds and shelter; the remaining large trees are removed later, perhaps in several cuts.

Table 7.8 Data from an experiment in Peru, where a strip of rain forest 30 m wide was cut down. The felled trees were removed

(**a**) Mean number of young trees per m^2 in the strip, recorded 14 months after felling of the previous forest.

Surviving from forest	0.23
Sprouts from tree stumps	0.46
New seedlings	1.79
Total	**2.48**

(**b**) Mean number of species among seeds caught in seed traps (each totalling 4 m^2, cleared twice a month).

Distance from forest (m)	Number of species
12.5	4.3
7.5	5.8
2.5	12.2
Within forest	20.0

Data of Gorchov *et al.* (1993).

Another alternative is that trees are first felled in small patches, creating gaps in which seedlings can establish; the gaps are gradually enlarged. So there is a variety of shelterwood systems practised within Europe.

It has been suggested that felling strips of tropical forest could retain some of the advantages of selection felling while reducing damage to unharvested trees. Experimental strips have been felled in two areas of rainforest in Peru, but so far only early regeneration has been reported (Hartshorn 1989; Gorchov *et al.* 1993). Table 7.8 shows that in one of these studies, just over a year after the felling the average abundance of young trees exceeded 2 per m^2, more than enough to form a new forest. Some of these were from small trees that had survived the felling operations, and some were sprouts from remaining tree stumps, but the majority were from seeds that germinated after the felling. Table 7.8 part (b) shows that the number of species in the seed rain decreased rapidly with increasing distance from the forest. So the width of the strip may have an important influence on the species diversity of the new forest that develops. However, Hartshorn (1989) found more than 200 species of young tree in two strips (20 and 50 m wide) cut in another area of Peru, 2–2.5 years after they had been cut.

Clear-felling large areas

In plantation timber production, following clear-felling of an area new trees are planted for the next rotation. This planting is time-consuming and therefore expensive, so why not leave it to natural regeneration? Figure 7.3 gives a clue to the answer. Although the forest is classed as even-aged, regeneration after the fire, and hence tree age, was spread over 70 years. Not all natural regeneration in conifer forests is as uneven as that. Among North American conifer forests which regenerated naturally after fire 50–100 years ago, some have canopy trees that are mostly within a 20-year age span (Day 1972; Johnson *et al.* 1994). To obtain forests more uniformly aged than that is likely to require planting, and foresters often consider it to be worthwhile. If, however, they decide to rely on natural regeneration, careful silvicultural management, requiring a high degree of skill, is often necessary to obtain sufficiently rapid and uniform establishment of young trees (Peterken 1993).

Conifer forests usually have only one or a few predominant species, and so management of the species composition should be straightforward. In temperate deciduous forests and tropical forests there are usually many more species; so, if we rely on natural regeneration after clear felling, will we get the right species composition? So far we have considered what happens after the felling of patches up to 2 ha (Fig. 7.8), and of strips (Table 7.8). At both these sites many new trees established rapidly, but the species composition was markedly different from that in the previous forest. We now consider what happens after even larger areas are clear-felled, asking whether it is still true that many new trees

After clear-felling, what tree species colonize?

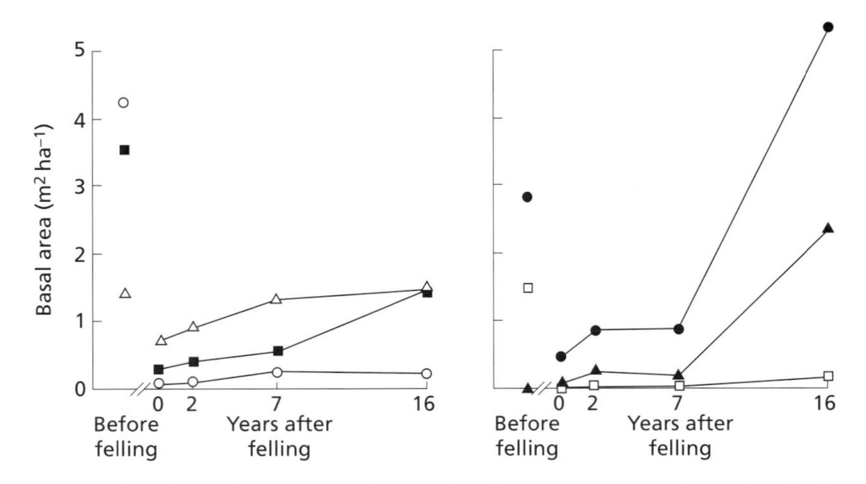

Fig. 7.9 Abundance of six species of tree in deciduous forest in North Carolina, before clear-felling of a large area and during subsequent natural regeneration. For meaning of basal area see text. ○ Hickory spp.; ■ red oak: □ chestnut oak; Δ red maple; ● yellow poplar; ▲ black locust. Data from Elliott *et al.* (1997).

will quickly establish, but not of the same species composition as before; and if so, how long would it take to regain the original forest composition? We are here concerned with what happens if trees are allowed to regenerate immediately after felling. The ability of forest to re-establish after a period of arable farming or grazing is considered in Chapter 11.

In 1974 a 59-ha watershed in the Appalachian Mountains in North Carolina, covered with mixed deciduous forest, was clear-felled. The vegetation was recorded before felling and on subsequent occasions. Abundance of each species was measured as basal area, which is the sum of the cross-sectional area of the trunks. By 16 years after felling the combined basal area of all woody species had recovered to near its prefelling value. However, the relative abundance the species was still very different from the preceding forest (Fig. 7.9). Yellow poplar and black locust were much more abundant than previously, and still increasing, whereas the oaks and hickories were still far below their original abundance. We do not have sufficiently long studies of regeneration following clear-felling of mixed deciduous forest to know how long it will take to return to its original composition and structure, but it is evidently substantially longer than 16 years.

Clear-felling a tropical forest

In the tropics clear-felling usually happens because the area is wanted for some other use: it is rare to clear-fell for timber and then allow regeneration immediately. This is primarily because the timber of many of the tree species is not commercially attractive. However, increasing ability to use these less attractive species, e.g. for paper pulp, may make tropical clear-felling commoner in future. One large felling operation started in 1973 in the Gogol Valley in Papua New Guinea. Before logging started the 880 km^2 valley was almost entirely covered by tropical rainforest. About 5000 people, who among them had 21 languages, lived in small isolated communities and practised shifting cultivation. Some rules were made about where strips and patches of forest were to be left by the logging company; felled areas were up to 3 km across, though most less than 1 km.

After removal of the timber the land was left to natural regeneration, and this was monitored (Saulei 1984, 1985; Saulei & Swaine 1988; Lamb 1990; see also Newman 1993, Table 5.5). Soon after the forest had been felled and the timber removed there were vast numbers of seeds of tree species in the soil, but most of them were of pioneer (i.e. early successional) species: only about 1% were from the non-pioneer primary forest species that had predominated in the forest that was felled. This is probably because seeds of pioneer species can remain dormant in soil for some time, whereas the seeds of non-pioneer species of the primary forest lose their viability within a few weeks (Ng 1978). The wide bands of felling may have prevented seed rain from the remaining forest reaching most of the felled area.

A year after felling there were on average more than two young trees per m^2, but few of them were primary forest species, and of those the majority were not from seed but were trees that had escaped felling or were coppice growth from remaining stumps. About 10 years after felling, among the young trees about one-sixth were of non-pioneer species, far less than the proportion in the original forest. Two positive messages from the results are:

1 Many young trees established quickly, which would help to prevent erosion.

2 If stumps are left they can be a source of non-pioneer species regeneration.

However, it seems likely that the large areas felled, and hence the large distances from remaining forest, reduced the number of primary forest species establishing because of reduced seed rain.

Clear-felling and wildlife

This chapter has suggested that clear-felling may be an appropriate system for conifer forests where previously fire occurred frequently enough to maintain large areas of even-aged forest. But does clear-felling simulate the effects of fire from the point of view of wild species? One might perhaps suppose that fire would kill more animals: felling operations would give them more warning and time to get away. However, animals often return to burned sites quite quickly. For example, 2 years after the great Yellowstone fires of 1988, browsing of aspen sprouts (mainly by moose) was as abundant in burned sites as in unburnt (Romme *et al.* 1995). The important question is whether the forests after fire and after felling provide similar habitats for wildlife.

Table 7.9 Occurrence of bird species in conifer forests in the Rocky Mountains, from Utah northwards. Forest sites were less than 10 years ('early') or 10–40 years ('mid') after burning or felling, or had not been burned or felled for a long time. Figures are percentage of sites at which the species was present

	After burning		After clear-felling		Forest not burned or felled
Species	Early	Mid	Early	Mid	
1. American robin	100	100	90	90	71–90
2. Yellow-rumped warbler	100	100	60	100	70–100
3. Golden-crowned kinglet	9	0	0	20	0–86
4. Mountain bluebird	91	60	80	70	0–12
5. House wren	26	40	20	50	0–40
6. Red-naped sapsucker	17	0	30	70	8–43
7. Cedar waxwing	9	0	10	40	0–4
8. Clark's nutcracker	65	60	10	0	0–48
9. Three-toed woodpecker	65	0	10	0	0–44
10. Black-backed woodpecker	78	0	0	0	0–12

Data from various studies, collated by Hutto (1995).

Response of birds and mammals to felling or burning

Table 7.9 shows the occurrence of some bird species in conifer forests in the Rocky Mountains of the northern USA and Canada. The response to burning and felling varies greatly between species. Some (species 1 and 2) occur in forests whether they have been burned, felled or neither; some (e.g. 3) are more abundant in some old forests than after burning or felling; whereas species 4 and 5 are more abundant after burning or felling than in most old forests. For all these species (1–5) clear-felling and burning seem to be approximately equivalent. However, species 6 and 7 are more frequently found after felling than after burning, whereas 8–10 are more frequent after burning than after felling. Two woodpecker species (9 and 10) are almost entirely confined to early regrowth after fire.

Thus for some bird species conditions after fire are evidently not adequately replicated by felling. After studying birds at forest sites in Wyoming and Montana following fires in 1988, Hutto (1995) concluded that standing dead trees left after fires are important (1) as a source of insects for woodpeckers and other insectivores, (2) for nest sites, and (3) as perches for aerial insectivores such as swallows. He also suggested that conifer seeds released from cones by the heat are important food for some birds.

Lindenmayer and Franklin (1997) compared the effects of felling and fire on native birds and mammals in *Eucalyptus regnans* forest in south-eastern Australia. Many eucalypt species can be considered as Australia's ecological analogues of the conifers of the northern hemisphere: in these eucalypt forests commonly only one or two species dominate, and fire has occurred for thousands of years. *E. regnans* normally regenerates after fire, and so may form even-aged stands. It grows into a very tall tree, attractive for logging. Lindenmayer and Franklin pointed out the importance to wild animals of large standing trees which are left after fire (either alive but fire-scarred or standing dead), and also of large fallen logs. Cavities in trees are important as dens or nesting sites for species of marsupials, bats and birds. Because there are no woodpeckers in Australia such cavities are formed only by termites and microbes, which takes a long time: large cavities are not formed until trees are 120–200 years old. Old trees are also important because of hanging 'streamers' of bark which provide habitats for invertebrates. Clear-felling of *E. regnans* is therefore not favourable for some animals because it is done when the trees are 80 years old or less, and no large trees, dead or alive, are left. Lindenmayer and Franklin suggested possible strategies more favourable to wildlife. These would need to involve leaving trees to grow on to a larger size, in large patches, small clumps, or as spaced individual trees.

Importance of patch size and spacing

Two other important questions, from the point of view of survival of wild species, are: does the size of area felled matter? Does the size of remaining patches of forest matter? The answer to both questions is yes. Chapter 10 (Conservation) considers in some detail the survival of species in patchy environments, where habitat patches suitable for them are separated by other, unsuitable habitat types. The species' survival is likely to depend on how large the habitat patches are, and hence how many individuals of the species each can support; and also on how easily they can migrate from one patch to another. Here I summarize one example of how it has been proposed that the size and spacing of felled areas might be planned to favour a wild species.

Conserving the northern spotted owl

The northern spotted owl inhabits conifer forest in the Pacific Northwest of the USA—Washington, Oregon and northern California. For its hunting and nesting it requires *old-growth forest*. This means forest more than 80 years old, which is the normal age for felling rotational forests in the area. This in practice means forests that have never been felled: they contain large trees which are attractive for felling in the future. Thus there is a conflict between timber extraction and conservation of the owl. The US Congress set up a Spotted Owl Scientific Committee to 'develop a scientifically credible conservation strategy for the Northern Spotted Owl'. Its report in 1990 was the basis for a management plan (Lamberson *et al.* 1992; Wilcove 1993; Noon & McKelvey 1996).

The basic plan was to leave patches of old-growth forest, separated by areas where the forest would be felled and allowed to regrow. The difficult questions were: in order to give the owl a good chance of surviving long term, how large should the patches of old growth be, and how far apart? The home range of a pair of the owls is fairly well established, by tracking of radiotagged owls: it is 5–40 km^2. However, having patches just that size would not necessarily be adequate. As Chapter 10 explains, in order for a species to have a high chance of surviving long term in a habitat patch, that patch needs to support a population (the *minimum viable population*) larger than one pair. The minimum viable population varies in size between species and is not easy to determine. For the northern spotted owl its size is unknown, and the committee had to make the best possible guess using information from other species, supplemented by modelling. They chose 20 pairs as the required population size, and therefore recommended patches of 200 km^2 (based on a mean home range per pair of 10 km^2). The distance between patches needs to be low enough to allow the spread of young owls to find new mates and territories. On the basis of radiotagging results this was set at 19 km.

Thus the committee made precise recommendations about the size and spacing of patches of old-growth forest. In reality these had to be modified to take account of the size and arrangement of old growth that actually remains today, but it was still a precise plan for conserving one species. Two major criticisms can be levelled against it.

1 It is based on inadequate knowledge of the ecology of the owl. For example, the requirement of 20 pairs per patch is open to dispute: some scientists have suggested that more would be better (Noon & McKelvey 1996), but a scheme for conserving the Californian subspecies of the spotted owl involved only a few pairs per patch (Andersen & Mahato 1995). A different type of model has been proposed, using 'cells' each large enough to support one pair of owls (Hof & Raphael 1997), but it still suffers from inadequate information about how survival, fecundity and dispersal are influenced by conditions in the cells and the arrangement of the cells.

2 There are many other wild species living in these old-growth forests, and this arrangement of patches will not necessarily suit them. There are, for example, salmon in the streams, which are affected by what sort of vegetation occurs near the streams (Wilcove 1993). A plan for their conservation would favour retaining forest strips along streams, not in patches 19 km apart. Using terms defined in Box 10.1, the northern spotted owl may have served well as a *flagship species*, attracting public attention to the need for planned management and conservation of old-growth forests, but it is not an effective *umbrella species*. Management plans for the Pacific Northwest are now turning more to considering the needs of many species, rather than one or a few flagship species (Noon & McKelvey 1996).

Logging and its related services are economically important activities in parts of the Pacific Northwest. It is understandable if people do not want to lose their livelihoods, and especially if this happens on the basis of management plans which appear not to have a sound scientific basis. The northern spotted owl and the forests of the Pacific Northwest provide a case study of conflict between timber production and wildlife conservation. Such conflicts are a challenge to applied ecologists, to devise the best management system in a situation where there is no ideal solution.

A question posed at the start of this chapter was: Is it possible to obtain our timber needs sustainably from seminatural forests while also conserving wildlife there? or should we in future get all our timber from plantations, and leave semi-natural and undisturbed forests for other functions, including wildlife? This chapter has provided scientific information upon which an answer to this can be based, but it has also explained why it is difficult to decide on the best answer. The same answer may not be best in all types of forest in all parts of the world. So I am not going to give a straight answer. What do you think?

Conclusions

- The world timber harvest per year is increasing, but is probably still below the total new wood growth per year in the world's unplanted forests.
- Loss of forest is occurring in many tropical countries, but in many temperate countries, including some major timber producers, there has been little change in forested area in recent decades.
- Most loss of forest occurs because the land is wanted for other uses, not because of timber harvesting.
- Following a large-scale disturbance new forest that establishes naturally will at first be approximately even-aged, but will gradually change to mixed-age.
- In the past fire was frequent enough in many northern conifer forests to keep large areas even-aged. This was probably not true of most temperate deciduous and tropical forests.
- Selection felling (removal of individual trees) has been widely practised in tropical forests. In the process, many unharvested trees are also killed. Some wild animal species tolerate such felling, though others are not favoured.
- Following clear-felling of large areas, natural re-establishment of trees can be rapid in temperate and tropical regions, but a return to the original species composition will take many years.
- The response of animals to clear-felling varies greatly between species. Their survival can be influenced by the size and spacing of the remaining forest patches, and whether any large trees are left in the cut-over areas.

Further reading

Forest ecology and management:
general: Perry (1994)
tropical: Whitmore (1998); Mabberley (1992); Gomez-Pompa, Whitmore & Hardley (1991)
temperate: Peterken (1993)

Plantation forestry:
tropical: Evans (1992)
temperate: Savill & Evans (1986)

Economics of forestry: Price (1989)

Chapter 8: Pest Control

Questions

- Can we prevent pests becoming resistant to pesticide chemicals, or delay that? How?
- Do pests and diseases ever have major effects in natural ecosystems? Are the host and its attacker always in a stable relationship, or are there sometimes outbursts of damage to the host? How does the host species manage to survive?
- Can pests in arable crops and forest plantations be controlled by altering the management? Can study of pests in natural ecosystems suggest ways of doing this?
- Can biological control provide effective long-term control of pests? Will a species that initially provides good control evolve to become less effective?
- Can we decide which species are likely to be effective biocontrol agents, before elaborate testing?
- Is biological control safe? How can we be sure it will not harm other, non-target species?

Basic science

- Predicting the rate at which resistance to a chemical pesticide will increase in a pest population.
- The status of insects and fungal pathogens in natural ecosystems: invasions, outbreaks, endemics.
- Mathematical models of host–parasite population dynamics.
- Predicting a critical host population size or density below which a parasite will not survive.
- Coevolution of a parasite and its host.

A pest is a species that we would like to get rid of, or at any rate reduce in abundance. This chapter is about microbial pests, which cause diseases in plants and animals; about plants which are weeds in cropland or pastureland; and about animal pests—insects that damage crops or forest trees, and also other damaging animals, e.g. rabbits.

There is no doubting the serious effects pests can have. As well as directly reducing food production, they have substantially influenced

Box 8.1. Methods that can be used for controlling pests.

1 *Chemical pesticides*. Fungicides, insecticides, herbicides, aimed at killing the pest. Also chemicals produced by insects which elicit a response in other insects can sometimes be used, e.g. sex pheromones can be used to disrupt mating or to attract insects to traps.
2 *Immunization* of mammals and birds against diseases.
3 *Genetic alterations* to animals and plants to make them more resistant to a disease or more tolerant of it, less palatable to a predator, or (crop) more competitive against weeds.
This can involve:
(a) selecting individuals with desired characteristics;
(b) conventional breeding;
(c) treatments to promote mutation;
(d) genetic engineering.
4 *Management*. Possibilities include:
(a) shifting cultivation, crop rotation;
(b) alter timing of soil cultivation, crop sowing, harvest (e.g. in relation to weed germination or insect arrival);
(c) grow mixtures of different plant species or varieties;
(d) allow other species around field margin.
5 *Biological control*. Using another species to control the pest. Principal methods:
(a) *Biological pesticides*. Release large numbers of a control species frequently, e.g. every year, not expecting it to persist. Can be either *inundation*, if control species is not native to the area, or *augmentation* where the aim is to increase the abundance of a species already present.
(b) *Classical biological control*. Introduce the control species once only. It is then expected to multiply and spread.

how farming has been carried out in the past. Pest control has been a major reason for carrying out shifting cultivation or leaving fields fallow, which reduces the area available each year for growing food. Cattle have been excluded from large parts of Africa because of trypanosomiasis, a disease caused by a protozoan which is dispersed by the tsetse fly (Jordan 1986).

The potential crop yield losses that can be caused by pests are rarely measured quantitatively, because it is so rare to grow crops with no pest control at all. Forcella, Eradat-Oskoui and Wagner (1993) grew soybean on a farm in Minnesota without any form of weed control. The reduction in soybean seed yield caused by weeds ranged from 0 to 80%, depending on weed density and the date the crop was sown. This is just one year's result, and does not take account of how the weeds' abundance might increase over some years if they were left unchecked.

Box 8.1 summarizes the main methods that have been used for controlling pests. These need not be seen as alternatives: it is often most effective to use two or more of them in combination, sometimes called *integrated pest management*.

Chemical pesticides

Chemical pesticides can be very successful

The control of agricultural pests by chemicals can be claimed as a great success story of the 20th century which contributed to the rapid increase in food production. Kogan (1986, p. 256) gives a list of insect pest species that have been virtually eliminated by the use of insecticides. Some arable weeds have become so rare that people are now concerned about their preservation (Potts 1991). However, not all pests have been successfully controlled. For example, there are still no adequate and consistent chemicals for the control of soilborne fungal pathogens of crops; and chemical control of insect pests of forests and grazing lands is in general not economic. The use of chemical pesticides can have two major disadvantages: the chemical may have harmful effects on non-target species, and the target pest species may become resistant to it. Chapter 9 (Pollution) gives examples of pesticides that are harmful to wild species. Pesticides can also indirectly increase pests, if a fungicide inhibits a mycoparasite, i.e. a fungus that was attacking a fungus pest (Cook & Baker 1983), or when an insecticide kills a parasitoid of an insect pest. Previously sparse fungi or insects can then increase to pest levels of abundance. An example of this is given by Waage (1989): when parathion was sprayed on apple trees in Pakistan to control a moth pest, one result was a marked reduction of two parasitoid species of a scale insect and a 20–50-fold increase in abundance of the scale. However, insecticide can also enhance the effect of a parasitoid: if the insecticide reduces the insect population before the time for parasitoid attack, then the number of parasitoids per host individual can be greater and the percentage mortality increased (Waage 1989).

. . . but they can have disadvantages, too

Evolution of resistance to chemical pesticides

After a chemical has been in use for some years individuals resistant to it may appear in the target species; in other words, some genetically different individuals can survive the chemical at a concentration that would previously have killed all of them. The proportion of resistants in the population may then increase. Resistance to pesticides began to appear substantially among arthropods during the 1940s–50s, among plant pathogens during the 1950s–60s, and among weeds not until the 1970s–80s (Cousens & Mortimer 1995). The number of resistant species has increased greatly in each group since then, and resistance to some types of pesticide is very widespread (Shaner 1995). This section describes some examples of pesticide resistance, and considers whether there are ways of preventing its appearance or slowing its spread. The rest of this chapter is about possible alternatives to chemicals for pest control.

Resistance to some pesticides has appeared more frequently than to others. For example, the oldest major type of selective herbicides, the phenoxyacetic acids (which include 2,4-D) have been in use since the 1940s, yet resistance to them is quite rare: Holt *et al.* (1993) reported that resistance to them was known in only six weed species. In contrast, resistance to the triazine herbicides is known from more than 1000 sites, in at least 57 species. If the causes of such differences were known, it could help in the development of pesticides to which resistance would evolve only slowly.

Mechanisms of resistance to pesticides

This has led to much research on the mechanisms of resistance to pesticides (Holt *et al.* 1993; Lemon 1994; McKenzie & Batterham 1994). One relevant question is whether the resistance arises by a single point mutation or by amplification (i.e. producing many copies) of an existing gene. There is evidence, for example, that gene amplification has led to increased esterase activity in the aphid *Myzus persicae,* which leads to increased ability to break down some insecticides (Devonshire & Field 1991). Such gene amplification may give only gradual increase in pesticide resistance, slower than from a single point mutation (Cousens & Mortimer 1995). An example of resistance due to a single point mutation is resistance to triazine herbicides, which is most often due to a single base change in the DNA of the chloroplast gene which codes for a binding site protein in the thylakoid membrane (Warwick 1991). This mutation occurs more frequently than many other mutations. However, because it occurs in the chloroplast DNA it is inherited only from the female parent, not carried by pollen, and one might expect this to slow down the rate of spread.

Where resistance is caused by a single point mutation it is important how common the resistant gene is in the population before the pesticide is first applied. This will in turn depend on how frequently the mutation occurs, and on whether there is selection against the resistant strain in the absence of the pesticide, and, if so, how strongly. Quantitative information about these is unfortunately very sparse, and this places a serious limitation on modelling the appearance of resistance and predicting how best to prevent it.

Rate at which resistance increases in the population

The rate at which resistant individuals increase in frequency can be predicted by mathematical models (Cousens & Mortimer 1995). Here I describe simple predictions for resistance conferred by a single point genetic difference. If the resistance gene is initially very rare in the population, the frequency of homozygous resistants will probably be negligibly small. So a resistance gene is only likely to be selected for, and become a problem, if it confers some resistance in the heterozygote, i.e. it is a dominant or partially dominant character. The proportion of individuals in the population that are resistant will be approximately equal to the proportional frequency of the gene. Let the fitness of the homozygous susceptibles and the heterozygous resistants be W_{ss} and W_{rs}, respectively. Then, if r_t is the frequency (proportion) of individuals that are resistant in year t, the next year its frequency will be approximately

$$r_{t+1} = r_t (W_{rs} / W_{ss}) \tag{8.1}$$

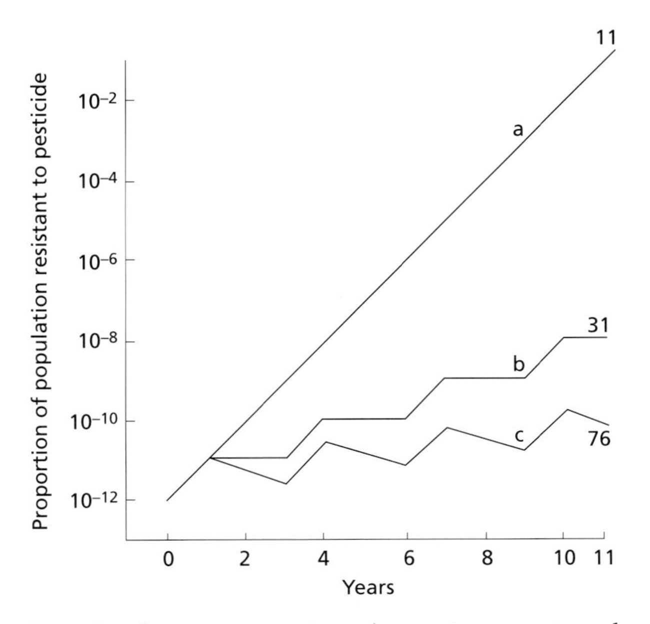

Fig. 8.1 Predicted proportion of individuals in a population that are resistant to pesticide, calculated from Equation 8.1. Line (a), pesticide applied every year; (b) and (c) applied every third year. Assumed relative fitness of pesticide resistant and susceptible individuals (W_{rs}/W_{ss}): years pesticide applied, 10; years pesticide not applied, 1 (line b) or 0.5 (line c). Figures at right: time (in years) taken for resistants to reach 0.1 (10%) of total population, starting from 10^{-12} in year 0.

and starting from time 0, after n generations (assuming continued pesticide application) it will be

$$r_n = r_0 (W_{rs} / W_{ss})^n \tag{8.2}$$

How long it will take for resistants to reach a significant frequency clearly depends on the frequency at the start (r_0), information which is not available. However, the ratio (W_{rs}/W_{ss}) is likely to be large, because pesticides are designed to kill the pest, and therefore even a modest ability to survive the pesticide is likely to give the mutant a substantial fitness advantage. If, for example, the fitness ratio (W_{rs} / W_{ss}) was only a modest 10, and the frequency of the mutant initially (r_0) was only 1 in 10^{12}, Equation 8.2 predicts that the resistants will reach significant abundance within 12 generations, which for many pests (e.g. annual weeds) is 12 years. This helps to emphasize the problem: even if there is only one individual insect or weed in a field that has the resistance gene, its increased ability to survive will soon make its progeny abundant. Figure 8.1, line (a) shows the time course of this 10-fold increase per year. It illustrates another problem: because the existence of resistants will not be noticed until they are a significant proportion of the total, say 1% or more, most of the increase in their abundance goes on unnoticed and unstudied, until finally they seem suddenly to burst into prominence.

The prediction that resistance can become serious within a matter of years or a few decades is supported by experience. In vineyards in northern France, after benzimidazole fungicides had been used for only three or four seasons they ceased to give effective control of important fungal pathogens such as *Botrytis cinerea* (Russell 1995). Cousens and Mortimer (1995, Table 83) give figures for herbicides where resistance has been detected after periods ranging from 3 to 25 years.

Table 8.1 Relative fitness of herbicide-resistant and susceptible strains of two weed species, assessed by reproductive capacity when the two strains were grown either separately or in competition in 50:50 mixture, in the absence of herbicide. Figures are ratio, for measured variable (see notes below), of resistant/susceptible

Species	Grown separately	Grown in mixture	Notes
Senecio vulgaris	0.57	0.20	1
Amaranthus hybridus	0.90	0.18	2

Notes. (1) Dry weight of reproductive parts. Data from Holt (1988).
(2) Weight of seed per plant. Data from Ahrens & Stoller (1983).

Delaying build-up of resistance

One approach to preventing resistance appearing in the pest population is to have fewer individuals of the pest within the whole treated area at the time the pesticide is first applied, in the hope that there will be no individual with a resistance gene. This requires the use of other control measures as well as the chemical, in other words integrated pest management. Because the frequency of resistants in the absence of pesticide application is not known, we have no sound basis for deciding whether this approach could be effective.

One way to slow down or stop the development of resistance is not to apply the pesticide every year. Some resistant strains are less fit than susceptibles when growing in the absence of the pesticide—their adaptation to survive the pesticide has been accompanied by some loss of ability to grow and reproduce in normal conditions. So, if the pesticide is applied only in some years, we should expect the resistants to decline in abundance in the 'off' years. However, if this is actually to prevent resistance developing, rather than just slow its development, the fitness advantage of susceptibles in 'off' years will need to be substantial, to counterbalance the substantial selective advantage of resistants in 'on' years. The relative fitness of resistants and susceptibles in the absence of pesticide is in general not known with any confidence. It is likely to depend on the conditions in the field situation. This can be illustrated by the experimental results in Table 8.1. In these two species of weed, strains were isolated that were susceptible or resistant to triazine herbicides but in other respects closely similar genetically. They were grown from seed in a greenhouse or an experimental plot, without herbicide application, the two strains either separate or in a mixture so they competed. Table 8.1 shows that when they competed the resistant strain had only about one-fifth the seed production of the susceptible, but when they grew separately the difference was much less: indeed, for *A. hybridus* the strains did not differ significantly. On a farm the weed would probably compete primarily with the crop plant, perhaps little with others of its own species.

Figure 8.1 illustrates the effect of having one year with a pesticide then two without it, comparing the situation where the resistants and susceptibles are equally fit in the 'off' years or where the susceptibles are twice as fit. It shows that a substantial component of the delay in resistance build-up (compared with pesticide applied every year) is simply due to there being no selection for resistance in the 'off' years (line b). However, if the susceptibles are more fit than the resistants in 'off' years this can further slow the build-up (line c); in the example given the time taken for resistance to become abundant is 76 years, which is longer than selective pesticides have been in existence. Graphs similar to Fig. 8.1 but based on more sophisticated models have been published (e.g. Gressel & Segel 1990).

If there is a pool of susceptibles that escape the pesticide treatment this can delay resistance build-up. For weeds this can be a bank of remaining seeds in the soil. For insects or microbial pathogens it could be a residual population in unsprayed field margins. It has also been suggested that the farmer might accept less control of the pest in order to leave some susceptibles. There is unfortunately little experimental evidence on how much effect these strategies can have in delaying the build-up of resistance. If there are several alternative pesticides these could be rotated, with each being used only once every few years. Provided the pesticides have different modes of action, one might expect that resistance to one of them would not increase resistance to another. Unfortunately, this has not always proved to be true: *cross-resistance* has been reported, meaning that individuals becoming resistant to one pesticide also become resistant to another, even though it is chemically unrelated (Holt *et al.* 1993).

Another suggestion is to use mixtures of pesticides which are all effective on the same pest. Then an individual which has resistance to one chemical would probably be killed by another, and so the probability of resistance being selected could be greatly reduced. However, if there is cross-resistance this does not work.

This section has outlined some of the problems with the development of pesticide resistance and suggestions for what to do about it. Most of the proposals are based mainly on models, but there is a shortage of key information to feed into the models. The main message is that there are ways of slowing down the development of resistance, thereby reducing the time before it becomes serious, but for many pesticides there is no clear way to ensure that resistance never develops.

Diseases and pest insects in natural ecosystems

If we aim to control pests and diseases on farms and in commercial forests, by means other than chemical pesticides, one approach is to ask whether we can learn by studying more natural ecosystems. Fungi and insects do not wipe out all the other species in those systems, so in some sense they must be under control. Maybe this can give us ideas on how to control them in our more artificial systems. This section describes some examples of insects and diseases that attack wild plants and animals, and considers whether there are any useful messages for pest control.

Chestnut blight

The first message is that epidemics *can* occur in near-natural ecosystems, even when they are quite species rich. An example is chestnut blight in North America, which is caused by the fungus *Cryphonectria parasitica* (Van Alfen 1982; Buck 1988; Newhouse 1990; Anagnostakis 1995). This spreads in the bark of stems, and once it has formed a girdle right round a stem all the branches above that point die. The fungus probably reached the United States in young chestnut trees imported from Asia. It was first noticed in 1904 in New York, and within the next 50 years had spread through the whole natural range of the native chestnut (*Castanea dentata*) in the eastern USA. Once a major tree of mixed deciduous forests in the Appalachians, the chestnut now occurs there only as regrowth coppice shoots from surviving stumps. These shoots rarely live long enough to set seed, so whether the species will survive indefinitely in the area is uncertain. The fungus also reached Mediterranean Europe, where it killed many of the native *Castanea sativa*. There, however, after about 15 years less damaging 'hypovirulent' strains of the fungus began to appear, which caused some symptoms but did not kill the tree. The character for hypovirulence is carried by a virus, which can transfer to virulent strains, reducing their virulence. This suggests an opportunity for biological control, which will be discussed later.

Phytophthora cinnamomi *in Australia*

Chestnut blight provides an example of a disease caused by a parasite which can attack plants only within a single genus (*Castanea*). In contrast, some other plant pathogens have a wide host range: one example is *Phytophthora cinnamomi*. This fungus infects the roots of many woody plant species, causing root rot, which can be followed by die-back of branches and ultimately the death of the whole plant. The fungus is confined to soil and roots and has no airborne spores, so it normally spreads slowly, in water or with eroding soil. Yet it occurs in every continent

Table 8.2 Records from a permanent plot in *Eucalyptus* woodland in the Brisbane Ranges, Victoria, Australia, in 1975 and 1985. *Phytophthora cinnamomi* was present from 1975 onwards. From Weste (1986)

	1975	1985
Bare ground (%)	10	40
Number of plant species*	18	24
Number of individuals in plot:		
Eucalyptus, all species*	75	41
Isopogon ceratophyllus (P)†	327	0
Xanthorrhoea australis (L)†	162	9
Banksia marginata (P)†	70	64
Hakea sericea (P)†	4	20
Lomandra filiformis (L)†	0	53

* in 630 m^2
† in 360 m^2
Plant family: L = Liliaceae, P = Proteaceae.

except Antarctica (Zentmyer 1985). It attacks commercially important plantation and orchard trees as well as species of natural forest. It has caused conspicuous die-back in eucalypt forests in several parts of Australia (Weste & Marks 1987). Table 8.2 shows changes in abundance of some of the plant species at a woodland site in Victoria, southeastern Australia. At the time of the first recording in 1975 *Phytophthora cinnamomi* had only recently invaded this area. Although bare ground increased substantially during the following 10 years, the number of species did not decrease. Some species that were very susceptible to the fungus decreased markedly in abundance, but other, less susceptible species increased. In a 'control' area where the disease had not yet reached, none of these species changed markedly in abundance between 1975 and 1985. So, the fungus caused major changes in the vegetation composition, as it has also in species-rich shrubland in Western Australia (Wills 1993). However, in the Victoria site some of the very susceptible species did not disappear altogether from study plots: even after 20 years a few individuals remained. These might be genetically different individuals with higher resistance to the fungus. Their survival could be aided by a decrease in abundance of the fungus, which happened in some areas after the main epidemic had passed. The cause of this decline could have messages for pest control. It could be a response to a reduced abundance of susceptible plants; but there is also evidence that in soils in some parts of Australia microbial populations build up which suppress *P. cinnamomi* (Malajczuk 1979). This same fungus is found in apparently undisturbed forests, including tropical rainforest, where it is presumably endemic. So far we have no evidence on whether it is having much effect on the vegetation there.

The previous examples involved invasions, which were started or at least aided by people transporting the fungus. However, there are native herbivorous insects and pathogens that show periodic outbursts: after remaining low in abundance for some time the population increases rapidly to a high and damaging density, then after a while decreases again to a low level. An example is the spruce budworm, *Choristoneura fumiferana* in eastern North America and *C. occidentalis* in the west. The larva feeds on the leaves and buds of spruce, fir and Douglas fir. When it is at its most abundant the insect causes severe defoliation over large areas of forest, leading to substantially reduced stem growth and, if the outbreak persists for several years, to the death of trees (Crawley 1983; Speight & Wainhouse 1989; MacLean *et al.* 1996). Figure 8.2 shows that in an area of New Brunswick budworm larvae were abundant during the 1950s, then sparse (though never completely absent) for a decade, but then increased again in the 1970s. That outbreak ended about 1987. There are written descriptions of two earlier outbreaks, about 1880 and 1912–20. Other outbreaks before that are indicated by changes in tree-ring width. Figure 8.2 shows that according to these estimates outbursts of the insect have occurred at fairly regular intervals of about 35 years. The western spruce budworm has also shown periodic outbreaks in

Outbreaks of spruce budworm in N. America

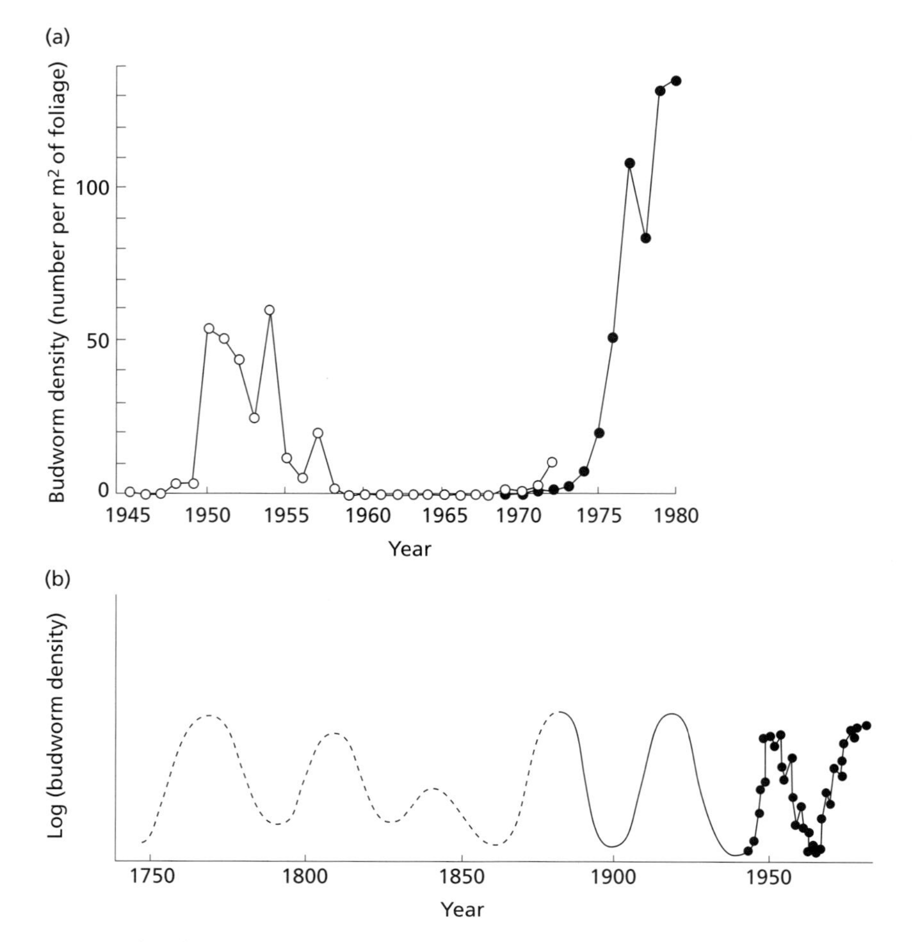

Fig. 8.2 Abundance of eastern spruce budworm (*Choristoneura fumiferana*) in conifer forest in New Brunswick. (a) 1945–80; ○ third- to fourth-instar larvae, ● eggs. (b) 1750–1980; – – – estimated from xylem ring widths in trees; —— based on written records; 1945 onwards data from part (a). From Royama (1984).

Colorado and New Mexico (Swetnam & Lynch 1993; Weber & Schweingruber 1995).

These outbreaks are of great practical importance, as these forests are commercial timber sources, but an understanding of their cause (or causes) would also have a wider significance, since it is highly desirable when instituting pest control, for example by biological means, to be able to predict whether it will be effective long term or whether it will allow periodic outbreaks of the pest. In spite of extensive research, the cause of the cycles in spruce budworm remain uncertain. Contributory factors that have been proposed include:

1 weather;

2 food abundance and quality, influenced by the frequency and age of different tree species and their nutrient status;

3 predation by birds;

4 parasitoids and microbial diseases (Crawley 1983; Royama 1984; Swetnam & Lynch 1993).

Speight and Wainhouse (1989) give information on six other species of forest insect that have undergone major outbreaks. Like spruce budworm, these species live in temperate or boreal forests where there are only one or a few abundant tree species. However, outbreaks can also occur in much more species-rich communities, including tropical forest. Wong *et al.* (1990) described conspicuous defoliation in 1985 of one tree species, *Quararibea asterolepis*, in Barro Colorado forest, Panama, by the larvae of two moth species, *Eulepidotis* spp. Among 460 *Q. asterolepis* trees in a 50-ha plot defoliation ranged from 0 to 100%, with about 100 of them suffering more than 90% defoliation. *Eulepidotis* larvae had not been abundant in the previous year, nor were they abundant in several years following the outbreak.

Endemic pests

In the examples given so far the disease or insect suddenly became much more abundant, and so its effects were conspicuous. However, we are just as interested in endemic pests. Are there insects and diseases whose abundance in natural systems is approximately constant from year to year? Do they have much or little effect on their hosts? What controls them? Could a pathogen or insect species be quite rare, yet have a major effect on the composition of the community? Study of the *Phytophthora cinnamomi* epidemic in Victoria, described earlier, suggests some answers. At a site where the fungus had invaded about 20 years earlier and had caused the local extinction of some plant species, it was no longer possible to isolate the fungus from the soil, yet plant species susceptible to the fungus remained absent or rare (Weste 1986). If the fungal invasion and its effects on the vegetation had not previously been observed, it would be very difficult to discover its continuing important influence.

A nematode in deer

The importance of an endemic insect or parasite can sometimes be inferred from field observations. An example in North America is a nematode worm *Parelaphostrongylus tenuis*, which infects some deer and a few related ungulates, including moose (Davidson *et al.* 1981; Nudds 1990). The worm has a complex lifecycle involving slugs or snails as an alternate host; it eventually ends up in the spinal cord or brain of the mammal. In moose the worm causes serious symptoms, including paralysis, usually leading to death. In contrast, in white-tailed deer it causes no harmful symptoms, and in many populations 50% or more of the deer are infected; so white-tailed deer act as carriers. Since the early 20th century the range of white-tailed deer has extended. When the deer extend into the range of moose, the moose usually become infected with the parasite and their population declines. This decline is probably due mainly to the parasite, though direct competition between deer and moose for food might also be involved (Price *et al.* 1988; Nudds 1990). Price *et al.* (1988) summarized several other examples where a parasite

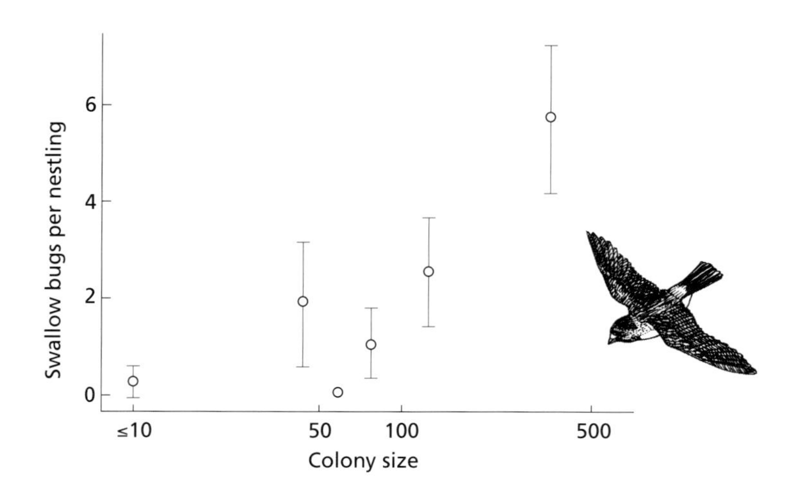

Fig. 8.3 Number of bugs per nestling in relation to colony size (number of active nests) for cliff swallows in Nebraska. Vertical bars are standard errors. From Brown & Brown (1986).

carried by one mammal is passed to another species, and is probably responsible for greatly reducing it. They also list several fungal diseases of plants, caused by rusts, which behave in a similar way.

A blood-sucking bug of swallows

Sometimes the abundance of a disease organism can be altered experimentally, in order to find out whether it is having much effect on the host population. An example, investigating myxomatosis in rabbits, is described later (see Fig. 8.12). Another example is a bug, *Oeciacus vicarius*, which feeds on the blood of nestlings and adults of the cliff swallow. This swallow, which is widespread in North America, builds nests out of mud under overhanging cliffs, ledges of buildings, or bridges. Usually there are several or many nests close together, forming a colony. Brown and Brown (1986) found that in Nebraska many nests and nestlings had the bugs; the mean number of bugs per nest and per nestling increased with increasing colony size (Fig. 8.3). To find out whether the bugs affect the nestlings they fumigated some nests: the fumigant killed the bugs but did not directly affect the birds. In fumigated nests the number of nestlings surviving to 10 days was usually higher, often substantially; they were also slightly larger (Table 8.3).

It is clear from these and many other examples that parasites and their hosts, and herbivorous insects and their food plants, can coexist long term (Grenfell & Dobson 1995). Whether or not the populations vary from year to year or are fairly constant, some sort of balance has been achieved which allows both species to continue to survive in the area. Box 8.2 summarizes the mechanisms that can be involved. In some of the examples we can see which of these is likely to be important. An example of mechanism (2) is the reduced virulence of chestnut blight in Europe owing to the fungus being attacked by a virus. The greater abundance of blood-sucking bugs in swallow colonies with more nests (Fig. 8.3)

Table 8.3 Effect of fumigation, to kill insects, on survival and growth of nestlings of cliff swallow in a colony in Nebraska. From Brown & Brown (1986)

	Nest fumigated	Nest not fumigated
Number of nestlings per nest surviving to 10 days	3.1	1.4*
Mean body weight per nestling (g)	23.7	20.3*

* Difference statistically significant ($P < 0.001$)

suggests that more widely spaced nests provide some control (mechanism 3). The decline of *Phytophthora cinnamomi* after invading Australian forests, and the survival of a few individuals of susceptible species, could be due to mechanisms (1), (2) or (3), or a combination: genetically more resistant individuals; an increase of microbial populations in the soil which are suppressive to the fungus; or the wide spacing of the few remaining susceptibles.

These examples can provide suggestions when we are considering how to control pests and diseases of useful plants and animals. Here I point out some questions that still remain.

1 Why do outbreaks of disease or insects occur in some ecosystems but not others? What prevents them in many ecosystems?

2 After an epidemic or outbreak the causal organism often declines to a low level, where it is much less damaging (e.g. *Phytophthora cinnamomi*, spruce budworm). What causes this decline?

3 Why do populations of some species fail to carry a parasite, even though it can infect them (e.g. moose and the nematode worm)?

An answer to any of these questions could be very helpful in planning control of pests or pathogens.

Box 8.2. Mechanisms by which a host species can coexist with a pest species in natural ecosystems.

1 Resistance characters in the host.
(a) Immune systems in vertebrates.
(b) Toxic chemicals in plants and invertebrates.
(c) Physical barriers, e.g. insect cuticle.
(d) Low resource quality, e.g. low protein concentration.

2 The pest species is itself attacked by a parasite or predator, which controls its abundance.

3 Low abundance and hence wide spacing of the host species interferes with spread of the pest, sufficiently to keep its abundance low. This only works if the pest is specific to the host, so there are no alternative hosts to support it.

Mathematical models of host–parasite population dynamics

These examples of parasites and herbivorous insects in quasi-natural ecosystems have led to useful questions and suggestions, but they have also indicated how difficult it is to understand fully what is going on. One way forward is to apply mathematical modelling. I start by outlining briefly a model which has substantially advanced our understanding of the relationship between animals and their parasites (Box 8.3). This can

Box 8.3. Simple models of the population biology of parasites and diseases and their hosts, and of herbivorous insects.

Based on Anderson & May (1979, 1986); Crawley (1983); Swinton & Anderson (1995). The letters are defined below.

(1) Diseases of animals; causal parasite transmitted from animal to animal by contact.

The rate of increase in the number of infected individual animals is

$$dI/dt = \beta SI - cI = I(\beta S - c) \quad (8.3)$$

If animals can become immune to the disease

$$c = b + \theta + \nu \quad (8.4)$$

otherwise

$$c = b + \theta \quad (8.5)$$

The disease will maintain itself in the population when

$dI/dt \geq 0$

i.e. when $\beta S \geq c$,

i.e. when $S \geq c/\beta$

Therefore, the critical number of susceptible host individuals above which disease will establish, is

$$S_T = c/\beta = (b + \theta)/\beta \quad \text{or} \quad (b + \theta + \nu)/\beta \quad (8.6)$$

(2) Parasite or insect can survive outside the host, and can in this way infect new hosts.

$$S_T = (c/\beta) \times (f/\lambda) \quad (8.7)$$

Meaning of letters:

b—death rate of host individuals from non-disease causes

c—rate at which infected individuals cease to be infected, by recovery or death

f—death rate of propagules

I—number of infected host individuals

S—number of uninfected, susceptible host individuals

S_T—critical number of susceptible individuals, below which the disease fails to persist

t—time

β—transmission coefficient of disease in the population

θ—death rate of infected individuals caused by disease (= virulence)

ν—rate at which infected individuals recover and become immune

λ—rate at which propagules produced

be applied, after some modifications, to diseases of plants and to insects eating plants.

In the model, at any time the total population of host individuals consists of S uninfected susceptibles, I infected, and, in some species (e.g. mammals), also some immune. The rate of transmission of the infection through the population depends on how many infected individuals there are, to provide a source of infection, and also on how many susceptibles there are ready to become infected. This rate of transmission thus equals βSI. The number of infected individuals *increases* by spread of the parasite to new individuals, but it *decreases* by the death of infected individuals and perhaps by the recovery of others; it is the balance between these increases and decreases that determines whether the parasite survives in the population. Equations 8.3–8.6 in Box 8.3 set this out, showing, by simple algebra, that there is a critical population size of uninfected susceptible individuals, S_T, below which the parasite will not survive and above which it will survive; this is defined by Equation 8.6. S_T can alternatively be expressed as a critical population *density*, if I and S in the preceding equations are densities, i.e. numbers of individuals per unit area.

Up to now we have considered diseases of animals which require contact between individuals for their transmission. Rabies is an example, considered later. Many diseases, however, have a stage that survives free of the host and is involved in transfer. For example, many fungal diseases of plants disperse by spores; viruses that attack insects may survive on plant surfaces until another host insect comes along; insects that attack other insects or plants may have a motile adult stage. Equation 8.7 modifies Equation 8.6 to take account of the formation and death of the propagule stage. This would be different, however, if the propagule is carried by a vector or alternate host.

Can the disease persist in the population?

The equations in Box 8.3 are very simple, but they can give us some important messages. The idea of a critical host population density, below which the disease will die out, is a very important one. Equations 8.6 and 8.7 show that a parasite can maintain itself in a sparser population if it is more effective at spreading (larger β), which is perhaps not surprising. Less obvious is that a disease that kills its host more quickly (higher virulence, larger θ) needs a denser population to survive. The quicker the animals die, the fewer infectives there are to spread the disease; or the more rapidly a herbivorous insect kills its host plants, the less chance that there will be other, live plants waiting for it to move on to. Earlier I described a nematode worm that can infect moose, and which usually kills them. Yet the worm is absent from most moose populations: only when white-tailed deer carry it into moose areas are the moose infected and killed. The explanation is probably that the worm kills moose too rapidly, so that in the absence of deer as a carrier there are soon few susceptibles left. In terms of Equation 8.6 θ is high so S_T is high, i.e. the worm can only maintain itself in a moose population if the density of animals is high. In deer, because $\theta = 0$, S_T is probably much lower.

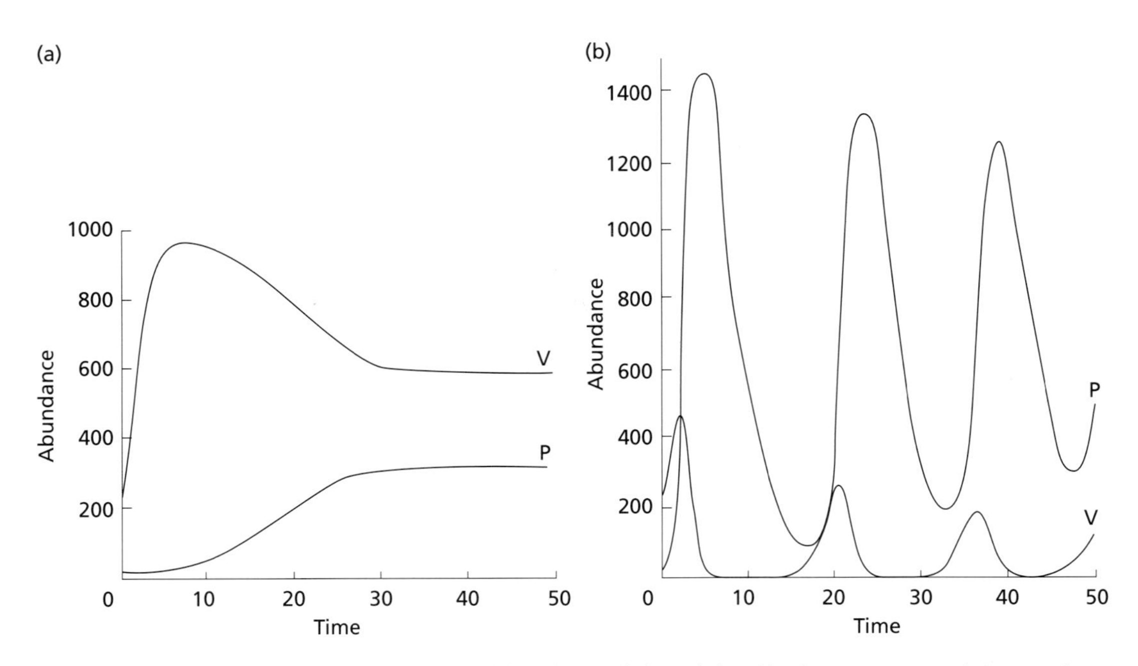

Fig. 8.4 Predictions of abundance of plants (V) and herbivorous insects (P). In graph (b) insect multiplication ability g 10 times as high as in (a). From Crawley (1983).

The models can be extended to predict how much the parasite or herbivore reduces its host's abundance, and also whether this abundance is stable or subject to cycles; but I do not present the equations here. Crawley (1983) presented predictions, from a whole family of models of this type, of the abundance of insect herbivores and their host plants. Figure 8.4 shows two examples. In the absence of the insects plant abundance would stabilize at 1000. The graphs show the predicted time-course of abundance of the insect and the plant, for two different values of g, which controls the rate of insect multiplication:

$$dP/dt = gPV - fP \tag{8.8}$$

where P is the number of insects, V is the abundance of plants (measured as numbers or biomass), and f is death rate of the insects.

A stable population or cycles?

In Fig. 8.4(a), with the lower value of g, plant abundance is reduced to a stable level, but only slightly below the level of 1000 attained in the absence of insects. If the insects have higher multiplication ability, g (part (b)), not surprisingly they increase in abundance compared with (a), and the plants are reduced more. However, the prediction is that both cycle in abundance. The insects increase so fast that they kill almost all the plants; many insects cannot then find a live plant to feed on, and so the insect population crashes; then the remaining few plants increase; and so on. This insect would probably not be considered an effective biological control agent. This has interesting similarities to real insect outbreaks, e.g. the spruce budworm (Fig. 8.2). However, in real cycles of

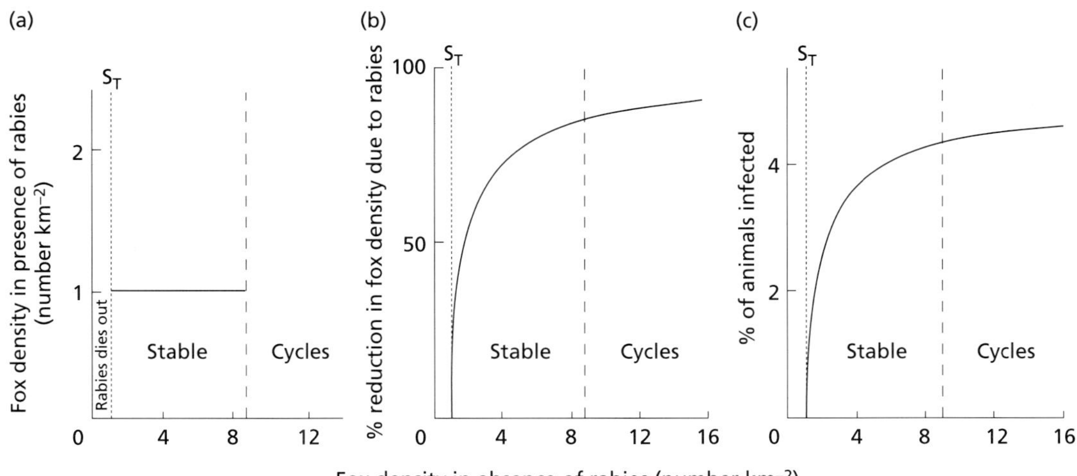

Fig. 8.5 Predictions by model of Anderson *et al.* (1981) of population dynamics of foxes and rabies. (a) Abundance of foxes when rabies present; (b) percentage reduction in fox abundance caused by rabies; (c) percentage of foxes infected at any time. S_T is the critical fox density below which rabies does not persist. Reprinted with permission; copyright Macmillan Magazines Limited.

herbivorous insects that have been investigated, the causes always seem to be more complex.

Rabies in foxes

A similar model has been applied to rabies in foxes in Europe (Anderson *et al.* 1981; Anderson 1982). The disease is caused by a virus, which spreads between animals by direct contact. It occurs in fox populations in much of central Europe and the USA and can be spread to humans, usually via domestic dogs and cats. In humans the disease has very unpleasant symptoms and is usually fatal. From research on foxes and rabies it is possible to derive values for the key variables in the model. Figure 8.5 shows predictions by the model of the percentage of foxes infected and the effect of the disease on fox numbers, for a range of initial (pre-rabies) fox densities. The model predicts that for any initial (disease-free) fox density within the range 1–9 per km^2, rabies will maintain the population near the critical density (S_T), which is predicted to be about 1 per km^2. Above a disease-free density of about 9 per km^2 the model predicts cycles, rather like those in Fig. 8.4(b). Outbreaks of rabies and consequent cycles of fox abundance, with peaks every 3–5 years, have been recorded in various parts of Europe and North America. Within the range where rabies and fox densities are stable, the percentage of foxes infected at any one time is predicted to be usually only 2–4% (Fig. 8.5(c)), which is in reasonable agreement with field observations that have shown the figure to be usually 3–7% where the disease is stable and endemic. In spite of this, the reduction in the fox population can be large (Fig. 8.5(b)). Rabies shows that a disease can be the most important single factor controlling the population density of its host, even if only a few per cent of the population are infected at any time.

Many more complex variants of these models have been presented and their predictions explored (Crawley 1983; Briggs *et al.* 1995). A principal aim has been to predict situations in which the host's abundance will be greatly reduced but in a stable manner. Although more complex models aim for greater realism, it is often difficult to know whether the predictions are in fact realistic.

Making species less prone to attack by pests

Paragraph 1 of Box 8.2 lists the main ways in which characteristics of wild animals and plants make them more resistant to attack by pests. Do these provide any suggestions for pest control?

Immunization

Immunization can be used on domestic animals, and has sometimes proved practicable for wild mammals. For example, an oral vaccine against rabies has been fed to foxes in many European countries. This started on a large scale in 1990, and was followed by a substantial decline in the number of cases of rabies in those countries (Stöhr & Meslin 1996; MacKenzie 1997). Models have been used to predict whether immunization or culling would be the more effective way to control a disease in a wild species, e.g. bovine tuberculosis in badgers and possums (Barlow 1996; Roberts 1996). The basic aim is to reduce the number of susceptible animals below the critical density (Equation 8.6); however, the proportion that would need to be culled may not be the same as the proportion that would need to be vaccinated, because the vaccinated animals will for a while remain alive, so the level of competition between animals is different. It will be more difficult to make culling effective if the species has the potential to reproduce rapidly to replace the culled animals.

Secondary chemicals in plants

Unlike vertebrate animals, plants do not have an immune system to protect them against diseases, but they produce *secondary chemicals* which are often poisonous to animals and microorganisms. There is an enormous diversity of these chemicals within the plant kingdom; major groups include phenolics, terpenes, alkaloids and non-protein amino acids (Harborne 1993; Bennett & Wallsgrove 1994; see also Chapter 10). They are one of the methods by which plants prevent animals eating all of them. However, these chemicals often taste nasty to humans, and some are poisonous to us. So, rather than breeding crops for increased secondary chemicals, to protect them against pests, breeding has often, in contrast, been aiming to reduce them. Many members of the cabbage family (Brassicaceae) contain glucosinolates (mustard oils). In vegetables such as cabbage and Brussels sprouts we value the glucosinolate for providing the characteristic flavour, but in the seed of oilseed rape the glucosinolate is not welcome to us. Oilseed rape was therefore selected for low glucosinolate concentration in all the tissues; however, these varieties were then found to be very susceptible to fungal infection (Bennett & Wallsgrove 1994). More recently varieties have been bred with glucosinolate concentration high in the vegetative tissues but low in the seeds. Another example is alkaloids in grasses (produced by fungi within the grass tissue), which provide protection against leaf-eating insects but harm cattle and sheep that eat the grass. It has been found that low-alkaloid perennial ryegrass will not persist in many parts of New Zealand, primarily because it is severely attacked by a weevil (Prestidge & Ball 1997). Research is aiming to produce grass varieties with alkaloids that are harmful to insects but not to cattle and sheep. Although such specificity has not yet been achieved for grass alkaloids, it is a feature of *Bacillus thuringiensis* toxins, as explained later.

Box 8.2 also lists two other sorts of mechanism by which a species can reduce the severity of attack by a pest: physical barriers and low resource quality. Both of these are likely to be associated with reduced value of plants as food for us and our domestic animals. Also, lower protein concentration in leaves is correlated with slower photosynthesis (see Chapter 2).

Management for control of diseases, weeds and insect pests

Management means the way a farm or forestry plantation is laid out and run. Can this help in the control of pests? Can the study of natural ecosystems provide any suggestions about this? Box 8.1, paragraph 4, lists some possibilities.

Long-term control of a weed

Viewed over a period of some years, a small change in the degree of control achieved can have a major effect on pest abundance. This can be illustrated by the corncockle (*Agrostemma githago*), a weed of cereal crops which originated in Mediterranean Europe but is now widespread in temperate regions. Its seeds are similar in size to those of the cereals and it is dispersed mainly as a seed contaminant. Improved seed cleaning during the 20th century greatly reduced its abundance. Firbank and Watkinson (1986) measured the growth and seed production of corncockle and wheat, grown separately and together at different sowing densities. They used the results as the basis of a model to predict the effect of different efficiencies of removal of corncockle seed from contaminated wheat seed. In the model, some of the seed harvested was resown the next year, with different proportions of the weed seed removed. The model predicted that if there was no removal at all of the weed seed the weed increased greatly within a few years, then levelled off to a stable abundance (top line of Fig. 8.6), which reduced the yield of wheat per hectare to only one-fifth of that in weed-free plots. Figure 8.6 shows that if about 80% or less of the corncockle seed was removed each year, the effect on its abundance was modest. The model predicts a critical efficiency of seed cleaning—removing about 90% of the weed seed—above which the weed declines over time to a very low abundance. The precise value of this critical cleaning efficiency would depend on the growing conditions,

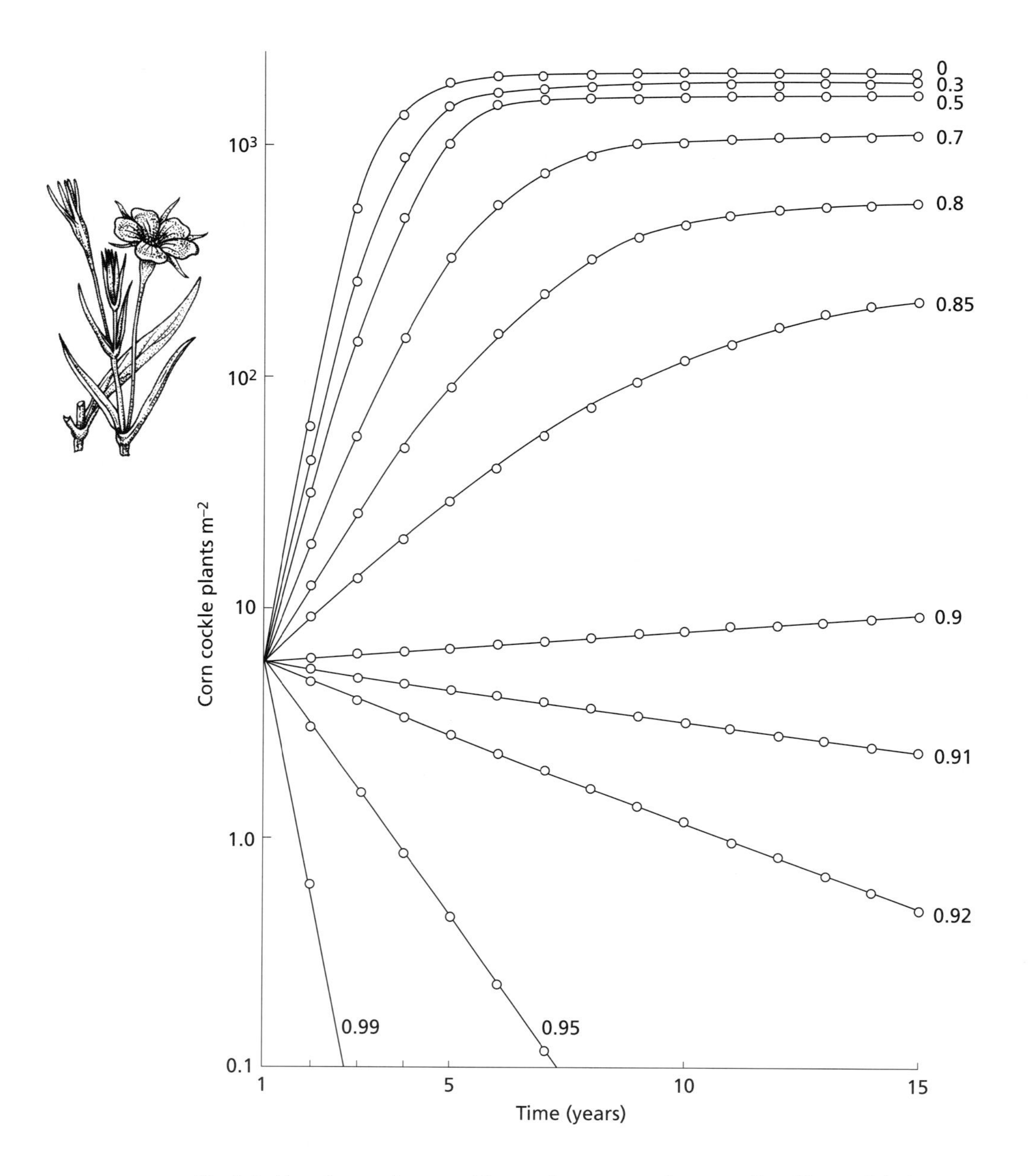

Fig. 8.6 Abundance of corncockle growing among wheat, predicted by model of Firbank and Watkinson (1986). Seeds sown per m^2 the first year: wheat 500, corncockle 10. After that part of the harvested wheat + corncockle seed was resown, 500 m^{-2} of wheat, with a proportion of the corncockle seed removed as shown by figures at right. Note that abundance is on a log scale.

but this model illustrates how it is important to think of pest control over a long timescale, not just 1 year.

Shifting cultivation and rotation

Shifting cultivation and rotation have already featured in Chapters 2 and 4. Among their functions is the control of weeds, pests and crop diseases. If shifting cultivation is being practised in a forested area, in the patch where crops are grown weeds, diseases and insect pests of the crop plants are likely to colonize and increase. This is one of the reasons why after a few years the patch is left for natural forest regeneration to occur, and the farmers clear another patch of forest. During the forest 'fallow' most crop pests decrease or disappear. The major disadvantage of shifting cultivation is that only a small proportion of the area is growing crops in any one year, and hence it can support only low densities of people (see Table 2.3). In Europe for many centuries the most common farming system involved a rotation, in which each field was left fallow every second or third year. During the fallow the field could be ploughed several times to reduce weeds; it also helped to control soilborne diseases and animal pests. But this system did result in only one-half to two-thirds of the potential arable land being under crops in any year, so the total food production of the farm was reduced (see Table 2.3).

Nowadays if rotations are used they more commonly have no bare fallow but rotate among different crops. However, a single year's break before a particular crop species returns is not necessarily enough to reduce its pests to acceptably low levels. In northern Germany at least 2 years of cereal are needed between each year of sugarbeet to control beet nematodes (Heitefuss 1989); Baker and Cook (1974) gave examples where an even longer break is recommended. In contrast, continued growth of the same crop year after year can lead to a decrease in some diseases (Campbell 1989; Whipps 1997): an example is take-all decline. Take-all disease of wheat is caused by the soilborne fungus *Gaeumannomyces graminis*. It has been observed in many countries that if wheat is grown every year on the same field take-all increases in severity for about 3 years, then declines again. This decline appears to be caused at least partly by living species in the soil, since fumigation, irradiation or heat treatment of the soil results in the disease increasing again (Baker & Cook 1974). Other examples of disease decline are shown by potato scab (*Streptomyces scabies*), *Fusarium oxysporum* wilt of melon and *Phymatotrichum* in cotton (Whipps 1997).

Effect of increasing plant diversity

It is often suggested that pest problems in modern farming and forestry are due partly to growing large areas of a single species, especially if it is a genetically uniform variety. How far could pests be controlled by increasing the species diversity? One message from the study of natural communities is that diversity is not a guarantee against major effects of pests. The drastic effects of the invading pathogenic fungi *Cryphonectria parasitica* and *Phytophthora cinnamomi* in species-rich forests showed this. Nevertheless, one mechanism by which pest species can be maintained at low abundance (Box 8.2) is by the host plant or animal being widely spaced. The model in Box 8.3 helps to explain how this works. Whether this could be relevant to farming and forestry depends on the critical density of the host. Figure 8.5 gives one example for an animal: fox density needs to be near 1 per km^2 for mortality from rabies to be low. A

Table 8.4 Summary of experiments in which the abundance of herbivorous insects was compared in 'more diverse' crops (i.e. mixed species, alternating rows of different species or crops + weeds), and less diverse crops. Data from 198 species of insect in all. The columns do not add up to 100% because some species had varying responses and have been omitted

Percentage of insect species that were:	Monophagous* species	Polyphagous species	All species
More abundant in more diverse crops	10	44	18
Little difference	11	4	9
Less abundant in more diverse crops	61	27	53

* i.e. insect ate only one of the plant species present.

Data of Risch, Andow & Altieri (1983).

plant–insect example is provided by the successful control of prickly pear in Australia (Munro 1967; Debach & Rosen 1991). These two introduced cactus species, *Opuntia stricta* and *O. inermis*, had in places become so abundant that the rangeland was made unusable. After the prickly pear had been greatly reduced in abundance by the introduced moth *Cactoblastis cactorum*, the plant and insect species appeared to be approximately in balance and the mean distance between surviving prickly pear plants ranged from 5 to 20 m. Study of the spread of rust fungus in cereal fields, from a single infected plant, showed that a plant needed to be somewhere between 0.2 and 1 m from the infection source in order to have only a low infection rate (Mundt & Leonard 1985). These and other examples indicate that to reduce the pest to an acceptably low level the distance between host individuals needs to be so large that this on its own is of little practical use for crops.

Pests in two-species mixtures

Nevertheless, growing plants in two-species mixtures has quite often been found to reduce pests. Risch *et al.* (1983) drew together results from a large number of investigations of abundance of herbivorous insects in crops, in which there was a comparison between pure stands of crops and mixtures of species. Table 8.4 shows that 53% of the insect species were less abundant in the mixed-species plots and only 18% were more abundant. The difference 18% vs. 53% is statistically significant ($P \ll 0.001$). However, it was only the monophagous insects that showed this difference. This suggests that insect abundance was reduced if they had difficulty in spreading from one food plant to another, because their food plants were more widely spaced or were obscured by other plants. Burdon and Chilvers (1982) reviewed research on plant diseases in relation to host density. Results on fungal diseases mostly showed more disease infection when the host plant density was greater. In contrast, the amount of infection of crops by aphid-borne virus diseases was often greater at lower crop densities; this could be either because the number of aphids per hectare was unchanged so that there were more aphids per

Table 8.5 Abundance of eggs of stem-borer insect *Chilo partellus* on maize and cassava, grown separately or in mixture, in Kenya. The eggs are laid in groups (masses). Figures are number of egg masses per 100 plants, counted 8 weeks after plant emergence. Data of Ampong-Nyarko, Seshu Reddy & Saxena (1994)

Species	Monoculture	Mixed crop
Maize	38.7	28.0
Cassava	0	11.0

plant, or because aphids are actually attracted when bare ground is visible between the plants.

In a two-species mixture the distance between each individual and its nearest conspecific neighbour will not be greatly increased, compared with monoculture; so we need to look for other possible mechanisms of reducing pest attack. One is that spores may be 'wasted' by landing on the wrong species, on which the fungus cannot develop, or an insect may lay its eggs on the wrong species. Table 8.5 shows an example. *Chilo partellus*, a stem-boring insect, is a serious pest of maize and some other cereals in Asia and Africa, but it does not attack cassava. In an experiment near the coast of Kenya, maize and cassava were grown in separate (monoculture) plots and also together in mixture. The plots were open to natural attack by the stem borer. Table 8.5 shows that if cassava grew on its own no eggs were laid on it, but if it was in the mixture some eggs were laid on it and the number laid on maize was reduced. Evidently the adults laid some on cassava 'by mistake'. The percentage of maize plants showing damage by the borer was substantially lower in the mixed crop than in monoculture.

Another possibility is that the second plant species provides a habitat for a species that can attack the pest. Dempster (1969) determined the mortality of Cabbage White butterfly caterpillars on Brussels sprouts plants grown at 90-cm spacing. If weeds were allowed to grow unchecked between the sprouts caterpillar mortality was twice as high as in plots where the weeds were removed (Table 8.6). This was probably caused mainly by a ground-living predatory beetle that climbed up the sprout plants at night. This was abundant among the weeds (Table 8.6), which provided a habitat for it to shelter in by day. However, in spite of the

Table 8.6 Results from an experiment in which plots of Brussels sprouts were either kept weed-free by hoeing or were left unweeded

	Unweeded	Weed-free
Percentage mortality of Cabbage White caterpillars	70.3	34.8
Number of *Harpalus rufipes* (beetles)*	69	13
Weight of Brussels sprouts produced (kg plant^{-1})	0.41	0.64

* Caught in pit-fall traps

From Dempster (1969).

reduced caterpillar attack the Brussels sprouts plants were smaller in the weedy plots, no doubt owing to direct competition with the weeds. Possibly some other plant species, or a non-living ground cover, could encourage the beetles without competing against the crop plants.

Influence of field margins

It is also possible that crops can benefit from diversity around the field rather than in it, for example beneficial species in strips of rough grassland or in hedges. For example, nettles can harbour a nettle aphid which in turn can support a parasitoid, *Aphidius ervi*, which also attacks the grain aphid *Sitobion avenae* (Wratten & Powell 1991). Thus patches of nettles in field margins might lead to reduced abundance of the grain aphid in cereal crops, though as far as I know this has not yet been demonstrated. However, hedges can also be a source of pests. For example, in northern Europe the disease fireblight, caused by the bacterium *Erwinia amylovora*, can spread from the common hedge shrub hawthorn to apple and pear trees (Billing 1981). As well as harbouring control species, weeds can also increase disease in crops. Cucumber mosaic virus infects lettuce and can cause discoloration of the leaves that makes the lettuces unmarketable. The virus is also carried by some British weed species, though without causing symptoms. The virus can be carried over from one year to the next in weeds that survive the winter, for example in the field margin. It can be transmitted from weeds to lettuce by aphids, and infected weeds are thus a potential source of infection (Tomlinson & Carter 1970, Tomlinson *et al.* 1970).

Thus there are various ways in which management can help to control pests. However, decisions need to be based on a knowledge of the particular crop and pest involved: there is little scope for general recommendations.

Biological control

Biological control involves using another species to control the pest. Box 8.1 defines the main biological control systems. In *classical biological control* the control species, once introduced, maintains itself and gives control of the pest long term. This can only happen if the control species was not previously present in the area: if it had already been present it would presumably already have been providing control. A *biopesticide*, on the other hand, is a species that provides short-term control of a pest but cannot maintain its abundance long term, and so has to be repeatedly applied in large numbers. It may have already been present in the area in low numbers, or be introduced from elsewhere. I now describe briefly a few successes of each type, before going on to consider how new biocontrol agents may be discovered and tested.

Case studies: biopesticides

Control of soil-borne plant pathogens

Much research has been applied to biological control of soil-borne plant pathogens (Whipps 1997). These pathogens, mostly fungi, attack the root systems of crop plants and can cause severe reductions in yield. Chemical fungicides have generally been ineffective in controlling them. It was explained earlier that in some areas, some of these diseases will decline if the same crop is grown in a field for many years, and that this seems to be due, at least in part, to living species antagonistic to the pathogen. There are in addition areas where the soil is suppressive to certain diseases, i.e. the severity of the disease is low even if many propagules of the fungus are present. Many microbial species have been isolated from suppressive and decline soils and tested for ability to reduce pathogen attack on the plants. Whipps (1997), in tables covering 17 pages, listed bacteria and fungi which have shown potential to provide biological control of various soilborne pathogens. Some were applied to seeds before planting, some to the soil, a few to cuttings or roots. He went on to list seven bacterial and 15 fungal species (or mixtures of species) which, at the time of writing, were available commercially or in the process of registration for commercial use.

The mechanism by which the soilborne pathogen is controlled varies between the control species. Some probably operate by competition, for example *Phialophora* spp. used to control the very similar fungus *Gaeumannomyces graminis*, cause of take-all disease in wheat. Other control species produce antibiotics. It is possible by genetic engineering to derive a bacterium or fungus that lacks a single gene for the production of a particular antibiotic, and to then show that this removes the ability of the species for biological control. For example, the biological control ability of some strains of the bacterium *Pseudomonas fluorescens* against several fungal pathogens has been shown to be particularly associated with the production of 2,4-diacetyl-phloroglucinol, whereas in another strain control ability is dependent on phenazine-1-carboxylic acid (Hokkanen & Lynch 1995, Chapters 12 and 13).

Bioinsecticide

The most widely used bioinsecticide is the bacterium *Bacillus thuringiensis*. This produces a protein which, when ingested by insects, is broken down in their guts to toxic polypeptides. Only insects are killed. Many strains of the bacterium are known, some of which kill only a limited range of insect species (Payne 1988). The bacterium is very widespread in soil but is a slow multiplier and spreader under field conditions. It is easy to grow in artificial media, and is marketed commercially as the bacterial spores or as the toxin itself in crystalline form, to be sprayed on crops or forests. It is widely used against two major North American forest pests, spruce budworm and Gypsy moth (Hokkanen & Lynch 1995, Chapter 17). Products are also available for the control of Colorado beetle and various lepidopteran larvae. Limitations to this application method are that (1) the toxin remains active on plant surfaces for only a few days, and (2) that as the insects have to ingest the spores or crystals the toxin is effective only against those that feed on the outside of the above-ground parts of plants. The protein toxin is controlled by a single gene, which has been successfully transferred to other species, where it is expressed. It has been transferred into soil bacteria, which can be effective in the control of root-attacking insects (Lindow *et al.* 1989), and has also been inserted into crop plants. Corn and cotton capable of producing the *B. thuringiensis* toxin were first planted commercially in 1996 (Wadman 1997). This gives them resistance against the corn borer and cotton boll worm. The gene has been inserted experimentally into many other plant species.

Case studies: classical biological control

Control of Klamath-weed in California

Klamath-weed (St John's wort, *Hypericum perforatum*) is a small herbaceous plant, a native of Europe, where it is not considered a problem. When introduced into the western USA and Australia it became a serious weed in pastureland. Cattle and sheep will not eat it, and nor will most invertebrates, probably because of a polyphenolic, hypericin, which it contains. The species was successfully controlled in California: three beetle species known to feed on it in Europe were introduced, but most of the reduction was caused by one of them, *Chrysolina quadrigemina*. Soon after the beetles arrived in an area the Klamath-weed population was almost wiped out, followed by a rapid decline in beetle numbers (Fig. 8.7), but in most areas both species persisted at low abundance. The beetles rarely lay eggs in shade, and Klamath-weed is now more common in shaded habitats than in the open. For further information see Huffaker & Kennett (1959); Harper (1977); Debach & Rosen (1991).

Some weeds have been successfully controlled by introduced fungi. An example is the control of skeleton weed, *Chondrilla juncea*, in Australia and western USA, by a rust fungus (Te Beest *et al.* 1992).

Control of rabbits by myxomatosis

The spread of the European rabbit and its subsequent control by the myxoma virus is described by Thompson and King (1994) and Williamson (1996). This rabbit (*Oryctolagus cuniculus*) was originally a native of Spain, where it has never been a pest. It has now spread across much of Europe. It was introduced into Britain in the 12th century and into Australia in the 19th century. Both these introductions aimed to provide a source of meat, but in both countries rabbits later became a pest, eating crops and pastureland. The myxoma virus occurred originally in forest rabbits (*Silvilagus brasiliensis*) in South America, where it has only slight effects, but in the European rabbit it causes the severe and often lethal disease myxomatosis. The virus was introduced into Australia in 1950–51 and began to spread rapidly. It reached Britain in 1953. In both countries the disease initially killed more than 99% of the rabbits, but it did not wipe them out completely. Subsequently the rabbit population has partially recovered; this recovery will be considered later.

A practical difference between biopesticides and classical biological control lies in the economics. As with chemical pesticides, the costs of development and testing have to be set against the likely income from sales before it can be decided whether the product is worth developing and

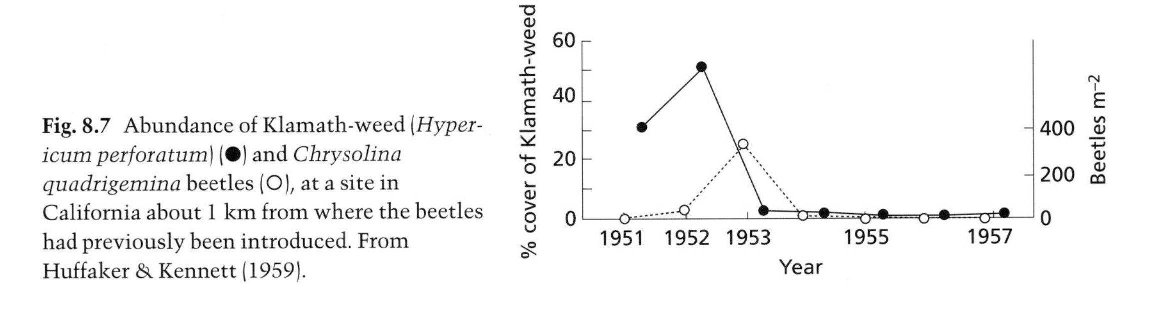

Fig. 8.7 Abundance of Klamath-weed (*Hypericum perforatum*) (●) and *Chrysolina quadrigemina* beetles (○), at a site in California about 1 km from where the beetles had previously been introduced. From Huffaker & Kennett (1959).

producing commercially. Biopesticides can, like chemical pesticides, go on being sold year after year, so they can be commercially attractive. But a classical biological control agent should need to be released only once in each area, so the opportunities for sales are much less. Their development has therefore usually been publicly funded. However, there has been economic assessment of the costs and benefits of some classical biological control programmes, and for some the benefits have greatly exceeded the costs (Hokkanen & Lynch 1995, Chapters 5 and 26). For example, the control of skeleton weed in Australia by a rust fungus was estimated to have cost US$3.1 million to develop, but the subsequent benefit was $13.9 million *per year* through increased rangeland production.

Finding and developing new biocontrol agents

These case studies were all success stories. This should not, however, blind us to the fact that many attempts at biological control have failed. Waage and Greathead (1988) summarized the success rate of insect species that were introduced for biological control of insect pests and weeds. Of the species that became established, only 40% of those aimed at insect control and 31% of those aimed at weed control were 'substantially successful'. This does not take account of species that were introduced but did not establish, for which records may not always be kept.

When assessing whether a species might be suitable as an agent of classical biological control, we need to answer three questions:

1 Will it establish and persist in the new area?
2 Will it sufficiently reduce the abundance of the pest?
3 Will it have undesired effects on non-target species?

The remainder of this chapter considers how we can obtain answers to those questions.

Will the control species establish in the new area?

Can we spot species that are good at invading?

This question leads us to ask: are there consistent characteristics of invader species, features that make a species good at establishing in a new area? There have been numerous published suggestions of such key characteristics, but these are usually followed by other writers pointing out exceptions where successful invaders do not have those characteristics. Williamson (1996) discussed five suggestions that have frequently been made: his discussion concerned all sorts of species, not just potential biological control agents. The suggested characteristics of invaders were:

1 ability to multiply rapidly;
2 genetic characteristics such as inbreeding;
3 a wide native range and high abundance in that range;
4 ecologically or taxonomically distinct from species in the area to be invaded;
5 native range has a similar climate to the area to be invaded.

Williamson concluded that none of these characteristics has been shown consistently to be a predictor of ability to invade.

A basic fact is that all species must have the ability to invade. All species vary in abundance through time, so a species that becomes rare must be able to subsequently increase in abundance. Metapopulation theory (see Chapter 10) shows that in any small area any species is at risk of becoming extinct sooner or later, and it therefore needs an ability to reinvade. It also follows that all communities must be invadable: because they are likely to lose species they must also be able to gain them, otherwise they will finally end up with none. This does not deny that some species may be better at invading than others, and some communities more easily invaded than others. Rather, it says that the search for consistent characters indicating species that can establish in new communities has proved unfruitful, probably for this fundamental reason; so this is not a good starting point to choose promising species for biological control. Whether a certain species can establish is likely to depend on particular features of that species and its relation to its host and to the habitat; there are no easy rules of thumb.

Sometimes a control agent has failed to establish at the first attempt, yet later it has succeeded in establishing and has provided effective pest control. An example is myxomatosis in rabbits (Thompson & King 1994). In the late 1930s there were several unsuccessful attempts to introduce myxomatosis in parts of Europe; and test releases in Australia in the 1930s and 1940s were considered a failure because the disease did not spread. Yet in the 1950s further attempts resulted in widespread establishment of the disease in both Europe and Australia.

How many individuals to release

One message from this is that if a new biocontrol agent fails to establish it may be worth making a second try. It also raises the question whether the number of individuals released may be important. As explained in Chapter 10 (Conservation), a small population is at greater risk of becoming extinct, both because of chance fluctuations in numbers and for genetic reasons. This will also apply to a small initial population of a species introduced intentionally (Williamson 1996). Figure 8.8 shows experimental evidence that initial inoculum size can be important for biological control agents. Thrips are being tested as a possible biocontrol

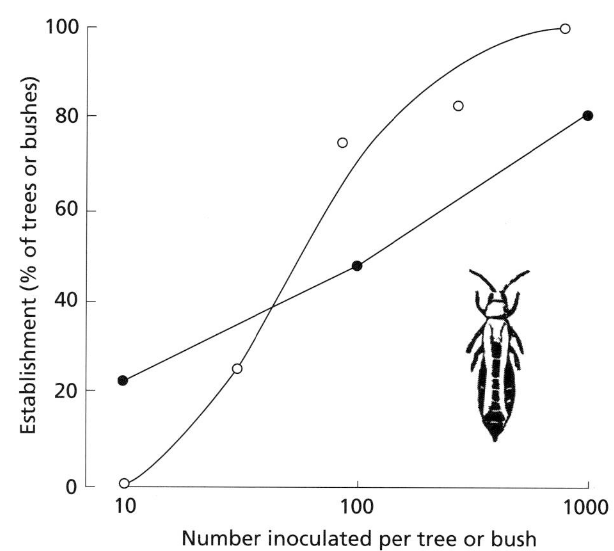

Fig. 8.8 Results of two experiments in which different numbers of insects for biocontrol were released per tree or per bush, and number of trees or bushes on which the biocontrol insect had persisted were later determined. ● Parasitoid of red scale released on to citrus trees in orchards in South Australia; data of Campbell (1976). ○ thrips (pictured) released on to gorse bushes in New Zealand; data of Memmott, Fowler and Hill (1998).

agent against gorse, which was introduced into New Zealand from Europe and has reached pest abundance in places. Figure 8.8 shows the percentage of bushes on which thrips were still present a year after being introduced. If thrips survive the first year on gorse they usually increase thereafter. The other example, the parasitoid *Aphytis melinus*, was introduced into Australia to control the Californian red scale (*Aonidiella aurantii*), which was infesting citrus trees. In this experiment the percentage occurrence of the parasite was recorded after 1 and 3 months. In both experiments a larger inoculum evidently increased the chance of the insect's establishing on that tree. It does not follow, however, that the larger the inoculum the better. If the total number of insects available to inoculate is fixed, set by our ability to catch or breed large numbers, then larger inoculum per tree means fewer trees inoculated. On the basis of their results (Fig. 8.8), Memmott *et al.* (1998), calculated that if a set total number of thrips was available for release, releasing about 90 insects per bush would result in the largest number of bushes with an established population.

Will the species control the pest?

Should we aim to wipe out the pest or just reduce it?

If the potential biocontrol species establishes on the pest, the next requirement is that it reduces the pest to an acceptably low level. A key decision to be made is whether it is better to aim for complete extinction of the pest or to reduce it to a low abundance. It might at first sight seem obvious that extinction is better, but whether this is the case will depend partly on whether the control organism is host specific, a major advantage of which is that this removes the danger of its attacking non-target species. A disadvantage is that if the pest species becomes extinct, so too will the control species. Then if the pest survives somewhere else there is no control against its reinvading later. Aiming for extinction can in practice result in subsequent periodic outbreaks of the pest.

If it is decided to aim for a stable pest abundance, clearly we want this abundance to be very low, so a key question is how much the control agent will reduce the pest. As explained earlier, models have been developed, following on from those of Box 8.3, which predict the stable abundance of the pest in the presence of a control agent, in relation to characteristics of the control agent. Figure 8.4 gives predictions from one such model, where the plant is the pest (i.e. a weed). The stable abundance of the pest cannot be below S_T of Equations 8.6 and 8.7, as the control species would then die out; it may be slightly above S_T or substantially above it. As explained earlier, models predict that a control species will maintain a low stable pest abundance if it has a high ability to spread (β) but intermediate virulence (θ); if θ is too high the control species kills the pest so rapidly that it hampers its own spread and survival. A rule of thumb that emerges from this type of model is: choose the control species that needs the lowest density of the host (pest) to just maintain itself to the next generation (Murdoch & Briggs 1996). A prediction for parasitoids is that if the pest insect suffers significant mortality from another cause besides the parasitoid, then a parasitoid that attacks the pest at an earlier stage of its life history will be more effective. This is because the density of the pest will be higher at that stage (Murdoch & Briggs 1996).

Intermediate virulence in the control agent may be the best

Parasitoids are promising for the biological control of insect pests, for example because of their high host specificity and hence low danger of harming beneficial species. Beddington *et al.* (1978) discussed various host–parasitoid models and how much reduction in host (pest) abundance they predict. For the simplest model, and several variants of it that they tried, the parasitoid could not reduce the host population by more than 70% if it was to maintain that level stably (compare Fig. 8.4). The basic reason for this is that in these models stability of the host population was dependent on interactions between host individuals, such as competition for food; if the host population became very sparse this density-dependent control became very weak. Host–parasitoid interaction does not involve a density-dependent control of this sort: if the host population grows rapidly it can outstrip its parasitoid population so that the control becomes less effective (at least for a time) just when a more effective control is needed. In four laboratory experiments parasitoids did, as predicted, reduce the host population by about two-thirds, but in six field studies cited by Beddington *et al.* (1978) the reduction was much greater, by one or two orders of magnitude. A family of models that predict a much greater, but stable, reduction in the host are those that introduce patchiness. For example, there may be a 'refuge' within the habitat where the host can escape attack by the parasitoid. An analogous situation may occur with Klamath-weed and *Chrysolina* beetles (see earlier): the ability

Can parasitoids provide effective control?

of some Klamath-weed plants to escape the beetles by growing in shade may be crucial for allowing the two species to coexist at low abundance. Another possible type of patchiness is if the parasitoid searches for new host individuals in a non-random way, for example searching preferentially in patches of high host density. These and later models of host–parasitoid population dynamics have strongly suggested that the choice of a parasitoid as an effective biological control agent will need to take into account 'patchy' behaviour of the host and the parasitoid. However, there are differences between models in their stability properties and the way that stability can be built in (Murdoch & Briggs 1996), so these predictions need more confirmation from observation and experiment.

Where to look for a control agent: . . . in the pest's home area?

These models have suggested various characteristics that we can expect an effective control species to possess. However, in real life we need to narrow down the list of species that we consider testing. Many species become pests only when they are introduced to a new area outside their native range. Their ability to reach pest-level abundance in their new home may be because they have left some of their natural control species behind. So a promising source of species for biological control is to look at what is attacking the pest species back in its original native area. This has been the basis of many of the success stories of classic biological control. An example already described is Klamath-weed, a native plant of Europe which was introduced into California and later controlled by introducing a beetle that feeds on it in Europe. Another example was the cottony-cushion scale insect (*Icerya purchasi*), which was accidentally introduced from Australia into California in about 1868 and became a serious pest of citrus. It was successfully controlled by two insects, a predatory ladybird (beetle) and a parasitic fly, introduced from Australia in 1888 (Thorarinsson 1990, Debach & Rosen 1991).

This last example illustrates how choosing the control species may not in practice be straightforward. When American scientists went to Australia to look for a species controlling cottony-cushion scale, they found that neither the scale nor any attackers were at all common there. When a species is providing effective control in its native range it may well be quite rare and so difficult to observe; this is illustrated by fox and rabies (see Fig. 8.5).

On the other hand, a species which is clearly providing control in one area may not work in another. As mentioned earlier, the cactus *Opuntia stricta* was very successfully controlled in Australia by introducing the moth *Cactoblastis cactorum*. Yet in South Africa the same insect was far less effective in controlling the same cactus, perhaps because many of its eggs were eaten by ants (Hoffmann *et al.* 1998). Another example is the virus, mentioned earlier, that reduces the virulence of the chestnut blight fungus, *Cryphonectria parasitica*. Hypovirulent strains of the fungus appeared in Italy, and when introduced into France the virus spread and provided effective control of the pathogenic fungus. However, in the

USA there has been little success with the virus as a control agent (Anagnostakis 1995). Native hypovirulent strains of the fungus occur in a few areas, e.g. in Michigan and Tennessee. When these or European strains were inserted into chestnut trees near a canker, the canker stopped growing. But the hypovirulence virus failed to spread from tree to tree, so it provided no large-scale control. The reason for this difference between Europe and the USA is still not known.

... or where the pest species is absent?

It can alternatively be argued that if the pest species and its attacker have been together in the native area for a long time they may have co-evolved, resulting in the attacker's being less damaging; and that it will therefore be more promising to look for a control species in an area where the pest species does not occur. Some successful biocontrol organisms have originated from other hosts. One example, described earlier, is myxoma virus, which occurs naturally in forest rabbits (*Silvilagus brasiliensis*) in South America but has been used to control European rabbits (*Oryctolagus cuniculus*). Another example is that the prickly pear cacti (*Opuntia* spp.) which became a serious pest in grazing land in Australia originated from the Gulf of Mexico, whereas the control insect *Cactoblastis cactorum* is native in South America, where it feeds on other species of cactus. Waage and Greathead (1988) analysed data from 441 introductions of insects (parasitoids or predators) aimed at the control of pest insects. Their data excluded introductions that failed to establish. The percentage of cases classed as giving completely or partially successful control of the pest were: if the pest and control species had been associated before elsewhere, 40%; if they had not, 34%. These two percentages were not significantly different statistically. So it is evident that either old or new associations can result in successful biological control, and there is no strong reason to think that one is more likely to prove successful than the other.

Will the biocontrol species evolve to become less effective?

As we have seen, a major problem with some chemical pesticides is genetic change in the pest which makes it less susceptible to the chemical. With biological control this could also happen, and there are known examples. The caterpillars of the diamondback moth, *Plutella xylostella*, are an important pest on oilseed crucifers in Canada. They have been effectively controlled by sprays of *Bacillus thuringiensis*, but in some areas resistance has now developed (Hokkanen & Lynch 1995, Chapter 23). On the other hand, resistance has sometimes failed to emerge when it might be expected. For example, the parasitoid *Aphidius ervi* was introduced into the United States in the 1960s to control the pea aphid *Acyrthosiphon pisum*. In a study in New York state, Henter and Via (1995) found a large amount of genetic variation among the aphids in susceptibility to the parasitoid, ranging from some clones where scarcely any of the parasitoid eggs that were laid developed further, to others where a parasite developed in about 80% of the aphids. Yet Henter and Via found no evidence that the frequency of resistant clones was increasing.

With biological control we have the additional complexity that the control agent also can (unlike a chemical) evolve. Selection might favour a parasite or predator evolving to become less damaging to its host, because this might increase the number of hosts available for it. In Fig. 8.4 the insect with the lower multiplication ability (part a) reaches a stable abundance (P), whereas a higher multiplication ability (part b) is predicted to result in wide fluctuations in the abundance of insect and plant, with the risk that the plant, and hence the insect, may become extinct in this area. So there are reasons for arguing that the control species will in time evolve to be less effective. But in reality things are not so simple. We observe that parasites vary greatly in their virulence, some being almost always fatal, e.g. many parasitoids. Evidently their evolutionary strategies vary.

Few reports of pests evolving resistance to biocontrol

Holt and Hochberg (1997) pointed out that, whereas the evolution of resistance to chemical pesticides has occurred often, there are only a few known examples of pests evolving resistance to biological control. They considered whether this apparent difference could be due simply to failure to report cases where biological control has become less effective. However, after careful consideration of the literature they concluded that the difference is real: resistance to biological control agents seems to evolve less frequently or less rapidly than to chemical pesticides. They suggest several possible reasons for this:

1 There may be trade-offs or costs: increased resistance by the pest may carry other, disadvantageous characters with it.

2 Selection pressure for increased resistance may be weak, e.g. because the pest has refuge habitats where it can escape the control agent. Whether this can happen at the same time as effective control of the pest is not clear.

3 Increased resistance in the pest may be counterbalanced by the evolution of increased effectiveness in the control species.

Rabbits and myxomatosis: genetic changes

Genetic changes in a pest and its control agent over several decades have been studied in rabbits and myxoma virus. As described earlier, the virus was introduced into Australia and Britain in the early 1950s, and initially the resultant disease, myxomatosis, killed more than 99% of the rabbits. However, a few survived, and the rabbit population in both countries has subsequently increased again. Precise data on rabbit numbers before myxomatosis and how they have changed since are unfortunately not available, but there are various indicators of the changes. Figure 8.9 shows how the numbers of rabbits shot on British game estates changed between 1961 and 1989. The main increase was from 1968 to 1977, with no consistent time trend after that. Of course the number of rabbits killed may reflect the number of person-hours devoted to shooting them, as well as the number of rabbits, but the trends in numbers are backed up by other surveys, e.g. signs of rabbit activity on farms (Trout *et al.* 1986).

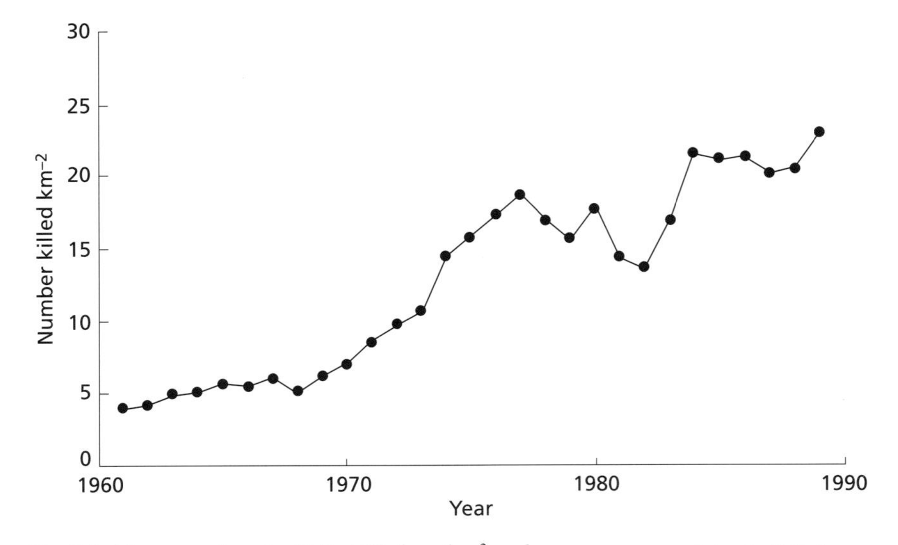

Fig. 8.9 Mean number of rabbits killed per km^2 each year on game estates in Britain. From Thompson & King (1994).

Rabbits became more resistant . . .

This recovery in rabbit numbers could be because they became more resistant to myxomatosis or because the virus became less virulent, or both. Figure 8.10 shows results from tests that were carried out on wild rabbits in an area of Australia over several years soon after the arrival of myxomatosis. Although they were all inoculated with a virus strain of the same virulence, the severity of the disease decreased with time; so the rabbits were becoming more resistant. The virus was changing, too.

. . . and the virus became less virulent . . .

The virus was classified into strains of five virulence grades, assessed by the percentage of infected rabbits that died in laboratory tests and how long it took them to die. The most virulent grade, I, caused more than 99% mortality, whereas the least virulent, V, caused less than 50% mortality. It is likely that the strain originally introduced into Australia was the highly virulent grade I. Figure 8.11 shows that this quickly became less abundant, as less virulent strains increased. Within a few years strains III and IV, of intermediate virulence, became predominant, and remained so for at least 25 years. In Britain, also, the most virulent strain rapidly became sparser, and by 1962 strains II, III and IV predominated (Thompson & King 1994, Table 7.4).

Thus within a decade of the arrival of the myxoma virus in Australia and Britain it had changed to become less virulent, and the rabbits had also become less susceptible. If we extrapolate this onwards we might predict that within a few decades the virus would cease to control the rabbits at all. Is that likely? In Australia the change in the rabbits' resistance

. . . but some control has been maintained

to myxomatosis shown in Fig. 8.10 subsequently became much slower: in two other areas of Australia rabbits were inoculated with standard strains of the virus from 1961 to 1981, but the percentage mortality

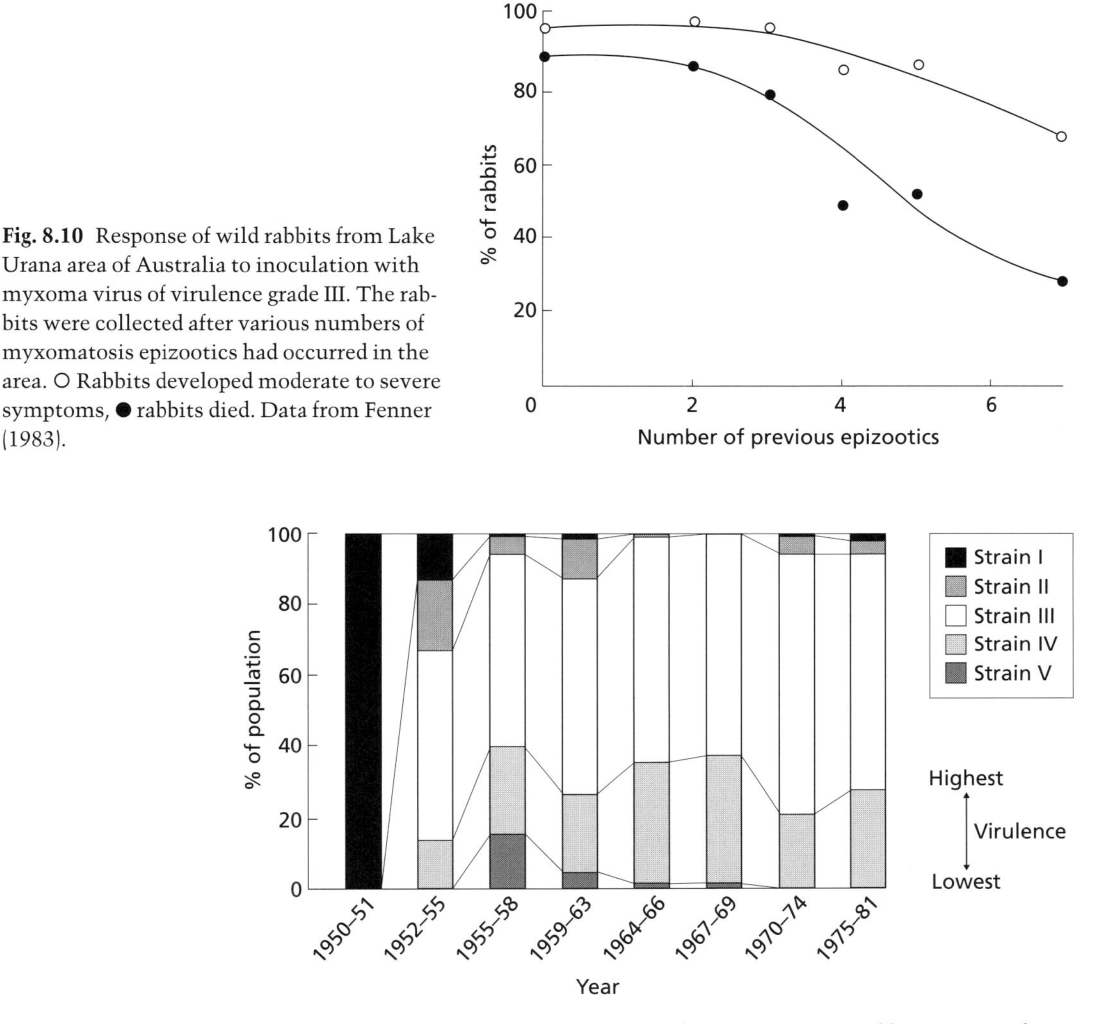

Fig. 8.10 Response of wild rabbits from Lake Urana area of Australia to inoculation with myxoma virus of virulence grade III. The rabbits were collected after various numbers of myxomatosis epizootics had occurred in the area. ○ Rabbits developed moderate to severe symptoms, ● rabbits died. Data from Fenner (1983).

Fig. 8.11 Relative abundance of five strains of myxoma virus in rabbits in Australia. Strains graded from I (most virulent, i.e. highest percentage mortality and most rapid death) to V (least virulent). Data from Fenner (1983).

decreased only slightly (Thompson & King 1994, Table 7.6). Figure 8.11 shows that the virus did not evolve towards eventual dominance by the least virulent strain, V; indeed, that strain decreased in percentage abundance after 1958. In Britain, also, the proportion of strain II increased somewhat from 1962 to 1981, whereas the proportion of strain V remained minute (Thompson & King 1994, Table 7.4).

This stabilizing of the host–parasite relationship happened because virulence, recovery and transmission interact in ways that tend to be balancing. The less virulent virus takes longer to kill its host, which favours the virus by giving it more time to spread; but more of the rabbits recover

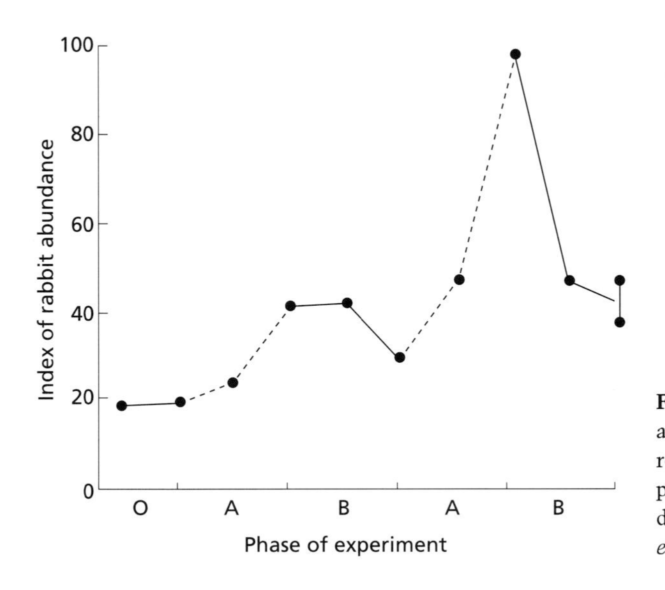

Fig. 8.12 Number of rabbits seen in March and April of each year on standard transect routes on a farm in southern England. During periods A and A myxomatosis was reduced, during B and B it increased again. From Trout *et al.* (1992).

and become immune, which slows the spread of the virus. Spread of the virus is by fleas and mosquitoes, which requires open lesions in the rabbit; these lesions are fewer if the virus is less virulent. Dwyer *et al.* (1990) developed a complex model to explore how these interactions would affect the coexistence of rabbits and myxomatosis in Australia. Their model was able to predict correctly that the rabbit–virus interaction would become dominated by grade III virus (Fig. 8.11). Their fundamental prediction is that if the rabbits evolve greater resistance, then the virus will evolve greater virulence. This is because a certain intermediate level of response in the rabbits (in terms of survival time, recovery percentage and lesion formation) is most favourable for the virus, and if the rabbit evolves to change this, selection will favour the virus changing to restore it. Dwyer *et al.* therefore predicted that the virus will remain effective in controlling the rabbits for some time to come.

Although it is not known for sure whether rabbits have returned to pre-myxomatosis numbers in parts of Australia or Britain, there is clear evidence from one site in Britain that myxomatosis was still reducing rabbit numbers 30 years after the virus first arrived. On a farm in southern England Trout *et al.* (1992) reduced myxomatosis experimentally over two 2-year periods (A in Fig. 8.12) by pumping an insecticide down rabbit burrows to kill fleas, the main transmitter of the myxoma virus in Britain. After the treatment stopped (periods marked B) flea numbers and myxoma infection increased again. Figure 8.12 shows that the abundance of rabbits in spring increased substantially when myxomatosis was reduced, but fell again when it increased.

I have described the relationship between rabbits and myxomatosis in some detail, because it is the example of a pest and its control agent where we know most about genetic changes over time. The coevolution of other pests and their control agents may well be different. Nevertheless, this example helps us to understand basic mechanisms that can allow biological control to remain effective for decades or longer.

Will the biocontrol agent have undesirable side-effects?

Before a species can be released for widespread use in biological control, we need to know not only that it will work, but also that it will not have harmful effects on non-target species. Clearly, if we want to kill weeds it is no use if we kill the crop as well. The use of some mycoherbicides has to be restricted for this reason. For example, one commercially available mycoherbicide contains a fungus that can control a leguminous weed, but it also attacks some legume crops (Hokkanen & Lynch 1995, Chapter 9).

Biocontrol agents can harm non-target species

A more widespread problem is that the biocontrol species might harm wild species of animal or plant. Species for biocontrol, and especially classical biocontrol, are potentially more dangerous in this respect than chemical pesticides, because they can spread and multiply. There have been some biocontrol disasters in the past. One that is often quoted concerns snails on Pacific islands (Williamson 1996). The giant African snail, *Achatina fulica*, has been introduced on to many Pacific islands and has become a pest in crops and gardens. To control it another snail, *Euglandina rosea*, a predator, was introduced. This attacks not only the giant African snail but many of the native snails, and has had very harmful effects on them, including the extinction of some endemic species. Another example of a generalist predator is the Indian mongoose, which was introduced into the West Indies and some Pacific islands to control rats, but has had a serious effect on some native birds (Simberloff & Stiling 1996).

The same examples of bad side-effects of biocontrol are cited in many books and reviews, which might suggest that the total list is short. One worry is that species may have been made extinct without our noticing, or without the cause being realized. There can also be indirect effects which are likely to be difficult to predict in advance. The great reduction in rabbits in Britain in 1953–55, following the introduction of myxomatosis, had many knock-on effects (Sumption & Flowerdew 1985). Predators for which rabbits had been a major item of diet, such as foxes and stoats, survived by altering their diet. The main effects were through changes in vegetation: grassland kept short by rabbits became tall grassland, shrubland or (in due course) forest. Tables 6.3 and 6.4 show how the cessation of grazing by sheep or cattle can result in the disappearance of some plant species; this was also true following the decline of rabbits. Some invertebrates are favoured by taller grassland (Fig. 6.13) but others disappear. The rarity of rabbits led to the disappearance from Britain of the large blue butterfly (*Maculinea arion*), because it is dependent on an ant which in turn requires short turf (Sumption & Flowerdew 1985).

The 'tens rule' states that about 10% of species that are introduced to a new region become established, and of those that establish about 10% become pests. Provided 'about 10%' is allowed to mean 5–20%, this rule has been found to apply to vertebrates, insects and flowering plants in various parts of the world, though there are also exceptions (Williamson 1996). This gives some indication of the risk of a new biocontrol agent establishing in a natural community and having undesirable effects. However, when a species does establish and have undesirable effects, they may be *very* undesirable, as the above examples show. So we cannot simply ignore this risk. Our aim should be to reduce the risk to a low level.

How can we test for side effects?

Whereas a new chemical can sometimes be tested in a limited area outdoors, it is rarely safe to perform a test introduction of a new species outdoors, since if it establishes it may be impossible subsequently to eradicate it. So, information has to come mainly from two sources: study of the species in its existing habitat, and tests in contained environments. In the existing habitat the main question of interest is how wide a host range the proposed control species has. If it has a wide range, this suggests a high risk that it would attack non-target species if introduced into a new area. Nowadays there is a reluctance to use generalist predators for control (mongoose and a snail were mentioned earlier), especially vertebrates; arthropod predators and herbivores need to be shown to have a narrow host range. Parasitoids often have a restricted host range.

A narrow host range in its home territory does not tell us definitely that there are no wild species the biocontrol agent could attack in its new range. So it is also useful to carry out tests, offering potential non-target hosts to the biocontrol agent (Hokkanen & Lynch 1995, Chapter 5). This is most often done in contained conditions, though an alternative is to introduce species to the home range of the biocontrol agent. It will not be practicable to test the response of the biocontrol agent to every wild species it might possibly come into contact with, so how do we choose which species to test it on? In the commonly used procedure for testing weed control agents, species closely related to the weed or other known hosts of the biocontrol species are considered most at risk; plants in the same genus are offered first in tests, and sometimes later plants from other genera in the same family. An alternative would be to test plants with similar secondary chemistry, i.e. similar protection systems.

There have been some cases where the biocontrol species did attack a non-target species in its new area, even though tests had indicated that it would not. Williamson (1996) cites the case of a gall wasp imported from Australia to South Africa to control an Australian wattle, *Acacia longifolia*. *Acacia melanoxylon* (blackwood) also occurs in Australia and South Africa; in Australia the gall wasp does not attack it in the field, nor did it in tests in South Africa, yet after its release in South Africa the wasp did attack blackwood. It is not clear whether this was due to different conditions in South Africa or to genetic changes in the wasp. The damage to blackwood was not serious, but this is nevertheless a warning of the difficulty of carrying out adequate tests to alert us to possible damage to non-target species.

It is thus clear that we can never state with complete confidence that the introduction of a species for biological control poses absolutely no risk to any non-target species. We cannot test the biocontrol agent on all species, in natural conditions, and we cannot wait to see if it evolves the ability to attack new species. So, decisions about biological control have to balance the benefits against the risks. This may involve comparing biological control of the pest with no control, or with some other control method, e.g. a chemical, which may well have its own risks of harmful side-effects. Agreeing how large the benefits and the risks are, and how to balance one against the other, can be difficult, as the published argument between Frank (1998) and Simberloff and Stiling (1998) illustrates. Chapter 10 discusses whether we can place a value on a wild species, and how much we should be prepared to sacrifice to prevent its extinction.

Integrated pest management

This chapter has considered most of the methods for pest control listed in Box 8.1, but it considered them one at a time. *Integrated pest management* was mentioned early in the chapter. To some people this seems to mean any way of reducing the amounts of chemical pesticides used; for example, it might include using the pesticide at the most effective time of year, and just enough for economically acceptable control of the pest. A more straightforward meaning for integrated pest management is that we use several of the available methods (Box 8.1), and in such a way that they complement and reinforce each other. Integrated pest management systems have been developed for some crops in some areas. Metcalf and Luckman (1994) provide chapters on its application to control of insect pests of cotton and apple. Cussans (1995) discusses its application to weed control.

Conclusions

- Resistance to chemical pesticides has become widespread among insects, plant pathogens and weeds. Techniques such as not using the chemical every year can slow down the build-up of resistance, but can rarely prevent it indefinitely.
- In natural ecosystems insects and fungal pathogens that attack plants and animals are widespread, but without wiping out their host species. Host survival may involve its having low abundance, its having the ability to resist attack, or the attacker being controlled by its own parasite or predator.
- Models can predict how much a parasite or herbivore will reduce its host's abundance, and whether the abundance will be stable or

subject to cycles. Biological control often aims for low, stable levels of pest abundance.

- A parasite that at any one time infects only a small percentage of the host population can nevertheless be a major controller of its abundance.
- Biopesticides involve the frequent release of large numbers of the control agent. They have been successful against some insects and soilborne fungal pathogens.
- Classical biological control requires the control agent to establish and multiply after its initial introduction. Models predict that parasites or predators of intermediate virulence will be most effective at maintaining a low, stable pest abundance.
- As an example of coevolution of host and parasite, the effectiveness of the myxoma virus in controlling rabbits has declined over several decades but has not ceased altogether.
- We can never be absolutely certain that a species introduced into a new area will not have undesirable side-effects. When species are introduced for biological control, the potential benefits and risks have to be weighed against each other.

Further reading

Biological control (general):
Hokkanen & Lynch (1995)

Invasions:
Williamson (1996)

Management and control:
Insects: Metcalf & Luckman (1994)
Pathogens: Grenfell & Dobson (1995); Whipps (1997)
Weeds: Cousens & Mortimer (1995); Te Beest, Yang & Cisar (1992)

Chapter 9: Pollution

Questions

- Are there simple, quick tests that can indicate whether a chemical will be harmful?
- What response of the organism should a test measure?
- How can short-term tests be used to predict the effects of long-term exposure to a chemical?
- Can one chemical enhance the harmful effect of another?
- If a chemical is to be released where there are many species, which of them should we test it on?
- Can tests on individual species adequately predict how the chemical will affect a whole community?
- What controls the concentration of pollutants in the tissues of living things?
- Do pollutants always increase in concentration as they pass up a food chain?
- Can measuring the concentration of a chemical in a plant or animal indicate whether the chemical has reached a toxic level?
- Can living things help us to get rid of pollutants? How can we promote this?
- Can we predict whether the breakdown of a new chemical by microbes will be slow or rapid?

Background science

- LC_{50}, LD_{50}, EC_{50} tests.
- Variations among species in response to pollutants.
- Methods of measuring the effects of exposure to a pollutant over many years.
- Changes in composition of multispecies communities exposed to a pollutant.
- Uptake of inorganic and organic pollutants by plants and animals, aquatic and terrestrial.
- Concentrations of pollutants in tissues of animals in relation to concentrations in their surroundings and their food.
- What factors limit the activity of pollutant-degrading microbes.
- Biochemical pathways by which complex organic chemicals are broken down.

What is a pollutant?

A pollutant is a substance that is potentially harmful, at least if it is in the wrong place at the wrong concentration. It may get into the environment by various means. It may have been made with the express aim of releasing it, for example a pesticide or an aerosol propellant. It may be an unwanted waste product dumped into the environment, for example, mine waste, partly treated sewage released into a river or sea, or gases from the burning of fossil fuel which are released into the atmosphere. Or the release may be unintended, e.g. CFCs leaking from an old refrigerator, nitrate leaching from farmland, oil from a damaged tanker.

Like the rest of this book, this chapter concentrates on biological aspects. Therefore it says only a little about sources of pollution and how pollutants are spread. The basic questions are how to determine whether a chemical is likely to be harmful, and if so what to do about it. The chapter is mainly about the effects of pollutants on wild species and communities: it says little about effects on people or domestic animals. So it covers what is often called *ecotoxicology*.

The chapter is not organized chemical by chemical, but in each section chooses whichever pollutant best illustrates the topic being considered. Box 9.1 lists pollutants that are mentioned in the chapter and gives a little information about each. Some topics relating to pollution are covered in other chapters: greenhouse gases in Chapter 2 (Energy), nitrate loss from farmland in Chapter 4 (Soil) and toxic materials in mine waste in Chapter 11 (Restoration).

Box 9.1. Pollutants mentioned in this chapter, with their principal sources.

Gases

Halocarbons, including CFCs (chlorofluorocarbons). Manufactured for use as refrigerants and aerosol propellants. See Chapter 2.
Nitrogen oxides. NO from burning fossil fuels and plant materials; subsequently oxidized to NO_2; see Box 9.4. N_2O: see Box 2.2.
Ozone. In the lower atmosphere produced principally by interaction of O_2 and NO_2 in sunlight.
Peroxyacyl nitrates (PANs). Formed by reactions involving ozone, NO_2 and hydrocarbon vapour in sunlight.
Sulphur dioxide. Mainly from burning fossil fuels; see Box 4.4.

Inorganic elements

Arsenic (As)	
Cadmium (Cd)* In P fertilizer.	Waste and contamination
Copper (Cu)*	from smelters, foundries,
Lead (Pb)*	mills. Mine waste.
Mercury (Hg)*	Industrial waste in sewage
Zinc (Zn)*	sludge
*Heavy metals	

Radioactive isotopes
^{134}Cs half-life = 2 years
^{137}Cs half-life = 30 years
From accidents at nuclear power stations (e.g. at Chernobyl, Ukraine, in 1986) and at nuclear waste processing works. From former atmospheric testing of nuclear weapons.

Organochlorine compounds
Insecticides, including DDT, lindane, aldrin, dieldrin, endrin. Production and use severely restricted in many countries.
Herbicides, including 2,4-D, 2,4,5-T, atrazine, dichlobenil, quintozene.
Polychlorinated biphenyls (PCBs). Examples in Fig. 9.10. Used for insulating materials, paints lubricants. Production restricted, but they are still abundant.
Chlorinated aliphatic compounds, including dichloroethene ($H_2C{:}CCl_2$) and carbon tetrachloride (CCl_4) are used as solvents.
Preservatives, e.g. pentachlorophenol.

Organophosphorus compounds
Insecticides, e.g. malathion
Herbicides, e.g. glyphosate.

Other organic pollutants
Natural pesticides, e.g. rotenone (Derris), a complex aromatic used as an insecticide. Pyrethroids are synthetic analogues of natural pyrethrin insecticides.
Polynuclear (polycyclic) aromatic hydrocarbons (PAH). Produced by burning fossil fuels or wood.
Crude oil. From oil wells, tankers, industrial waste. Most of the total release is from many small, unpublicized events.

Hellawell (1986, Table 7.22) and Freedman (1989, Table 8.1) give much longer lists of synthetic pesticides, with their full chemical names.

Measuring how toxic a chemical is

It is rarely possible to state simply that a chemical is harmful or harmless. Even apparently innocuous substances such as carbon dioxide or common salt (NaCl) can be harmful in high enough amounts. Usually the key question is: above what concentration or amount is it harmful? If the chemical is to be released where there are wild animals and plants, the question becomes: what concentration or amount can we be sure will not harm any of the species there? There are many new chemicals synthesized every year whose safety we need to assess. For example, in the USA more than 2000 chemicals are submitted each year to the Environmental Protection Agency for safety assessment (Alexander 1994). Do we test each of these in many concentrations? And do we test it on every species with which it might possibly come into contact?

Some of these problems are elaborated later. The key point here is that to perform every possible test with every chemical that might be released would be quite unrealistic. Nor is it realistic to say that the release of any possibly harmful chemical should be banned: that would, for example, mean stopping the burning of any fuel, wood-based or fossil. Human excreta can be harmful, too, in the wrong place. This section discusses how we can make the best assessment of the risk of using and releasing a chemical without unrealistic demands on time and expense. First I describe simple tests that can be carried out in a laboratory in one or a few days, and then go on to consider various steps that can be involved to scaling up from there to predict a safe concentration for the chemical in real ecosystems outdoors. Boxes 9.2 and 9.3 list methods that will be described and discussed in the text, and some problems that need to be addressed.

Simple, quick tests of toxicity

Almost always a chemical has to get into the tissue of an animal or plant before it can affect it. The route by which this happens can have an important influence on how the organism responds to a pollutant in its environment. The pollutant may enter the organism from a surrounding solution through wet membranes. For example, pollutants often enter algae, earthworms and aquatic amphibia through their whole external surface, or plants through their roots, fish through their gills. For such species, the *concentration* of pollutant in the external solution is likely to be important. On the other hand, many terrestrial animals (vertebrate and invertebrate) acquire most of the pollutant through their food; this is also true of marine mammals and birds that eat fish (Walker *et al.* 1996). For these species the *dose* they eat, rather than the concentration in it, may be a more useful measure. There is more about how pollutants enter and leave organisms later in the chapter.

Concentration or dose

A very common procedure for measuring the toxicity of a chemical is to determine what dose or concentration of it is required to kill individuals of a test species. Figure 9.1 shows an example in which five sets of aphids were sprayed with a solution of an insecticide, each at a different concentration. From the graph it is possible to determine the concentration that would kill 50% of the individuals: it is 4.9 mg l^{-1}. This is termed the LC_{50} (LC standing for *lethal concentration*). If the chemical is fed by mouth, e.g. to rats, the dose (amount per animal) sufficient to kill 50% of the animals is the LD_{50} (LD = *lethal dose*).

LC_{50}

LD_{50}

The LC_{50} and LD_{50} can be very useful for comparing both chemicals and species. If populations of the same test species are treated with several chemicals, an LC_{50} or LD_{50} value will be obtained for each chemical which can show which are the more toxic ones and give some indication

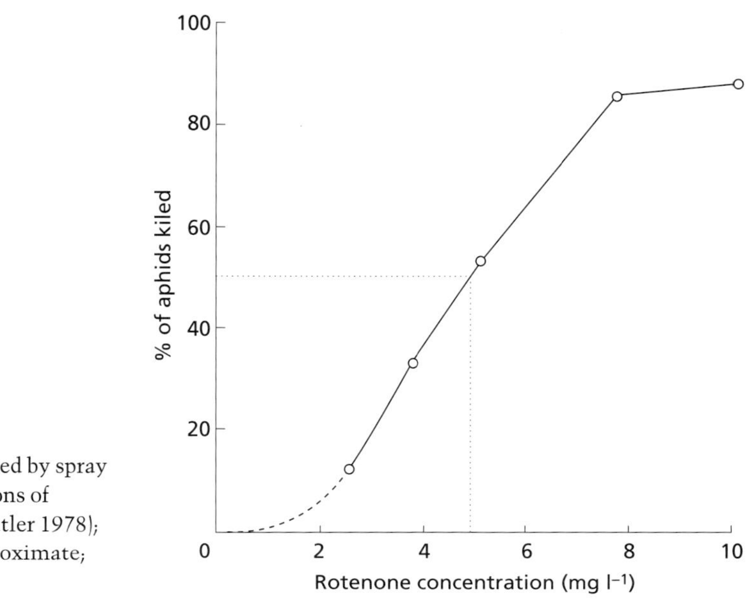

Fig. 9.1 Percentage of aphids killed by spray containing different concentrations of rotenone. O measured values (Butler 1978); - - - - expected continuation, approximate; · · · · indicates LC_{50}.

of how much more toxic one is than another (e.g. see Fig. 9.4). If many species are treated with the same chemical this can show their relative sensitivity (e.g. Table 9.1). However, the LC_{50} is clearly not a safe concentration to have outdoors in the real world. We would not, for example, wish to have in a lake a concentration of pollutant that killed half the fish. We now consider possible ways of moving on from an LC_{50} to being able to predict a safe concentration or amount to release into a field situation.

One suggestion would be to use an experiment similar to that of Fig. 9.1, but to extend the range of concentrations downwards to find out what concentration would kill no animals. The difficulty is that the line is likely to curve (e.g. as shown in Fig. 9.1, bottom left) and to reach the bottom axis only gradually; so the highest concentration causing no deaths is difficult to determine precisely. Nevertheless, provided we can accept some margin of uncertainty, such *'no observable effect'* measurements can be made.

'No observable effect'

We may reasonably ask whether death is the best measure of response to a chemical. One could argue that harm may be done to an animal or plant without it being killed, and that we should be measuring other responses which are more sensitive indicators that a species has been affected. An indicator commonly used for land plants is growth. Figure 9.2 shows the shoot weight of barley plants grown for 11 weeks in different concentrations of SO_2 under near-natural conditions in open-topped chambers. At 270 nl l^{-1} SO_2 reduced growth by half. This is the EC_{50}, where EC means *'effective concentration'*; EC is analogous to LC, but something other than death is the response measured. The results show that under these conditions barley growth begins to be reduced when SO_2

Effect on growth

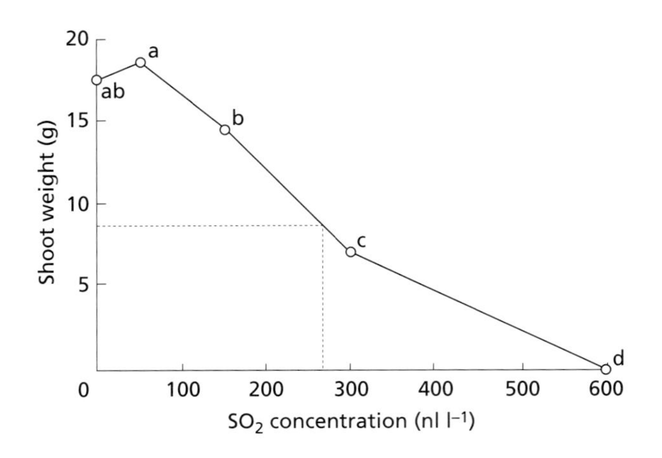

Fig. 9.2 Final shoot dry weight of barley plants grown for 79 days in an atmosphere containing different concentrations of SO_2. Points not bearing the same letter are significantly different at $P = 0.05$. Dashed lines indicate EC_{50}. From Murray & Wilson (1990).

is above about 100 nl l^{-1}, and this is likely to be more useful information than the EC_{50} for planning controls on SO_2 emissions.

In Fig. 9.2 there is a suggestion of a slight promotion of growth by SO_2 at 50 nl l^{-1}. It was not statistically significant in that experiment. However, when Ashenden (1993) treated 43 plant species with SO_2 at 50 nl l^{-1} for 8 weeks, 10 of them showed significant growth enhancement. Sulphur is in fact an essential element for all living things. Zinc and copper are two other examples of substances that are beneficial at low concentrations but harmful at higher concentrations. This serves to emphasize the importance of not simply labelling a chemical as toxic, but studying how different concentrations and doses affect living things.

Other responses that can be measured

Box 9.2 lists other test species responses that can be measured. All can be measured within hours or days in some species, but the choice of what to measure will clearly be affected by what test organisms are being used. Changes in animal respiration rates are difficult to interpret: the initial response to a harmful chemical is sometimes an increase in respiration rate. A decline in high-energy compounds, in RNA/DNA ratio or in protein synthesis is a more reliable indicator of a harmful effect on metabolism. Respiration is, however, often used to assess whether a pollutant is having a harmful effect on soil (e.g. Somerville & Greaves 1987). Provided the pollutant is allowed to act over several days, this can provide an indication of its effects on population size and the overall metabolic activity of the soil community. However, large changes in the species composition in soil could go undetected: if some species died and were replaced by other (pollutant-tolerant) species, the respiration rate might remain the same. That is why other, more specific measurements, such as rate of N fixation and nitrification, can be useful.

Length of the test affects the results

These tests are kept as short as possible, so that many species and chemicals can be tested and results obtained quickly. In real situations the organisms may be subjected to the pollutant for much longer. The toxic effect may increase with time, for example if the chemical is taken up

Box 9.2 Measuring how toxic a chemical is.

Short-term tests on individual species in artificial conditions: responses by the species that are commonly measured.

1 Death.

2 Growth rate. Measured by weight increase or linear extension (e.g. by plant roots).

3 Population increase. Can be measured within hours or a few days on bacteria, unicellular algae and some small invertebrates (e.g. *Daphnia*, collembolans).

4 Measures of metabolic state or activity:

(a) Respiration rate, by CO_2 production or oxygen uptake. Useful to measure microbial activity, e.g. in soil or decomposing litter; less useful for animals (see text).

(b) Other fairly simple measures of metabolic activity in microbes include nitrate production, acetylene reduction as a measure of N_2 fixation, activity of some common enzymes such as dehydrogenases.

(c) Photosynthesis by plants.

(d) RNA:DNA ratio.

(e) Amounts of high-energy compounds, e.g. lipids, glycogen.

(f) ^{14}C-aminoacid incorporation into proteins.

(g) The Microtox test. Uses the natural luminescence of the marine bacterium *Photobacterium phosphoreum*.

Further information: Walker *et al.* (1996)

slowly, so its concentration within the test individuals increases slowly. Figure 9.3 shows an example of how the measured LC_{50} can vary depending on how long the animals are subjected to the chemical. The organophosphorus compound diazinon was mixed into three soils at various concentrations. Earthworms were put in the soil, and at various times up to 3 weeks the proportion dead was recorded. Figure 9.3 shows the results expressed as $1/LC_{50}$; a higher $1/LC_{50}$ indicates greater toxicity. In the clay soil the measured $1/LC_{50}$ stabilized after 48 hours, but in the sand it was still increasing after 21 days. If the measurements had been made at 24 hours only, a quite different conclusion would have been reached about the toxicity of diazinon in the loam and sandy soils. Why the three soils gave such different results was not investigated. It may be that most of the diazinon was quickly adsorbed by the clay soil, and so uptake by the worms stopped, but in the other soils it remained available for uptake longer.

Relationship between short-term and long-term effects

The question now arises whether responses to 'acute' (i.e. short) exposure to a chemical can predict the effect of 'chronic' (i.e. long-term) exposure to the same chemical. Figure 9.4 compares the acute and chronic toxicities of 50 substances, including organic pesticides and also salts of heavy metals such as lead, copper and mercury. Each chemical was tested on a fish species (either fathead minnow or rainbow trout) and on the crustacean *Daphnia magna*, and the results used from whichever species

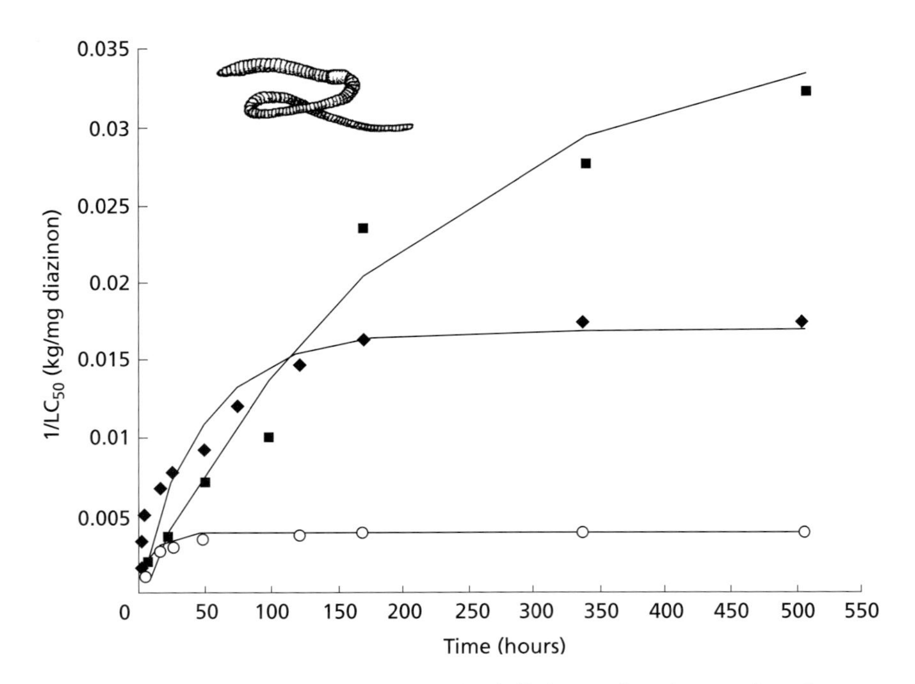

Fig. 9.3 Concentration of diazinon in soil that killed 50% of earthworms (*Lumbricus terrestris*) when they were in the soil for various lengths of time. Concentration expressed as $1/LC_{50}$, i.e. kg soil per mg diazinon. ○ clay soil, ◆ loam, ■ sandy soil. From Lanno, Stephenson & Wren (1997). Reprinted with permission from Elsevier Science.

proved to be the more sensitive. The horizontal axis ('acute') in Fig. 9.4 is the 96 h LC_{50}; the vertical axis ('chronic') gives the concentration to which the species could be exposed 'indefinitely' without suffering any observable adverse effect on survival, growth or reproduction. (Some chemicals affect egg and sperm production by fish even at concentrations too low to have a detectable effect on growth or mortality (Kime 1999).) As we should expect, the chronic concentration is always lower than the acute concentration: if a concentration is just low enough not to affect any of the fish when acting over a long time, it will not kill half of them in 96 hours. In Fig. 9.4 there is clearly a strong correlation between the two measures of toxicity: the correlation coefficient is 0.919 ($P \ll 0.001$). The regression line fits closely to: chronic level = (acute level)/14.8.

This suggests that a measurement of LC_{50} made in 4 days can be used to predict a safe concentration to which the species can be exposed long term, simply by dividing by 14.8. Notice, however, that some individual points diverge widely from the regression line. The equation above predicts that if the 96 h LC_{50} is 414 $\mu g\ l^{-1}$, any concentration below about 28 $\mu g\ l^{-1}$ should have no observable effect long term. Yet one point on the graph (circled) shows that cadmium, which gave a 96 h LC_{50} value of 414 $\mu g\ l^{-1}$, can have long-term effects at a concentration as low as 2.4 $\mu g\ l^{-1}$; the prediction was wrong by more than a factor of 10. So, LC_{50} values

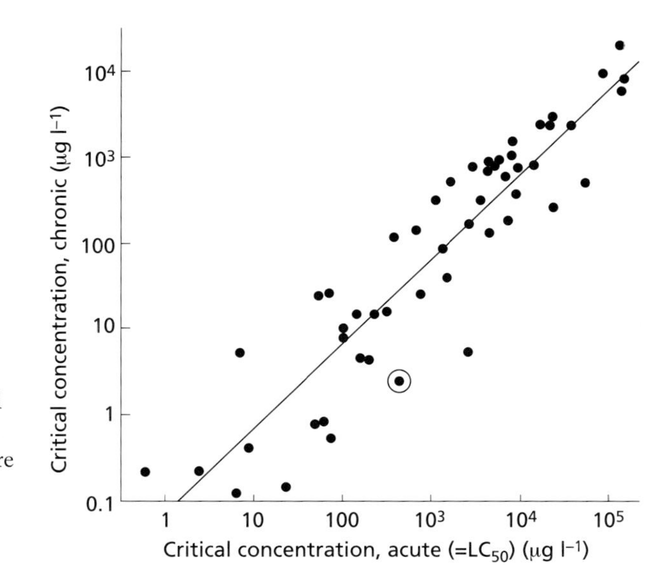

Fig. 9.4 Relationship between critical concentrations of 50 chemicals in acute (short-term) and chronic (longer) toxicity tests on freshwater animals. The meaning of 'critical concentration' is explained in the text. Each point is for a different chemical. Both axes are log scales. The straight line is the linear regression of chronic on acute. Data from Giesy & Graney (1989).

determined over a few days are by no means worthless in providing an indication of the long-term toxicity of chemicals, but there could be a wide margin of error in particular cases. Relationships between acute and chronic toxicities have been determined for other chemicals and other species; the equation above is not intended to be definitive, only an example. There has been a tendency to divide the measured LC_{50} or EC_{50} by an arbitrary number, most often 10, 100 or 1000, which is intended to incorporate a safety margin (Calow 1992).

Which species to test?

If a chemical is to be released into the environment it will probably come into contact with many species of animal, plant and microorganism. It is not feasible to test the response of every one of these. So, how many species do we need to use to obtain a reliable indication of toxicity, and which should we choose? There is a tendency for the test species to be chosen at least partly for convenience, because they are easy to breed and grow in laboratory conditions. This has been a major reason why, among invertebrates, *Daphnia* (freshwater) and *Artemia* (marine) have been so widely used. There is a danger that species robust enough to be cultured easily may tend to be less susceptible than average to toxic chemicals. Obviously there are differences between species in their response to at least some chemicals: herbicides such as atrazine that inhibit photosynthesis have little direct effect on animals, though they can kill non-target plants such as algae. Hellawell (1986) provided tables of LC_{50} values for many organic and inorganic chemicals on freshwater fish and invertebrates. Table 9.1 summarizes a few of the results. Even though only a few species have been tested, the toxicity of a chemical to fish can

Variation between species in their sensitivity

Table 9.1 Range of response among species of fish and aquatic invertebrates to four potential pollutants

	Number of species tested		96 h LC_{50} ($\mu g\ l^{-1}$)	
Substance	Fish	Invertebrates	Fish	Invertebrates
Cu^{2+} in hard water	10		300–10200	
Dichlobenil (herbicide)	4	3	4200–8000	8500–13000
Endrin (organochlorine insecticide)	6	4	0.27–1.96	0.25–5.0
Malathion (organophosphorus insecticide)	7	3	120–20000	0.76–50

Data collated by Hellawell (1986).

vary 10-fold or even 100-fold, and to invertebrates nearly as much. Invertebrates are not consistently more sensitive or less sensitive than fish, and the range of sensitivities within the two groups is at least as large as the differences between them. A common practice for potential pollutants of fresh water has been to test them on three species, of which one is a fish, one an invertebrate and one an alga; but Table 9.1 shows that this may be no more informative than testing, for example, two fish species.

Can we predict which species will be most sensitive?

One possibility is to choose for the tests species that play a key role in the ecosystem that will receive the chemical (see keystone species in Chapter 10). Another is to choose particularly sensitive species. In practice it is difficult to decide which species fulfil either of these criteria (Calow 1992). The species that is most sensitive to one chemical is not necessarily the most sensitive to another. This was shown by measuring the LC_{50} of 12 aquatic macroinvertebrate species in solutions of 15 different chemicals (two inorganic and 13 organic). Table 9.2 shows a small selection of the results. For each of the three chemicals a different species was the most sensitive, i.e. had the lowest LC_{50}.

It may be that if species are classified by their ecological strategies rather than taxonomically, this can indicate species that are likely to be more sensitive to pollutants. Grime (1977) proposed that plants can usefully be classified according to their tolerance of disturbance and of stress. The stress-tolerators tend to have slower growth rates, even in favourable conditions. When 41 herbaceous species were assessed for their sensitivity to SO_2, it was found that slower-growing species that had previously been rated as more stress-tolerant (e.g. towards drought and nutrient deficiency) were also more tolerant of SO_2, as measured by percentage growth reduction (Ashenden *et al.* 1996). In another piece of research slower-growing herbaceous species were also found to be the ones least harmed by ozone (Reiling & Davison 1993). Whether these stress-tolerator plants are also less sensitive to non-gaseous pollutants has not yet been investigated.

Testing on 'representative' species

Instead of asking which species to test our new chemical on, another approach is to test it on a representative selection and to use the results

Table 9.2 48 h LC_{50} values ($mg\ l^{-1}$) for three invertebrate species exposed to three different chemicals

Species	$HgCl_2$	Penta-chlorophenol	Benzene
Chironomus (midge)	0.55	0.11	100
Cloeon dipterum (mayfly)	0.05	5.9	34
Ischnura elegans (damselfly)	10.3	42	10

From Sloof (1983).

to estimate the range of responses likely to occur among all the species present, including those not tested. Kooijman (1987) proposed a method for calculating the concentration of a chemical that would be harmless to most of the species in a community, when the only information we have is the short-term LC_{50} values for a sample of the species. There are three basic steps to arriving at his formula.

1 The LC_{50}s were measured over a short time; they would be lower if the test had gone on for longer (see Figs 9.3 and 9.4). This is allowed for by assuming that the change in LC_{50} with length of exposure to the chemical occurs because the chemical is still being taken up into the organism, and Kooijman made assumptions about the time-course of this uptake.

2 The concentration causing no harm to an individual species is lower than the LC_{50}. Kooijman made assumptions about the shape of the curve (e.g. Fig. 9.1) relating harm to the organism and concentration of the chemical.

3 LC_{50} was measured on only some of the species, but we want to use a concentration that harms no species. In fact, Kooijman requires us to choose some probability (or risk) that the most sensitive species will be harmed, and uses this combined with the variability among the LC_{50}s of the measured species as a basis for the calculation.

Using these steps, Kooijman proposed a way to calculate HCS, a 'hazardous concentration for sensitive species':

$$HCS = X_m / T$$

where X_m is the geometric mean of the LC_{50} values of the measured species, and T is an 'application factor' calculated from the standard deviation of the measured LC_{50}s (i.e. the amount of variation in sensitivity between the species), and other parameters which take into account the number of species measured, the total number of species in the community and the acceptable risk (decided by the user) that some species will in fact be harmed.

Kooijman applied these calculations to real data for the response of an aquatic community to seven chemicals. The calculated value of the application factor, T, varied among the chemicals by a factor of 1 000 000, from 60 to 7×10^7. The number of species was the same in all cases, so the variation in T was due solely to variation between chemicals in the standard deviation of the LC_{50}s.

This method assumes that each species acquires the chemical from its environment, not through its food. The model could therefore be applied to many aquatic communities. It has also been used to estimate the minimum concentration of SO_2 in the atmosphere that will harm heathland plants in the Netherlands (Dueck *et al.* 1992). The calculation is based on a number of assumptions, and the predictions have not, so far as I know, been rigorously tested. Nevertheless, it has a more scientific basis than simply dividing the LC_{50} value by some arbitrary safety factor such as 1000.

Effects of pollutants in the environment

The aim of all the tests described so far is to predict the responses of individual species and whole ecosystems to pollution in the real world outdoors. This presents several sorts of difficulty (Box 9.3, section 2). One is

Box 9.3 Scaling up from short-term LC_{50} measurements to predicting safe concentration in real ecosystems.
For further explanation, see text.

Measure responses other than death (see Box 9.2)

Measure 'no observable effect' rather than effect on 50% of the population

Relate concentration producing response in short test ('acute') to that producing response when treatment prolonged ('chronic'), e.g. Fig. 9.4.

Use tests on a few species to predict safe concentration for may species.
Test: ecologically most important species?
species most sensitive to the chemical?
a representative range of species?

Problems with scaling up
If environmental conditions during test different from field conditions, may affect toxicity.

Effects on animal behaviour may be difficult to detect with captive animals.

In a mixture, chemicals may enhance each other's toxicity.

Observations and experiments in the real world
Long-term records, e.g. of growth, showing when a response first occurred, can help to indicate a pollutant responsible.

Conduct tests on multispecies microcosms or mesocosms.

Use species composition at a site as an indicator of amount of pollution there.

Measure concentration of pollutant within the organism.

that environmental conditions may affect the toxicity of particular chemicals. For example, Table 9.1 shows the LC_{50} of copper to fish species in hard water, meaning for those experiments $CaCO_3$ 220–360 mg l^{-1}. Two of the species were also tested in softer water ($CaCO_3$ 31 and 42 mg l^{-1}); their LC_{50} was lower by a factor of more than 10: in other words, copper was then much more toxic to them.

Effects on behaviour

Another type of problem in scaling up from a laboratory test to the field is that some chemicals, notably organochlorine and organophosphorus insecticides, can affect the behaviour of mammals and birds at concentrations that do not affect their growth or survival, and it may be impossible to detect this on animals in cages. One example was provided by Grue *et al.* (1982), who carried out an experiment on starlings that were nesting outdoors. Some adults were given a single dose of an organophosphorus insecticide; others were undosed as controls. During the next 24 hours the weight change of the dosed and control birds did not differ significantly. However, at the end of the 24 hours the dosed birds had only half the concentration of cholinesterase in their brain tissue that the control birds had; this could have resulted in impaired functioning of the nervous system, which might reduce their ability to find food. During the 24 hours after being treated the dosed birds visited their nestlings to feed them less frequently than before, and the nestlings lost weight. If such behaviour continued for some days it would probably result in the death of nestlings. Thus this change in behaviour, which could not have been detected in birds in a cage, could influence the abundance of the species in future years.

Effects of mixtures of chemicals

Industrial waste often contains several harmful chemicals, so we have the additional problem of predicting how a plant or animal will respond to such a mixture. Some research has found the effects of several pollutants to be approximately additive. Alabaster *et al.* (1972) studied the relationship between industrial pollution and occurrence of fish at 73 points on rivers in central England: the amount of pollution varied greatly between the points. They calculated a single toxicity value for each point, from the measured concentrations of ammonia, cyanide and heavy metals. The method of calculation took into account the individual toxicity of each chemical, as shown by its LC_{50} to trout, but otherwise the effects were assumed to be additive. The method was quite successful: fish were found to be absent from most of the rivers where the calculated toxicity was above a critical value.

Sometimes, however, one chemical can greatly increase the toxicity of another. This can happen if the organism has a system for rapidly detoxifying one chemical, but this system is inhibited by another chemical. For example, some insects can detoxify pyrethroid insecticides by monooxygenase enzymes. Any chemical that inhibits monooxygenases will thus increase the toxicity of pyrethroids to these insects. This is thought to be how EBI fungicides (ergesterol biosynthesis inhibitors) increase the toxicity of pyrethroids to bees by a factor of 10 or more (Walker *et al.* 1996).

Box 9.4. Summary of reactions leading to formation of sulphuric and nitric acids in cloud water and rainwater.

Combustion	S*	+	O_2	→	SO_2
In solution	SO_2	+	H_2O	→	$HSO_3^- + H^+$
In solution	HSO_3^-	+	O_3	→	$HSO_4^- + O_2$

* In fossil fuels

Heat from combustion	N_2	+	O_2	→	2NO
Gas Phase	NO	+	O_3	→	$NO_2 + O_2$
Gas phase	$2NO_2$	+	O_3	→	$N_2O_5 + O_2$
In solution	N_2O_5	+	H_2O	→	$2NO_3^- + 2H^+$

Further information: Wellburn (1994)

Mixtures of gaseous pollutants

There are important interactions between gaseous pollutants which increase their harmful effect. SO_2 and NO are products of combustion which dissolve in cloud droplets and rain. In this form they are of limited toxicity, but if they are oxidized they can form sulphuric and nitric acids, the principal constituents of acid rain, which can be very damaging. This oxidation is brought about by other pollutants, for example ozone, hydrogen peroxide or peroxyacyl nitrates (PANs). Box 9.4 summarizes the main reaction steps involved.

Nitrogen oxides, SO_2 and acid rain also provide an example of how the toxicity of chemicals can be influenced by environmental conditions. There has been concern since the 1960s about 'die-back' of red spruce, a tree which is widespread in the mountains of the northeastern USA. Needles become mottled or completely brown; later the terminal bud and then the whole branch dies, and many trees have died. Commonly browning of needles was first seen in early spring. The symptoms are less common further south in the Appalachians. This led to the suggestion that cold weather, including being covered with mist (clouds) for long periods, was the cause; but this did not explain why the problem apparently started in the 1960s. Experimental research has provided strong evidence that sulphate dissolved in cloud water makes the foliage more susceptible to damage by low temperature (Cape *et al.* 1991; Johnson 1992; Sheppard 1994). So the damage evidently originates from SO_2 blowing in from industry elsewhere. But a test of SO_2 effects, such as in Fig. 9.2, which did not take into account the cloud and low temperatures that occur in the spruce's natural habitat, would have failed to show this.

Long-term responses: trees

Trees can provide a long-term record of growth rate, in the annual rings in their trunks, which can be important for pinning down the effect of pollutants. The previous example, red spruce, also illustrates this. Figure 9.5(a) shows annual xylem radial growth in spruce at a stand in the Adirondacks where many trees have died since the 1960s. There was a steady increase in growth rate until the early 1960s, then a steady and

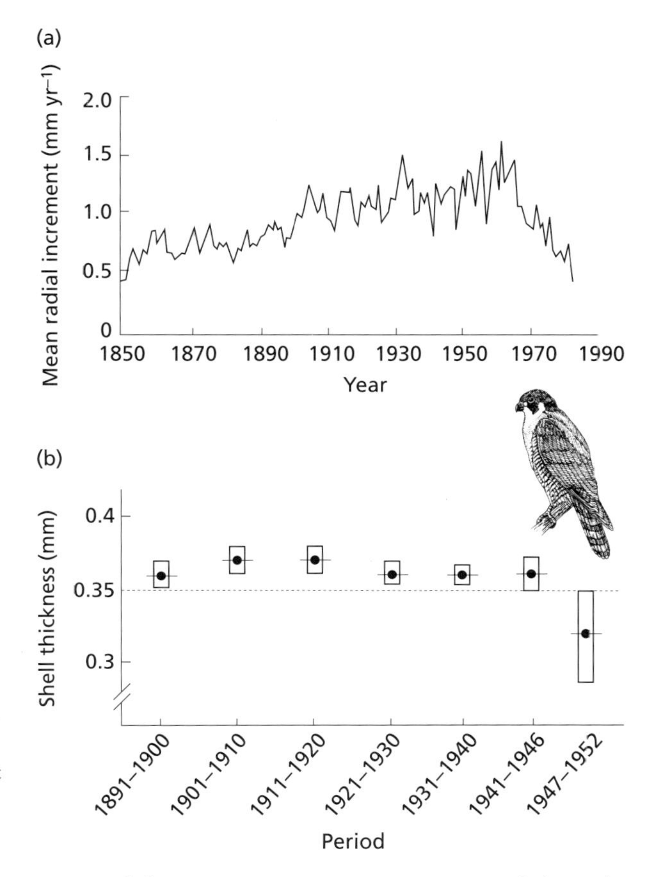

Fig. 9.5 Two examples of long-term records, where sudden changes helped to indicate involvement of a pollutant. (a) Annual radial xylem growth in trunks of red spruce trees at 1100 m altitude on Whiteface Mountain, Adirondacks, New York state. From McLaughlin *et al.* (1987). (b) Mean shell thickness, with 95% confidence limits, of eggs of peregrine falcon collected in California from 1891 to 1952. From Hickey & Anderson (1968). Reprinted with permission from Science volume 162 page 272. Copyright American Association for the Advancement of Science.

substantial decline. Many of the trees were young in 1850, and the subsequent increase in growth rate was probably age related. However, the later decline in growth cannot have been due to the trees reaching a critical age, because many other stands of spruce, of various ages, all showed a decline starting at about the same time. A detailed study of how spruce growth related to weather conditions over the period 1900–60 gave strong evidence that the decline in growth after 1960 was not simply a response to unfavourable weather. These measurements thus directed attention to whether pollutants might be involved.

. . . birds of prey

Figure 9.5(b) provides another, unrelated example of how a time-course can provide crucial evidence for pinpointing a pollutant effect. During the late 1950s and the 1960s there was concern about declines in the numbers of birds of prey, including peregrine falcon, in North America and western Europe. Various possible causes were considered. An important step forward was the discovery that many eggs were breaking in the nests, resulting in fewer young being successfully reared. It was possible to show that eggshells had become thinner, and to show when this had started, by measuring eggs collected at various dates and kept in museums and private collections. Figure 9.5(b) shows that the eggs of peregrines in California had maintained a nearly constant mean shell thickness from 1891 to 1946, but that the mean for 1947–52 was significantly lower. Other species suffered eggshell thinning at about the same time: bald eagle and osprey in the USA, golden eagle, merlin, sparrowhawk and shag in Britain (Hickey & Anderson 1968; Ratcliffe 1970). These findings, along with other research, helped to direct attention to DDT and other organochlorine insecticides as the likely cause: they had come into widespread use during the late 1940s and the 1950s. After the use of organochlorine insecticides was reduced, peregrine eggshells got thicker and population size recovered (Ratcliffe 1980; Newton *et al.* 1989; Newman 1993, pp 217–221).

Multispecies systems

Microcosms

So far we have been considering one species at a time, and how a pollutant affects it. We should now consider more complex systems where many species, perhaps at several trophic levels, interact with each other. It may be possible to simulate these in *microcosms*. A simple example of a microcosm is a soil sample. Even 1 g of soil will contain many species, some competing with others, some eating others. To measure an 'ecosystem process' such as the overall respiration rate of the soil sample, without any information on individual species or even groups, is by no means useless but far from ideal. A more useful example of an ecosystem process to measure is rate of litter breakdown, as this will affect rates of nutrient cycling and the development of soil organic matter. It can be measured by putting weighed litter samples in mesh bags, which can be placed at test sites outdoors, and weighing them again later. Berg *et al.* (1991) used this method to study the rate of decomposition of Scots pine needles at different distances from a mill and a smelter in Sweden that were sources of heavy metal pollution. At positions less than 1 km from the source decomposition was markedly retarded; beyond that distance the effect was small. Unfortunately the results do not show in detail how decomposition rate related to the heavy metal content of the needles.

Mesocosms

Lampert *et al.* (1989) compared the effect of the herbicide atrazine on the freshwater crustacean *Daphnia* in a simple laboratory microcosm and in a 'mesocosm', an enclosed volume of water in a lake in northern Germany. Table 9.3 summarizes the results. The top two lines show the concentration of atrazine needed to affect the population growth and mortality of the *Daphnia* if the atrazine was in the water containing the *Daphnia*, but its food, the alga *Scenedesmus*, was grown elsewhere with no atrazine. If instead (line 3), atrazine was added to the water in which the alga was growing, and then standard volumes of algal suspension were supplied to the *Daphnia*'s container, atrazine affected the *Daphnia* at a much lower concentration, by reducing its food supply. As atrazine is an inhibitor of photosynthesis this is perhaps not surprising. More surprising is the fact that when *Daphnia* was in the mesocosm, in the natural environmental conditions and among the full plankton species

Table 9.3 Response of *Daphnia* to the herbicide atrazine in its surrounding water. 'Effective concentration' is the lowest concentration at which an effect was observed.

Where experiment was conducted	Species treated with atrazine	What was measured*	Effective concentration of atrazine ($\mu g\ l^{-1}$)
Laboratory	*Daphnia*	Death (48 h LC_{50})	10000
Laboratory	*Daphnia*	Growth and reproduction	2000
Laboratory	*Scenedesmus* (green alga)	Population biomass	50–100
Lake, enclosure	Phytoplankton + zooplankton, including *Daphnia**	Population numbers	0.1–1

*All measurements were on *Daphnia*.

From Lampert *et al.* (1989).

complement of a lake (although fish were excluded), it was more sensitive still: its population was reduced by atrazine two orders of magnitude more dilute than was effective in the laboratory microcosm. The reason for this marked difference between microcosm and lake is unknown, but this is a clear indication of the difficulties of making predictions of safe concentrations outdoors based solely on tests in simple systems in the laboratory.

Insects in a stream

Figure 9.6 gives results from an experiment in Ohio which provided information on the response of the normal insect population of a stream to one pollutant. Over a period of 3 years copper salt was added at one point, at a rate which maintained an approximately constant Cu concentration near that point. The concentration declined with distance downstream, but after 2.6 km was still about twice that in the unpolluted stream above the contamination point (Fig. 9.6(a)). The abundance of bottom-living insects decreased greatly where the copper concentration was highest, and recovery in abundance downstream closely mirrored the decline in copper. The insect species composition also changed (Fig. 9.6(b)). In particular, mayfly larvae almost disappeared from the most Cu-rich sites, whereas the larvae of chironomids (midges) increased. The results show that a mean Cu concentration of 40 $\mu g\ l^{-1}$ was sufficient to have a marked effect on the insects. Although this is a useful result, it would be unwise to assume that 40 $\mu g\ l^{-1}$ would be the critical concentration in other streams, where species and conditions might be different. For example, as already mentioned, the $CaCO_3$ concentration in the water can have a marked effect on copper toxicity.

Invertebrates as indicators of pollution

The change in species composition in response to copper shown in Fig. 9.6 raises the question whether the abundance of particular species or groups of organisms can be an effective way of monitoring for pollution. This has been the basis of methods for monitoring water quality, especially of streams and rivers (Metcalfe 1989; Mason 1996). Living things

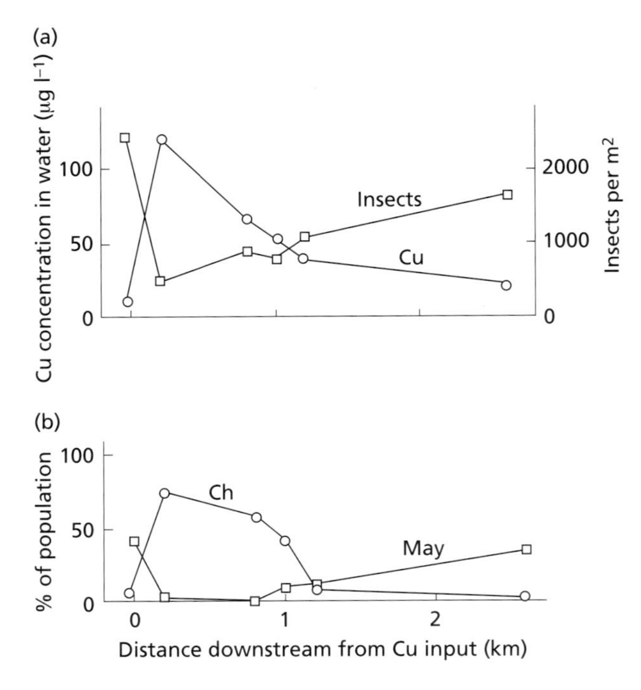

Fig. 9.6 Results from an experiment in a stream in Ohio, in which copper was added at one point. (a) Mean Cu concentration in water and number of insect individuals in stream bottom. (b) Percentage of total insects that were chironomids (Ch) and mayflies (May). Data of Winner, Boesel & Farrell (1980).

can, if fairly long-lived, provide an indication of the environmental conditions integrated over time. This is an advantage over measurements of a chemical: if it is discharged intermittently, regular sampling may miss the periods of highest concentration. Collecting river species usually requires only simple equipment, but sorting and identification can require a lot of time, as well as taxonomic skills. Identification has been simplified in the widely used Biological Monitoring Working Party (BMWP) system. Benthic (i.e. bottom-living) large invertebrates are collected by a standard procedure and are identified to family only. The number of individuals need not be counted. Each family represented contributes a score, from 1 to 10, depending on the sensitivity to pollution of that family. The least tolerant score 10, the very tolerant score low. The site's pollution rating can be expressed either as the total score for all the families present, or as that score divided by the number of families present, i.e. the mean score per family. The lower the score, the more polluted the site. The scoring system is based on the sensitivity of each family to a high input of organic matter, causing high activity of saprophytic microbes, which use oxygen and hence create low-oxygen or anaerobic conditions; so that is the type of pollution the system aims to measure. It is unlikely that each family would rate the same score for sensitivity to every other pollutant. Table 9.2 shows some differences between species of different invertebrate families in sensitivity to different chemicals; this is a small part of a data set for more species and chemicals (Sloof 1983). We can note that chironomids score 2 on the BMWP scale, i.e. very tolerant of organic pollution, and many mayfly families score 10 (very intolerant); so the tolerance of these two groups to organic pollution parallels their response to copper shown in Fig. 9.6 and to $HgCl_2$ in Table 9.2, but not their relative response to pentachlorophenol.

Overall species diversity of invertebrates has also been used as an indicator of pollution of fresh waters (Metcalfe 1989). This is based on the assumption that the least polluted sites will have the highest diversity, because pollution will kill off some species and there will be fewer new species tolerant of the pollutant. This assumption has not been rigorously tested, but seems reasonable. A practical objection to this alternative method of biological monitoring is that it requires more detailed identification, and commonly used diversity indices also require counts of the number of individuals of each taxon. So it requires more samples and more identification work than does the BMWP system.

Pollutant concentrations within living organisms

So far this chapter has been mostly about the responses of living things to chemicals outside them or in their food. But it is the amount that gets into their tissues that is critical in determining the effects. The concentration in plant and animal tissues can, as we shall see, be either higher or lower than the concentration in their surroundings or their food. As well as determining whether there is a harmful effect on that individual, the concentration in it will determine how much is passed on to an animal that eats it. This section presents information on what controls the concentration of pollutants in organisms, and whether this helps us to predict the toxicity to the species.

Aquatic species

If uptake and loss are through external surfaces

Aquatic species all have surfaces across which they exchange respiratory gases with the surrounding water—gills in fishes, other respiratory surfaces in many invertebrates, the whole outer surface in microscopic species. These surfaces allow many chemicals to enter the tissues, but they can also be the site of losses, depending on concentration differences between inside and outside. We should expect that if a chemical can enter and leave the individual only by this route, then an equilibrium concentration should in due course establish within its tissues. Figure 9.7 shows an example of this, for three species placed in a dilute solution of the insecticide dieldrin. The experiment was so designed that the two animal species could not obtain any dieldrin through food eaten. In all three species the internal concentration reached a plateau. However, this took longer the larger the organism. Also, the equilibrium concentration was not the same among the three species, and not the same as in the external solution, which was only 0.003–0.005 $\mu g\ g^{-1}$ water.

The concentration of the pollutant in the organism does not necessarily reach a plateau: it is possible for uptake to be so slow that its

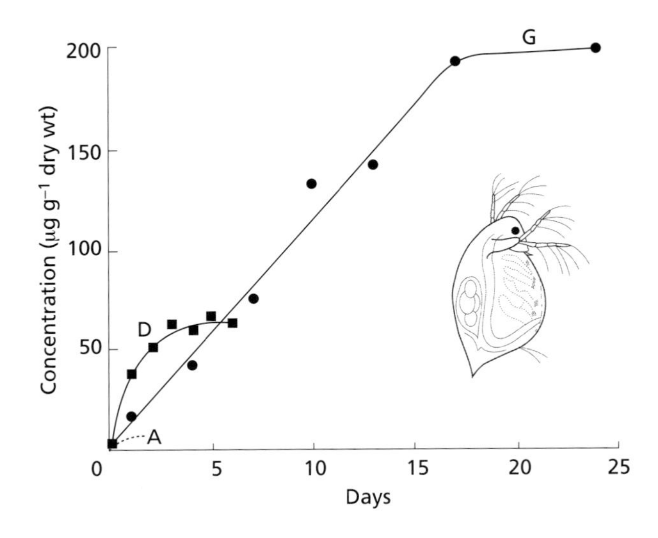

Fig. 9.7 Concentration of dieldrin in an alga *Scenedesmus obliquus* (A), a cladoceran *Daphnia magna* (D) and guppy (G) (a small fish). The species were immersed, separately, in dieldrin at 3–5 μg l^{-1}. The *Daphnia* were not fed during the test, the fish were fed on uncontaminated *Daphnia*. Data of Reinert (1972).

concentration in the tissues goes on increasing throughout the life of the individual. This happened, for example, with DDT and dieldrin uptake by lake trout in Lake Michigan in the early 1970s (Kogan 1986, Table 10.10).

Solubility in fat and water

One reason why the equilibrium concentration of a chemical may be higher in the organism than in its surrounding water is that the organism contains a lot of fat. If oil, water and a small amount of a chemical are shaken together, many organic chemicals will end up more concentrated in the oil than in the water, because they have a higher solubility in oil than in water (Clark *et al.* 1988). A very simplified 'model' for an aquatic animal would be to consider it as a lump of fat, surrounded by water, with a solute capable of diffusing from one to the other. Given enough time for equilibration, the solute should reach a steady concentration in the animal which depends on the relative solubility of the solute in fat and water. This relative solubility is usually measured by the octanol:water partition coefficient. Octanol, water and the chemical are shaken up together, left for the octanol and water to separate into two layers, and the concentration of the chemical in each layer then determined. Figure 9.8 shows that the octanol:water partition coefficient can sometimes be a good predictor of the concentration of a chemical that will build up in a fish. This in turn has led to the use of the octanol:water partition coefficient as a predictor of the toxicity of chemicals, on the grounds that, other things being equal, the more concentrated the chemical becomes in the animal's tissues the more toxic it will be. This may sound an excessive oversimplification, but Table 9.4 shows that it can work quite well. The toxicity of chemicals was measured on five quite unrelated species. The first four listed could acquire the pollutants from the surrounding water through membranes. For these species the octanol:water partition coefficient correlated strongly with the measured toxicity. (The

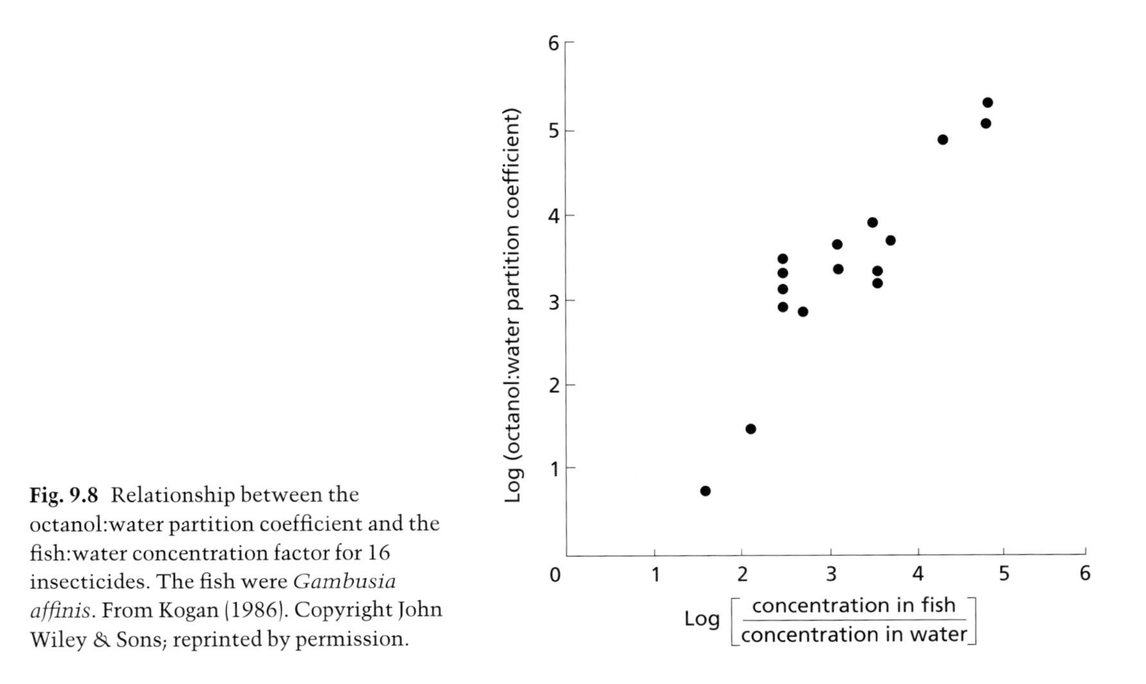

Fig. 9.8 Relationship between the octanol:water partition coefficient and the fish:water concentration factor for 16 insecticides. The fish were *Gambusia affinis*. From Kogan (1986). Copyright John Wiley & Sons; reprinted by permission.

correlations are negative because chemicals that are more soluble in octanol give lower EC_{50}, LD_{50} and LC_{50} values, i.e. they are more toxic.) However, the correlation coefficients were far enough from 1 to show that the prediction of toxicity may not be fully precise or reliable. In contrast to these species, rats obtain most pollutants through their food, and for them the method was very poor at predicting toxicity.

Considering an animal as just a lump of fat is clearly a drastic simplification. More complex and sophisticated models of pollutant entry into fish have been proposed which take into account food as a source, as well as the surrounding water. The model of Clark *et al.* (1990) showed that if

Table 9.4 Correlation coefficients between log (octanol:water partition coefficient) of 54 or 59 organic pollutants and their toxicity to various species

Species	How chemical applied	What measured	Period	Correlation coefficient
Lettuce	In solution culture, roots immersed	Growth, EC_{50}	16–21 days	– 0.76*
Fathead minnow	Animals immersed	Death, LC_{50}	96 h	– 0.66*
Daphnia magna	Animals immersed	Death, LC_{50}	24 h	– 0.59*
Photobacterium phosphoreum	Bacteria immersed	Luminenscence, EC_{50}	30 min	– 0.71*
Rat	Fed by mouth	Death, LD_{50}	96 h	– 0.23 NS

Statistical significance: *$P < 0.001$; NS, not significant ($P > 0.05$). From Hulzebos *et al.* (1991) and Kaiser & Esterby (1991).

an uncontaminated fish is placed in polluted water and starts to eat polluted food, it will initially gain pollutant from both sources. But later, as its internal concentration nears a steady state, it may be gaining pollutant through its food but losing it through its gills. So the concentration in the food would be expected to have an effect on the concentration in the fish. However, most experimental studies with aquatic animals have found that food is a minor source of pollutant intake compared to absorption from surrounding water (Moriarty 1988). In one experiment guppy (small fish) were kept in water containing dieldrin at 0.8–2.3 $\mu g\ l^{-1}$ and fed with uncontaminated *Daphnia*. Other *Daphnia* were kept in water with a similar concentration of dieldrin, and were then fed to guppy that were in uncontaminated water. After the concentration of dieldrin had reached a plateau in both sets of fish, the concentration was about 10 times as high in those that had received dieldrin from the surrounding water as in those that had received it via their food (Reinert 1972).

Terrestrial species

Uptake of inorganic ions by plants from soil

Although land plants can have pollutants deposited on their leaf surfaces, most of their uptake is through their roots. The uptake of organic substances is influenced by their relative solubilities in fat and water (e.g. lettuce in Table 9.4). However, the uptake of many inorganic ions is active and selective: in other words, metabolic energy is used to achieve a much higher concentration of the element inside than outside, but this applies only to certain elements. Some ions which are essential to the plant in small amounts and are taken up actively can reach toxic concentrations in the plant if the concentration outside is high. Examples are zinc and copper (see Fig. 9.9). Other toxic elements are chemically similar to essential elements and are taken up by the same carriers. For example, arsenic is taken up by the same mechanism as phosphorus. Caesium and potassium are another similar pair. The radioisotopes ^{134}Cs and ^{137}Cs were released into the atmosphere by a serious accident at a nuclear power station at Chernobyl in the Ukraine in 1986, and were deposited on various parts of Europe. Some of this has subsequently been taken up by plants. The half-life of ^{137}Cs is 30 years, and as Cs is leached only slowly from most soils (Szerbin *et al.* 1999), plants in these areas will be able to take up radioactive Cs throughout much of the 21st century.

Uptake of essential elements from soil by plants has been the subject of research for more than a century. One practical aim has been to predict whether growth rates of crops on a particular soil would be restricted by the deficiency of a particular nutrient element, and, if so, how much fertilizer should be added. In soil most elements are in a variety of chemical forms which differ in their availability to the plant. Uptake will also be affected by features of the plant, such as root depth and morphology, mycorrhizal associates, and ability to excrete acids. More information on this is given by Wild (1988) and Marschner (1995). In spite of these complexities, some useful prediction of likely uptake of cations can be made by measuring the amount of the ion extracted from the soil by a fairly strong salt or acid (e.g. 1 M (molar) ammonium nitrate). This has led to 'rules of thumb' used by agricultural advisers in Britain, quoted by MacNicol and Beckett (1985), e.g. if 0.5 M acetic acid extracts 100 or more μg zinc per g of soil, Zn is likely to be toxic to plants.

Moist-skinned animals living in soil can take in chemicals through their skin; for example earthworms (see Fig. 9.3). However, most land animals take in pollutants mainly from their food. This adds to the complexity of predicting a safe concentration of a pollutant, since these animals are not responding directly to the pollutant in the physical environment, but rather to the pollutant in their food.

Biomagnification up food chains?

There has been great interest in *biomagnification*, or increase in concentration along food chains. In other words, is it true that the concentration of a pollutant chemical is higher in the tissues of herbivores than in plants, higher in carnivores than in herbivores, and higher still in top carnivores? Some older ecology textbooks treat this as a universal truth. For example Collier *et al.* (1973) state: 'This process of biological concentration or magnification of materials is a general property of food chains'. Subsequent research has shown that things are not so simple.

If all the species in a food chain get their pollutant mainly from the external medium through their outer surfaces, there is no basic reason why species higher up the food chain should have higher concentrations. In Fig. 9.7 in fact there was such an increase up the food chain, which in that experiment was nothing to do with pollutant in food. However, the opposite can also happen in aquatic systems. For example, Moriarty (1988, Table 6.8) gives data from samples taken from the Mediterranean in which the concentration of PCBs in the phytoplankton and herbivorous zooplankton was about 10 times higher than in carnivorous shrimps.

If an animal gets the pollutant mainly from its food the situation is more complex; but it is certainly not true that biomagnification is universal, for organic or inorganic pollutants. When hens were fed diets ranging 1000-fold in their DDT concentration, the concentration in their eggs was always very close to that in their food (Moriarty 1988, Fig. 5.15). Examples involving inorganics are given later. Much of the food eaten by any animal is made of C, H and O, and most of the C, H and O that passes into the animal's mouth will not remain long term in its body: some will pass out in faeces, some in urine and some as CO_2 and water vapour. The key determinant of whether biomagnification of a pollutant occurs is whether the animal is more or less effective at getting rid of the pollutant than the other materials in its food. To be more precise, we have to consider what proportion of the pollutant in the food is retained in the animal's body and what proportion of the other constituents of its food are retained.

Thus a key determinant of whether biomagnification occurs is the ability of the animal to get rid of the pollutant. This loss may occur as dissolved substances in urine, or in bile which passes into the gut. In either case the excreted chemical needs to be water soluble, and the metabolic ability of the animal to convert a fat-soluble toxic compound into a water-soluble compound may be the limiting step in its ability to excrete it.

^{137}Cs in rabbits

Table 9.5 shows an example where excretion of a pollutant was fast enough to have a major effect on the concentration of pollutant in the animal. Alfalfa grown in northern Italy in 1986 became contaminated with ^{137}Cs from the Chernobyl explosion. Some of this alfalfa was made into meal and used to feed rabbits during a 6-week experiment. The ^{137}Cs content of whole animals was determined, so that average whole-body concentrations can be calculated. Table 9.5 shows that if rabbits were fed a diet high in ^{137}Cs, within 3 weeks the concentration had reached a plateau: it increased little more after 3 further weeks of feeding the high-^{137}Cs diet. If the animals were switched back to low-contamination food, within 3 weeks their ^{137}Cs concentration was back to normal. These rapid adjustments were possible because large proportions of the ingested ^{137}Cs were lost in faeces and urine (Table 9.5(b)). However, 2.4% was retained in rabbit tissue during weeks 4–6. Because the animals were growing quite rapidly, there was about enough new tissue to accommodate the extra isotope without increasing the whole-body concentration. In contrast, when rainbow trout ate food contaminated with ^{137}Cs they eliminated only about 1% of the body content per day, too slow to have much effect on the total body content (Cocchio *et al.* 1995).

Pollutant sequestered in harmless form

An alternative strategy is that the animal retains the toxic chemical, but in a part of its body where it can do little harm. For example, woodlice feeding on litter contaminated with the heavy metals copper (Cu), cadmium (Cd), zinc (Zn) and lead (Pb) sequester much of it in their hepatopancreas, blind-ending tubes off the main digestive system. Table 9.6 shows results from an experiment in which woodlice and dead leaves were collected from a wood in southern England; it borders on a busy

Table 9.5 ^{137}Cs concentration and balance of rabbits fed with alfalfa that had been contaminated during growth by fallout from the Chernobyl explosion

(a) Whole-body concentration of ^{137}Cs in rabbits (Bq kg^{-1})

Feeding regime	Day 1	Day 21	Day 42
High-^{137}Cs food days 1–42	27	76	81
High-^{137}Cs food days 1–21, then low-^{137}Cs	27	20	
Low-^{137}Cs food days 1–42	27	16	

(b) ^{137}Cs balance, days 21–42, for rabbits fed high-^{137}Cs throughout

	Bq per animal	% of total
Lost in faeces	2060	78.4
Lost in urine	507	19.3
Retained in tissue	62	2.4

Data from Battiston *et al.* (1991).

Table 9.6 Concentrations ($\mu g\ g^{-1}$ dry weight) of heavy metals in leaves of a tree species (*Acer campestre*) collected from the litter layer of a polluted wood in England, and in woodlice (*Porcellio scaber*) that were fed on the leaves for 20 weeks in controlled conditions

	Cu	Cd	Zn	Pb
Leaves	52	26	1430	908
Woodlice main tissues	66	3.7	135	11.3
whole body	1130	73	1370	132
% in hepatopancreas	92	93	85	52

Main tissues = whole body except gut and hepatopancreas.
% in hepatopancreas = (wt of element in hepatopancreas/wt in whole body) × 100
From Hopkin (1990).

motorway, and is 3 km downwind from a large smelting works. The woodlice were fed on the leaves (and nothing else) for 20 weeks. Although the hepatopancreas was only 5% of the dry weight of the animals, it contained half the total body Pb and an even higher proportion of the other heavy metals. These heavy metals in the hepatopancreas are kept away from the metabolic machinery of the woodlouse, but are ingested by anything that eats it.

Heavy metals in food chains

Table 9.6 compares the concentrations of the heavy metals in the woodlice with those in their food, all on a dry weight basis. Using the whole-body concentration (which is what a carnivore eating them would ingest), Cu and Cd were more concentrated in the woodlice than in their food, Zn was about equal and Pb was less concentrated in woodlice than in food. So the generalization that pollutants become more concentrated up the food chain is not consistently supported here. From the point of view of the woodlice, sequestering much of the heavy metal in the hepatopancreas allows the concentrations in the rest of the tissues, where the main metabolism takes place, to be kept well below the food concentration for three of the elements.

A detailed study of amounts of Cu and Cd in soil, plants and animals, involving many samples through a 12-month period, was made by Hunter *et al.* (1987a, b, c). Their samples were from three areas of rough grassland in Merseyside, England, at sites which differed greatly in heavy metal pollution: one close to a copper refinery, another 1 km away and a third in a much less polluted area. Table 9.7 shows results for the only carnivorous mammal, the common shrew. At all three sites the concentration of Cu in the shrew's tissue was lower than in its food. The Cu concentration in its diet increased more than 10-fold between the least polluted and the most polluted site, but its body concentration increased only twofold. In contrast, the Cd concentration in the shrew was slightly higher than in its food, rising as the food Cd rose. There were also two herbivorous small mammals at the sites, which had diets lower in Cu and Cd but, like the shrew, they were able to maintain their internal Cu

Table 9.7 Copper and cadmium concentrations (μg g^{-1} dry wt) in common shrew (*Sorex araneus*, a carnivore) and in its food, at three sites in northern England associated with a refinery

	Cu			Cd		
	Least polluted	1 km from refinery	Near refinery	Least polluted	1 km from refinery	Near refinery
Concentration in food	52	104	652	2	16	55
Concentration in shrew	13	17	29	4	19	71

From Hunter *et al.* (1987c).

concentration nearly constant as Cu in their food increased, whereas their body Cd concentration was always close to that of their food. These results show a strong ability of these mammals to control their body Cu even in heavily polluted areas, but not to control their Cd. Cu is an essential element for mammals, since it is a constituent of the enzyme cytochrome oxidase, whereas Cd is not beneficial even at low concentrations. The better ability of these mammals to adjust their body Cu concentrations than their Cd concentrations may relate to this.

Organochlorines in food chains

Table 9.8 shows an example of the transfer of some organic pollutants in a food chain—a carnivorous bird and two fish species that it eats. Each chemical was more concentrated in the gull than in its food, but the concentration factor (i.e. concentration in gull/concentration in fish) varied greatly between chemicals. DDT and dieldrin had almost identical concentrations in the fish, but were substantially different in the gull. Another example of organochlorine insecticides reaching birds through fish they eat relates back to the eggshell thinning in the 1950s (Fig. 9.5(b)). One bird that suffered serious thinning was the shag, and its eggs contained substantial concentrations of organochlorine residues (Ratcliffe 1970; Newman 1993, Fig. 7.10). The shag's diet is ocean fish. This illustrates the important discovery that pollutants washed into fresh waters or even the oceans cannot be assumed to be safely lost, in spite of being greatly diluted.

The conclusion, on both organic and inorganic pollutants, must be that they are not consistently more concentrated as they pass up food chains.

Table 9.8 Concentrations (μg g^{-1}) of some organochlorine chemicals in whole bodies of herring gulls and two fish species in Lake Ontario

	DDT	Dieldrin	DDE	PCBs
Gull	0.029–0.15	0.18–0.53	14–30	100–200
Fish:				
alewife	0.015	0.017	0.16	0.90
smelt	0.014	0.016	0.12	0.46
Concentration in gull/ concentration in fish	2–11	11–33	88–250	111–435

Data from Clark *et al.* (1988).

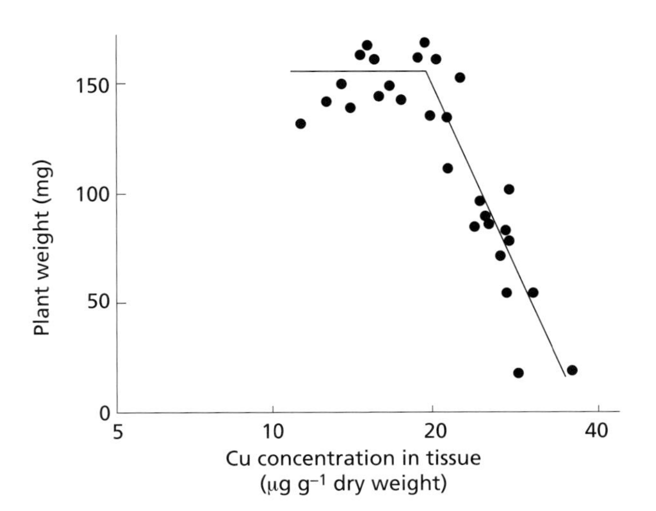

Fig. 9.9 Relationship between shoot dry weight and shoot Cu concentration of young barley plants grown with a range of Cu supplies. From Beckett & Davis (1977).

Sometimes animals contain higher concentrations in their bodies than was in their food, but other animals do not. Clearly, this makes it more difficult to predict what concentration there will be in a particular species, and hence how it is likely to respond to a particular chemical released into the environment.

Concentrations in animals and plants as indicators of pollution

The previous section showed that we now have some ability to predict the concentrations of toxic chemicals that will build up in plants and animals living on land and in water; but it also pointed out the difficulties and limitations. An alternative approach would be to measure concentrations within plants or animals in the field and to use these as indicators of whether amounts of particular pollutants have reached harmful levels at the site. One problem is that any chemical is likely to vary in concentration between different parts of an individual plant or different organs of an animal, and different parts are likely to have different sensitivities. Beckett and Davis (1977) minimized such variation by growing plants of a single species, barley, to the five-leaf stage and then analysing the shoot material, which would be mainly young leaf tissue. They grew the plants with various amounts of Cu, nickel (Ni), Zn or Cd in the nutrient solution, and found that for each element the shape of the relationship between growth rate (as measured by final dry weight) and heavy metal concentration in the shoot was like that shown in Fig. 9.9. There was a range of concentrations that gave a plateau weight, but above a critical concentration the weight decreased. If they plotted the concentration on a log scale (as in Fig. 9.9) the right-hand, declining region formed a straight line, and it was possible to decide the position of the corner fairly precisely.

Critical concentrations in plants?

Can this method be applied to a range of species growing under a range of conditions outdoors? MacNicol and Beckett (1985) collected together data from a variety of crop species grown in soil, peat or solution culture. There were 28 sets of results that allowed determination of the critical concentration for Cu, in other words the corner in Fig. 9.9 where Cu begins to cause a reduction in growth. These 28 values of critical concentration varied by more than an order of magnitude, from 5 to > 64 $\mu g\ g^{-1}$ dry weight. The results for other elements were more variable: among 46 data sets, critical concentrations for Cd ranged from 4 to 200 $\mu g\ g^{-1}$.

... in animals?

Graphs similar to Fig. 9.9 have been obtained for animals, provided a single species is treated with a single chemical. For example, in Fig. 8.11 of Walker *et al.* (1996) the log of tributyltin concentration in mussel vs. its 'scope for growth' (i.e. energy available for growth) showed a sharp corner similar to that in Fig. 9.9. But with animals, as with plants, once we move away from single-species, single-chemical situations the critical concentration becomes much less precise. McCarty and Mackay (1993) summarized the toxic concentrations in fish of major categories of chemicals, for example respiratory uncouplers and acetylcholinesterase inhibitors. For some types of chemical the critical body concentrations ranged over more than two orders of magnitude, though for others they varied only about fivefold.

It seems that, with our present knowledge, we cannot specify a precise concentration of a pollutant, or group of pollutants, in tissues of organisms and say that above that is toxic, below it is not. We may, however, be able to suggest a concentration below which harm is unlikely, provided we include a safety margin. For example, for Cu in plant tissues (see earlier) it would lie somewhere below 5 $\mu g\ g^{-1}$. Part of the problem is that many pollutants act only in certain organs, tissues or 'target sites'; their concentration elsewhere in the animal may not be important. For example, the concentration of a heavy metal in the whole woodlice of Table 9.6 is of little significance in assessing the toxicity to the woodlice, because much of it is sequestered in the hepatopancreas.

Removing or degrading pollutants

We have covered in some detail how to decide whether a chemical is present in large enough amounts or concentrations to be dangerous. If a chemical is considered to present a danger, then something needs to be done about it. Some common methods of dealing with toxic pollutants include:

1 Discharge it into a river or ocean, where it will (hopefully) be dilute enough to be harmless.
2 Discharge it as gas into the atmosphere, e.g. by heating polluted material or blowing air through it.
3 Burn it.
4 Put it in a large hole in the ground and cover it with topsoil.
5 Enclose it, e.g. in cement.

Of these, burning can convert a pollutant containing only C, H and O to pure CO_2 and H_2O, provided the temperature is high enough and the burning gases are adequately mixed; but if the pollutant contains other elements such as N, S or a halogen the resulting gases may still include pollutants. The other methods do not get rid of the pollutant, merely put it somewhere where people hope it will not be harmful. To be sure that it will really remain harmless in the long term may be difficult. One example is the difficulty of deciding how to dispose of radioactive wastes with half-lives of thousands of years. Another problem is that some organic pollutants (e.g. organochlorines from insecticides) that are very dilute in the sea can nevertheless be much concentrated in the bodies of fish, which are in turn eaten by birds.

Bioremediation

Sometimes living things can help with disposing of pollutants or making them harmless. This is called *bioremediation*, and is the subject of this section.

Sewage treatment

Sewage from towns contains a large amount of organic matter intermingled with a variety of toxic chemicals. A primary aim of sewage treatment is to reduce the amount of suspended organic matter. If it were discharged raw into a river it would support large populations of heterotrophic microorganisms, whose respiration could produce a severe biological oxygen deficit. Sewage treatment has first to remove large suspended and floating objects, and grit. Then the primary sedimentation stage removes organic particles by allowing them to settle out. The secondary treatment could be classed as bioremediation. The liquid sewage trickles down through gravel filters or is put into actively aerated tanks, where complex mixtures of bacteria, fungi and protozoa (also algae on filters, rotifers in tanks) oxidize much of the remaining organic matter. The primary and secondary treatments both produce sludge, which can be further treated by microorganisms anaerobically, to form a dried sludge which may be suitable for use as an organic manure (see Chapter 4).

Removing heavy metals

This standard sewage treatment is not good at coping with large amounts of toxic inorganics, such as heavy metals, which may occur in modern sewage. Some methods of removing these involve microorganisms. If a dissolved metal can be precipitated out, the precipitate can then be removed. Under anaerobic conditions, bacteria which reduce sulphate to sulphide can form insoluble precipitates of metal sulphides, for example CuS, PbS, ZnS. Another option is to use microbial material—algae, fungi or bacteria—to absorb the metals. Since much of this removal is adsorption by extracellular wall material, the cells can be either dead or alive, or a mixture. This allows waste biomass, produced cheaply in large quantities, to be used, e.g. yeast waste from breweries, *Penicillium* from antibiotic production, sludge from sewage farms. When the biomass is fully charged with the metals, it is either treated as waste to be dumped in a hole in the ground, or it may be possible to elute the metals with an acid solution and use them again. For more information on sewage treatment and heavy metal removal see McEldowney *et al.* (1993), Harrison (1996).

When we turn to considering organic pollutants, there is also the possibility that they can be converted to less toxic chemicals. Higher plants can take up a variety of organic chemicals from soil, which may then be (1) translocated to the upper parts and lost as vapour, (2) sequestered in vacuoles or as insoluble compounds, or (3) converted to non-toxic chemicals (Salt *et al.* 1998). However, compared to microorganisms plants appear to have a much less extensive range of chemicals that they can degrade to simple, non-toxic compounds; so more research has been devoted to biodegradation by microorganisms and how to promote it.

Biodegradation by microbes may happen naturally and rapidly, in fresh water, oceans and soil. However, some synthetic organic chemicals break down only very slowly under natural conditions, and may take years to disappear. After sewage sludge containing polynuclear aromatic hydrocarbons was mixed into soil at a site in England, 20 years later substantial proportions of some of them were still there (Wild *et al.* 1991). This section considers ways of promoting the decomposition of organic chemicals, especially those that normally decompose very slowly.

Organic pollutants: which are more readily degraded?

The ability of microorganisms to break down organic chemicals under laboratory conditions has been much investigated, and some generalizations can be made about which sorts of chemicals are more easily degraded. So it is possible, just from the formula and structure of a chemical, to give some indication as to how rapidly it will be broken down, which in turn is one indicator of how harmful it will be in the environment. Howard *et al.* (1991) classified 235 organic chemicals as degrading quickly or slowly under aerobic conditions. From this they were able to allot a score to each component of the molecule, indicating whether it tended to slow down or speed up breakdown. Table 9.9 gives as examples the score allotted to a few of the components. A larger positive score means a faster degrader; a larger negative means slower. The presence of Cl instead of OH makes the chemical more likely to be a slow degrader, but for NH_2 this is true only for aromatic compounds. In this way it can be predicted, for other organic chemicals, whether they will be fast or slow degraders.

A limitation of Howard *et al.*'s model is that it does not take into

Table 9.9 Influence of some chemical characteristics on biodegradability, according to model of Howard *et al.* (1991). The more negative the number the stronger the prediction that the chemical will degrade slowly

	Carbon skeleton	
Attached	Straight chain (aliphatic)	Ring (aromatic)
OH	+ 0.08	+ 0.06
NH_2	+ 0.09	− 0.26
Cl	− 0.18	− 0.21

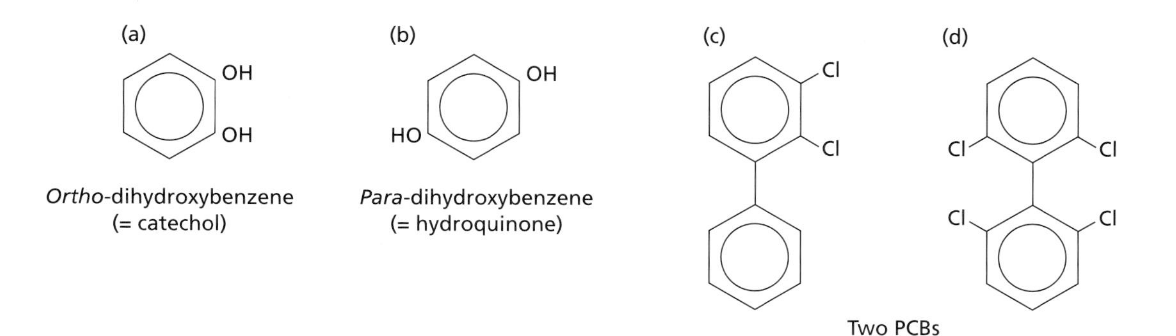

Fig. 9.10 Examples of aromatic chemicals. (b) is more difficult to degrade than (a), (d) more difficult than (c); 'difficult' here means that fewer species can accomplish it, and it takes longer. PCB = polychlorinated biphenyl.

account how many atoms of Cl (for example) are in the molecule or where they are positioned. In the breakdown of aromatic compounds the splitting of the benzene ring is often the limiting step. Whether a particular enzyme or a particular bacterial species can accomplish this depends on what side groups are attached to the ring and in what positions (Betts 1991). For example, more species can split a ring if there are two OH groups ortho to each other than if they are para (Fig. 9.10). If Cl is present, the position it occupies is important. Chapter 2 of Betts (1991) considers in detail the degradation of chlorine-containing aromatic compounds. Some enzymes can remove a Cl from a ring, replacing it with OH or with H, but others can split the ring before the Cl is removed. PCBs have the two-ring biphenyl framework shown in Fig. 9.10(c,d), with various numbers of Cl attached in various positions. What is used in industry is always a mixture of several different PCB compounds. In general the more Cl atoms in the molecule the slower the breakdown, and molecules with more than five are extremely resistant to breakdown. Degradation is faster if one of the two rings has no Cl, or failing that at least the ortho and meta positions of one of the rings are free. Figure 9.10 shows an example of a more-easily and a less-easily degraded PCB. However, the rate of breakdown of different organochlorines can respond differently to environmental conditions. Among aliphatics, compounds with a few chlorines per molecule, e.g. $H_2C{:}CCl_2$, are more readily degraded by aerobic organisms than by anaerobes; but those with more chlorines, e.g. $Cl_2C{:}CCl_2$, are degraded more rapidly under anaerobic conditions by reductive dechlorination (Lee *et al.* 1998).

Ways of promoting pollutant breakdown

If a toxic chemical released into the environment breaks down only very slowly or not at all, this may be due to either of two basic limitations.

1 It may be that there are no microbial species present that can metabolize the chemical. If so, we need to find a suitable species and inoculate with it.

2 Suitable species may be present, but in low numbers and activity because of unsuitable environmental conditions or lack of some requirement. This requirement might be oxygen, water, mineral nutrients or an organic substrate. If we can supply the requirement or modify the physical conditions, microbial breakdown of the chemical may become rapid without need for inoculation.

Oil breakdown . . .

An example of how breakdown can be speeded up by supplying a requirement for the microorganisms is in the treatment of crude oil spills. Oil pollution receives increased publicity after a major spill from a tanker accident, but much more oil is released each year from many small individual sources, such as leaks, industrial waste and natural seeps. Crude oil is a mixture of various aliphatic and aromatic hydrocarbons, some of which can be rapidly lost by evaporation. The less volatile components are attacked by a range of microbial species, including bacteria, streptomycetes, yeasts and fungi. These can be found sparsely in water of the open oceans and in coastal mud, so if there is an oil spill a natural inoculum of suitable microbes is normally present, which can then multiply. Ward *et al.* (1980) reported on the abundance of bacteria that could use hydrocarbons as substrate, in coastal muds in Brittany a year after the wreck of the tanker *Amoco Cadiz* in 1978 caused a very large oil spill there. At unoiled control sites there were a few hundred of these bacteria per gram of mud, but at heavily oil-polluted sites numbers ranged from tens of thousands to tens of millions per gram. Chaîneau *et al.* (1999) found that soil from agricultural land in France that had not been contaminated with oil contained 12 species of bacteria and seven species of fungi that could break down hydrocarbons. After oil waste was spread on the soil the number of oil-degrader bacteria increased greatly, though the number of species did not. Inoculation with oil-degrader species has sometimes been carried out at oil-polluted sites, but there is little evidence that it does speed up oil breakdown (Prince 1992). More useful is to try to promote the multiplication and metabolic activity of suitable microbial species already present.

. . . often suitable microbes are present

. . . but they need water, oxygen, mineral nutrients

Two common limitations to oil breakdown are the availability of water and oxygen, as microbial breakdown of the hydrocarbons under anaerobic conditions is very slow (Ward *et al.* 1980). Much of the oil may be out of contact with water and oxygen if it is forming a thick layer on the ocean surface, or is filling the pores in mud. Breakdown of thick oil films can be speeded up by promoting fairly stable emulsions of oil and aerated water (Hughes & McKenzie 1975). Mineral nutrient supply is another possible limiting factor. Following the large *Exxon Valdez* spill in 1989, oil spread over the shoreline in southern Alaska. Because the shores were coarse gravel and shingle, and were wetted by tides and waves, it was thought that water and oxygen would not be limiting; so remediation efforts concentrated on adding N + P fertilizer. To prevent it being quickly washed away it was applied as an oil-soluble liquid plus slow-release granules. Bragg *et al.* (1994) describe tests in 1990, a year after the spill, at three sites which had not been treated previously. It was estimated that 38–67% of the oil originally deposited had been lost during the year (without any treatment to speed up loss). This was partly by vaporization and washing away; the natural biodegradation rate was estimated to have been 4–46% per year. The NP fertilizers were then added to part of each site. At the site where breakdown had previously been slow, it was 6–9 times faster in the section where the fertilizer was added.

Adding another organic chemical

Sometimes a bacterium can degrade a synthetic chemical but does not obtain metabolic energy in the process. Therefore it may be necessary to add another organic chemical that it can use as an energy source. Sometimes this needs to be a chemical related to the pollutant. An example is the addition of biphenyl to soil, which can greatly enhance the breakdown of PCBs. (Biphenyl is the two-ring framework without any Cl attached; see Fig. 9.11.) Brunner *et al.* (1985) found that if a PCB was added to a soil only 2% was broken down in 10 weeks, but if biphenyl was added as well about 17% of the PCB was broken down. There was a lag of about 3 weeks before PCB breakdown became rapid, presumably while the degrader bacterial population multiplied.

In other situations an organic chemical unrelated to the pollutant may promote its breakdown. As mentioned earlier, some microorganisms can carry out reductive dehalogenation of organohalogens under anaerobic conditions. For this they require an electron donor substance, which can be a simple organic substrate such as lactate, acetate or ethanol (Lee *et al.* 1998). In one example (Semprini *et al.* 1992), carbon tetrachloride and some other chlorinated aliphatics were being injected down a well into a layer of sand and gravel 5 m down. Conditions there were anaerobic and little or no biodegradation was occurring. As an experiment, pulses of acetate were also put down the well. After several weeks the amounts of carbon tetrachloride and some of the other chlorinated compounds began to decline, and by 2 months had reached very low concentrations. The breakdown did not occur at the bottom of the well where acetate was being injected, but some distance from it. It may be that another microbial species was converting the acetate to another compound, which was then the actual substrate for the dechlorinator species.

In these examples, microorganisms capable of degrading the pollutant must have already been present, at least in small numbers, so that once their requirements were supplied they could multiply and speed up the degradation. In other situations this is evidently not the case, and degradation can only occur if a suitable species can be introduced. An example of this operating on a field scale is described by Lamar *et al.* (1994).

Inoculating with a degrader microbe

At a site where wood had been treated with preservative over a period of 40 years, soil had become mixed with a 'sludge' of creosote and pentachlorophenol. A lignin-degrading fungus, *Phanerochaete sordida*, was grown on a grain–sawdust mixture to which nutrient medium had been added, and this inoculum was mixed into the contaminated soil at the site. During 20 weeks following the inoculation pentachlorophenol decreased by 64%, whereas in uninoculated control areas there was little decrease. So the inoculation was evidently effective. Nevertheless, the decrease in pentachlorophenol was slower than had been obtained with the same fungus in earlier small-scale trials. This was probably due to much of the fungus being killed during the inoculation in the field, which involved mixing with the large volume of soil. It is quite common for a system for biodegradation of a particular chemical to be developed in the laboratory and to work well there, but to fail when tried in the real situation outdoors. Alexander (1994) discusses possible reasons for this. They include: the physical conditions are different (e.g. temperature lower); inadequate ability of the inoculated species to spread through the large volume of contaminated soil or water; predators (e.g. protozoa) which attack the inoculum.

If inoculation is required, clearly we first need to find and culture a microorganism capable of degrading the pollutant. This often involves inoculating soil or water, collected from a large number of sites, into a culture medium favourable for microbial growth but containing the pollutant chemical, and then measuring whether the chemical has been degraded. Sometimes a suitable species can be found at the polluted site. For example, a site in northern England where gas had formerly been made from coal was left heavily contaminated with coal tar (which consists mainly of polynuclear aromatic hydrocarbons) plus phenols; these were disappearing only very slowly (Bewley *et al.* 1989). It was thought that suitable degrader microbes might be present, but in small numbers. A medium consisting of coal tar from the site, mineral nutrients and surfactants was inoculated with soil from the site. Using this method several microorganisms were isolated and multiplied, some of which were more effective than others at breaking down the constituents of the tar. In a small-scale field trial it was found that when one of these was mixed into contaminated soil, with mineral nutrients and surfactant, coal tar constituents decreased over 8 weeks, whereas when only mineral nutrients and surfactant were mixed in there was no decrease. Finally, this species plus nutrients plus surfactant were used to treat the whole site, which was then virtually cleared of coal tar and phenols in 15 months. It may seem surprising that this was effective using a species that had been originally found on the site. Presumably multiplication of the degrader microbes from their initial population was extremely slow, unless this was carried out under favourable culture conditions.

Inoculation with a single degrader species may not fully solve the problem, because it may convert the pollutant chemical to another chemical which is still toxic. In the example described earlier of carbon tetrachloride in a sandy layer, as the CCl_4 decreased chloroform ($CHCl_3$) increased. Evidently microbes capable of breaking down chloroform were not present, or were not able to degrade it as fast as it was formed. One solution to

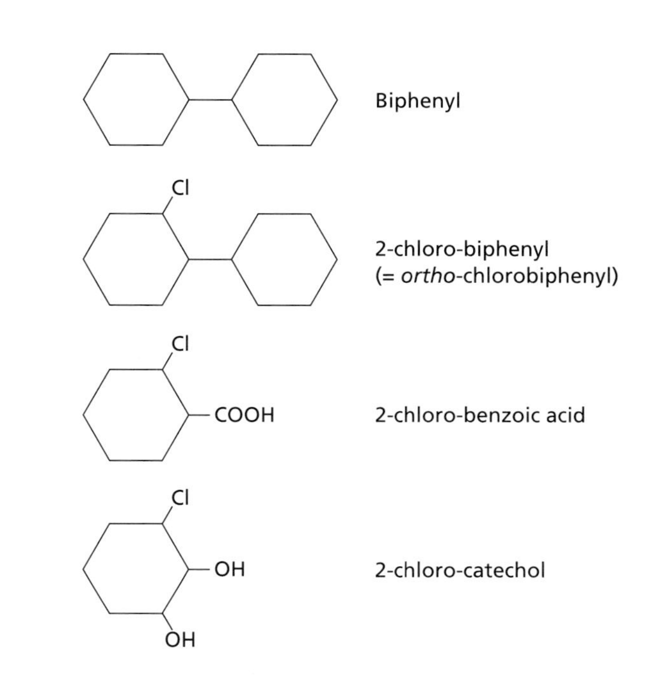

Fig. 9.11 Formulae of chemicals referred to in text, when describing the development of a bacterial strain that can degrade chlorobiphenyls.

Producing new, more effective microbes

such a problem involves inoculating with several microbial species, each of which degrades the product of the previous one in the chain, until harmless substances are produced. Alternatively, it may be possible to produce a new strain that incorporates two or more degradation steps within it, which previously only existed in different species. This may be achieved by natural genetic exchange, or by genetic engineering. As mentioned earlier, polychlorinated biphenyls (PCBs) are particularly slowly degraded if they have one or more Cls at the ortho (2-) position (see Fig. 9.10). Some bacteria can break the bond between the two benzene rings, to form chlorobenzoic acid or chlorobenzoates (Fig. 9.11), but these are still toxic and need to be further degraded. Havel and Reineke (1991) were able to combine genes for three stages of the degradation process into a single *Pseudomonas* species, by encouraging natural conjugation between strains in mixed cultures. Genes for chlorocatechol degradation were first combined with benzoate oxidation capabilities to give a strain that could degrade chlorobenzoate, and this was then combined with ability to use biphenyl as a carbon source. The result was a bacterium able to degrade Cl-biphenyls with the Cl in any position, to produce solely CO_2, H_2O and Cl^-.

Conclusions

- Various rapid, simple tests of whether a chemical is harmful to a species have been widely used.
- Using these tests to predict whether a chemical can be safely released

into the environment presents difficulties, which have so far been only partially overcome.

- These difficulties include:
  the relationship between short-term and long-term response varies between pollutants;
  species may differ substantially in their response to a chemical, and in practice only a few species can be tested;
  chemicals may interact;
  interactions between species may affect their response.
- The concentration of a pollutant in an organism depends on several factors, including:
  whether the chemical enters from the external solution or from food;
  its solubilities in water and fat;
  what mechanisms the organism has for getting rid of the chemical.
- Pollutants, organic or inorganic, are not consistently more concentrated at higher levels of food chains.
- We have some ability to predict, from its chemical structure, whether or not an organic chemical will be readily broken down by microorganisms.
- Breakdown of organic pollutants can sometimes be speeded up by supplying a requirement, e.g. inorganic nutrients or an organic substrate, for existing microbes. In other situations the inoculation of a microbial species from elsewhere has led to breakdown.

Further reading

Pollution, general:
Walker *et al.* (1996)
Harrison (1996)
McEldowney *et al.* (1993)

Air pollution:
Wellburn (1994)

Freshwater pollution:
Mason (1996)

Bioremediation:
Alexander (1994)

Chapter 10: Conservation and Management of Wild Species

Questions

- What should be our aims in conservation?
- Can we realistically aim that no species should ever become extinct? If not, how do we decide which species should have higher priority in conservation?
- If a large area of natural vegetation is reduced to small, separated fragments, how will this affect the plant and animal species living in it?
- Why can particular species not survive in habitat patches smaller than a certain size?
- Is ability to migrate between patches important? Can corridors promote it? What features make a corridor effective?
- Which will preserve more species, a few large habitat patches or many small ones?
- Can we alter conditions to promote high biodiversity? How?

Background science

- Keystone species in communities.
- Minimum viable population. How to predict it. Population viability analysis.
- Risks to small populations: genetic problems; fluctuations in population size.
- Metapopulations.
- Species–area relationships. Why do there tend to be more species the larger the area?
- Mechanisms that allow many species to coexist. What prevents a few species ousting all the others?

What does 'wild' mean?

The title of this chapter may seem to contain a contradiction: if we manage a species, how can it be wild? In this chapter I adopt a broad meaning for 'wild', to cover any species that we are not deliberately growing or keeping. So a species is not wild if we grow it on a farm to provide food or in a plantation for timber, if we plant it to decorate our gardens or keep it as a pet. However, birds and butterflies in suburban gardens are considered as wild, and so are plants and animals in a roadside verge or hedge.

'Semi-natural' is the term used to include areas of vegetation that have been much influenced by people. Examples are rough, unsown pasture-land, even if grazed by domestic animals; and mixed-species forests, even if cut down in the past and then left to regrow. Arable crops, sown single-species grassland and single-species tree plantations are not considered natural or semi-natural, though they may have wild species within them.

How to decide aims in conservation

'Conservation' implies trying to keep things the way they are. This chapter is about how to preserve landscapes, ecosystems, communities and species. The next chapter (Restoration) considers ways of re-establishing ecosystems in areas where they have been lost, and of reintroducing species.

A recurring theme of this book is that things cannot remain unchanged everywhere, because there is an increasing human population in a world of fixed size (Chapter 1). This is generating strong pressures to convert some areas of natural vegetation to other uses, such as farmland, forestry plantations, roads or towns. There are pressures to change the management of cropland, pasture and forest in ways that can increase production but may harm wild species (see Chapters 6, 7 and 8). And other by-products of human activity—such as climate change and pollution—will also influence wild species (see Chapters 2 and 9). It is therefore inevitable that there will be conflicts between wildlife conservation and other demands and desires of people. So we need to consider carefully what should be the aims and priorities for conservation.

Aims for conservation: preserve ecosystem processes . . .

Box 10.1 lists some aims that have been suggested for conservation. The first—preserve ecosystem processes—often has high priority. For example, cutting down forest and converting it to other vegetation can alter climate (see Chapter 3) and increase soil erosion (see Chapter 4), changes that we usually want to avoid. However, the ecosystem processes listed in Box 10.1 are likely to be primarily influenced by characteristics of the most abundant plant species, and much less by sparser plants and by animals (Grime 1998). Reduction of biodiversity is not always harmful to ecosystem processes. For example, the replacement of a mixed-species hardwood forest in North Carolina by a pine plantation reduced water run-off, and so presumably reduced the risk of serious soil erosion (Swank & Douglass 1974). Plantations of mahogany or pine in Puerto Rico had higher above-ground net primary productivity than nearby mixed-species regrowth forests of about the same age (Lugo 1992). Many people consider that preserving ecosystem processes is not in itself a sufficient aim for conservation (Simberloff 1998).

. . . maintain genetic diversity . . .

The second aim, maintaining genetic diversity, is also highly desirable. Within each species and population there is genetic diversity which is important to the species, as an insurance against future environmental change (e.g. climate change) and against sudden catastrophes (e.g. disease

Box 10.1. Some possible aims for conservation.

1 Preserve ecosystem processes
e.g. primary productivity, water balance, energy balance, nutrient cycling, slow soil erosion.
2 Maintain all the genetic diversity within living things.
3 Prevent any species from going extinct.
4 Preserve particular species.
(a) Flagship species. Species with strong public appeal.
(b) Umbrella species. Preserving one of these is likely to result in preservation of many others as well.
(c) Keystone species. It plays a key role in the community, so if it becomes extinct many others will too.
(d) Endangered species. Species considered to be in danger of becoming extinct soon, e.g. because they are rare or declining.

Further information: Simberloff (1998).

epidemic). It may also be important for people: a gene may carry a useful character which we can later insert into a farm crop or animal, or the ability to produce a useful chemical such as an antibiotic. However, it is not realistic to say that we must never destroy any genetic diversity. When you boil some water in a kettle you kill bacteria, one of which might, for all you know, contain a unique form of a gene. So the maintenance of genetic diversity should be one of our aims, but it cannot be the sole decider of what we conserve.

. . . prevent species from going extinct

The third aim—prevent any species from becoming extinct—has been supported by many people. According to this view, for people to make a species extinct is wrong, full stop. The preservation of any species is an absolute, which must take priority over anything else. This is approximately the attitude of the US Endangered Species Act 1973 (see Caughley & Gunn 1996; Schemske *et al.* 1994). A species is classed as endangered if it is in imminent danger of extinction throughout all or part of its range. The act requires that federal government agencies ensure that none of their activities, or activities they authorize or fund, threatens any endangered species. It also requires that a recovery plan be developed for each endangered species, setting out how to ensure its recovery. The basic argument for preserving every species is that each species is unique, and once it is lost it has gone for ever. However, there are practical difficulties, arising from the large number of species in the world. Table 10.1 shows figures for the number of known species in the world, in various groups of living things and in total. The figures for mammals and birds are probably not far from complete (to the nearest thousand), but for most invertebrate groups and for microorganisms there must be many more species yet to be discovered. The total number of insect species probably lies between 2 and 10 million (Gaston & Hudson 1994). The world total

Table 10.1 Approximate number of known species in major groups

Group	Thousands
Mammals	4
Birds	9
Reptiles	6
Amphibians	4
Teleost fishes	19
Insects	900
Molluscs	50
Nematodes	15
Angiosperms	270
Gymnosperms	1
Bryophytes	17
Fungi	70
Bacteria	4
Total*	1.5–1.8 million

* Includes other groups not listed above.

Figures from Holdgate (1991), May (1992).

of all species has been variously estimated to be between 3 and 30 million (May 1992). There are many habitats in which most of the species are still unknown. To take one example, a detailed study by Fenchel (1992) of the sandy sediment in a bay on the coast of Denmark enabled him to state that in a single core 10 cm deep × 1 cm^2 cross-sectional area there are typically 30 species of ciliates. In addition there are numerous species of flagellates, algae, bacteria and cyanobacteria, but many of them are not yet described and named. So how can we make a rule that no species must be made extinct when the existence of most of them is unknown to us, and we do not even know, to within a factor of 10, how many there are in the world?

Giving priority to particular species

This leads to the idea that we should give some species priority in conservation. Box 10.1 lists four possible criteria. The first suggests that we should concentrate on conserving species that have strong and widespread public appeal—*flagship species*. Because there will be widespread support for conserving areas where they live, many other wild species will also be saved. Examples of flagship species are the Florida panther, the northern spotted owl (see Chapter 7), the giant panda, the African elephant, and in New Zealand the takahe (a flightless bird). It seems that the species which command high concern among many people are, like us, warm-blooded and care for their young. People are concerned about dolphins but not about herring. In addition, butterflies are often actively conserved but most other insects are not. Among plants there is concern

Flagship species

for large trees and for plants with pretty flowers, but much less for other species. At the other end of the scale the extinction of the smallpox virus was considered an event to celebrate: there were, as far as I know, no protests from conservation organizations.

Finding out what people value

I can cite no references to back up the statements made in the previous paragraph. Is there any more rigorous basis for discovering what people value among wild species and natural communities? There have been numerous attempts to express this in monetary terms, and Box 10.2 summarizes the most-used methods. None is ideal. One conclusion is that many people attach value even to some species they have never seen (except perhaps on television), and would be willing to make some monetary sacrifice to ensure their continued existence. For example, Americans who were questioned said they would be willing to pay, on average, $40–64 for the continued existence of humpback whales, but only $27 for the continued existence of the Grand Canyon (Pearce & Moran 1994).

When deciding how to manage the countryside, one difficulty is that timescales are long—decisions taken now may affect what is there in a century's time—and that opinions change. Gardens provide visible evidence of changing attitudes. Famous formal gardens of Europe—such as Versailles (near Paris), the Generalife at Granada and the Boboli Garden

Box 10.2. Ways of putting a monetary value on wild species, natural communities, countryside.

1 Potential future value of products, e.g. medicines. Can be estimated from number of new products discovered per year × mean value per product. See Brown (1994).
2 How much people are willing to pay to hunt or fish.
3 Entry charges to national parks, country parks.
Limitation: entry to most countryside is free.
4 How much people pay to support conservation organizations.
5 Travel cost method. Find out how far people have travelled to reach the site, calculate their travel costs and value of their time spent travelling.
Limitations include: they may enjoy the journey, and it may enable them to visit more than one site.
6 House price method. Compare prices of similar houses close to and far from countryside.
Limitation: it may be difficult to exclude effects of other differences, e.g. cost of travel to work.
7 Willingness to pay, willingness to accept. Ask people how much they would be willing to pay to preserve a particular species or a piece of countryside; or how much they would require to be paid to allow it to be destroyed.
Limitation: the people know the payment will not really happen.

Further information: Turner (1993); Pearce & Moran (1994); Simberloff (1998).

in Florence—with their straight lines, enclosed spaces and vistas towards the palace, indicate owners who saw the garden as a safe haven from the dangers beyond, as an extension of the palace. This contrasts strongly with 18th-century English landscape gardens and parks—such as Blenheim Palace and Stourhead—whose designers saw the garden as an extension of the countryside. In AD 2100 what will people value most?

A serious limitation of using flagship species as the centre of conservation planning is that there is no guarantee that preserving them will preserve many other species. An example of this is the northern spotted owl in the forests of the US Pacific Northwest (Chapter 7). The plan for preserving forest patches was specifically geared to the needs of the owl, and was clearly not suited to other species, for example some fish and amphibians. This has led to the suggestion of giving priority to the preservation of *umbrella* species, which are chosen because in order to preserve them we shall also preserve suitable habitat for many other species. The difficulty is that we know too little about the ecology of most of the species in any community to have a sound basis for choosing umbrella species (Simberloff 1998). If we want to preserve a whole community it would be better to concentrate on that; or if we want to preserve as many species as possible, there are guidelines about how to manage for high biodiversity (see later).

Umbrella species

Keystone species

A third suggestion is to identify and conserve *keystone species*. A keystone is at the top of an arch: if it is removed the whole arch will fall down. If a keystone species disappears many other species will be affected and will perhaps disappear too. The converse is *redundant species*, whose disappearance has little or no effect on the remaining species. It has been suggested that keystone species should have high priority in conservation and redundant species low priority. One definition of a keystone species is 'a species whose impacts on its community or ecosystem are large, and much larger than would be expected from its abundance' (Power & Mills 1995). This distinguishes between keystones and dominants. The sudden disappearance of a dominant plant species, e.g. the principal tree species in a boreal conifer forest or the principal grass in a savanna, would probably have a major effect on many other species. However, the loss of such dominant species is unlikely and need not concern us here.

Aquatic keystone species

Early studies of keystone species particularly emphasized aquatic habitats. Pimm (1980) summarized 19 studies of communities of rocky shores and freshwater ponds in which one animal species, either herbivore or carnivore, had been removed experimentally or by a natural event. In 15 of these studies at least one other species disappeared; in some cases removal of the animal allowed a species it had formerly eaten to increase greatly and many other species disappeared. For example, in several experiments a sea urchin species was removed. A large alga it had fed on then increased greatly, and so did some species associated with it, whereas other algal species were outcompeted and disappeared.

Grazers as keystone species

On land the disappearance of a mammalian grazer species can have a major effect on the vegetation. Chapter 6 provides examples where the exclusion of cattle or sheep had a marked effect on the abundance of plant species, and led to the disappearance of some (see Tables 6.3, 6.4, Figs 6.9, 6.10). An example of the near-extinction of a naturalized grazer is the sudden killing of most rabbits in Britain by myxomatosis in the mid-1950s (see Chapter 8). Rabbits had been abundant in many lowland grasslands; after their near-disappearance many of these areas became colonized by shrubs and trees, and some of them are now forests. So, the loss of one species caused a major change in the species composition, plant and animal.

The elephant is an example of a species that is sometimes endangered and which can have a major effect on savanna vegetation: by destroying trees elephants can influence the balance between trees and grassland (Owen-Smith 1989). Elephant could certainly be classed as a keystone species, and a flagship species too.

The influence of a keystone species need not be through what it eats: for example, beavers influence many other species by building dams and so creating ponds.

Invertebrate grazers may also be keystone species. An example from biological control is the Klamath-weed/*Chrysolina* beetle story (see Chapter 8). Klamath-weed and *Chrysolina* are both sparse in California at present. If the beetle becomes extinct there Klamath-weed will presumably return to its previous status, extremely abundant in rangeland, reducing many other plant species and presumably their associated invertebrates. Chapter 8 gives more information on how a pathogen or herbivorous insect can control the abundance of its host. There may be many such associations in natural ecosystems, where inconspicuous species, perhaps of low abundance, are having major effects on more visible species.

Disappearance of carnivores

The disappearance of a carnivore species may be expected to lead to an increase in the herbivores it ate. This could in turn have a knock-on effect on the abundance of plant species. An example of this in North America concerns white-tailed deer, whose predators, including wolf and lynx, have disappeared from much of the deer's range. White-tailed deer have become much more abundant and widespread in North America during the last 100 years. We have no basis for accurate estimates of deer population numbers several thousand years ago, but it does seem likely that their present abundance in many forested areas of the northeastern USA is higher than it was then (Alverson *et al.* 1988; Whitney 1990). If the deer become too abundant their browsing on tree seedlings and saplings can be intense enough to prevent the regeneration of important tree species. Tilghman (1989) set up large fenced areas (13 or 26 ha) in four hardwood forests in northern Pennsylvania and placed white-tailed deer in them to give a range of densities. Figure 10.1 shows that after 5 years deer had reduced the number of tree seedlings, but had not eliminated them.

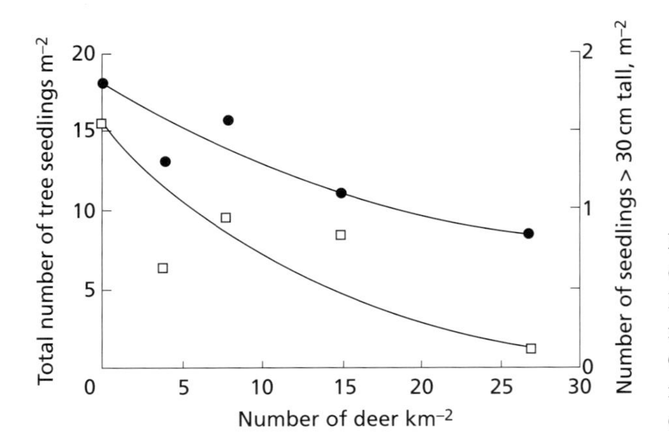

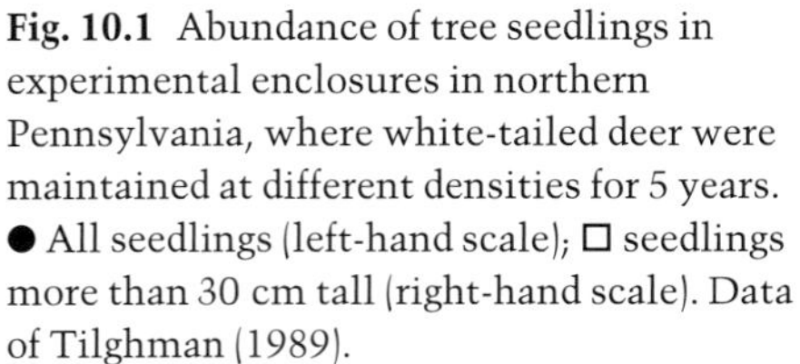
Fig. 10.1 Abundance of tree seedlings in experimental enclosures in northern Pennsylvania, where white-tailed deer were maintained at different densities for 5 years. ● All seedlings (left-hand scale); □ seedlings more than 30 cm tall (right-hand scale). Data of Tilghman (1989).

However, most of these seedlings were very small: if only seedlings more than 30 cm tall are counted, deer were reducing the numbers to a point where regeneration of trees could be affected. More important than the total numbers may be the species composition: in the highest deer density some major forest tree species, including two of the dominants in the canopy, sugar maple and red maple, had no seedlings at all more than 30 cm tall. In that treatment about half the sample plots were dominated by a single species, black cherry. So this experiment indicates that although deer over this range of density would not prevent tree regeneration altogether, they would probably prevent some species from regenerating and so alter the composition of the forest. Exclosures set up in other forests to keep out deer and other large grazers have confirmed this (Alverson *et al.* 1988). The highest deer density used by Tilghman is probably rare, but densities of 10–15 per km^2 are common.

Are any plants keystone species?

Although we now know of many examples of herbivorous and carnivorous animals that are keystone species, there are no convincing examples among plants. Some insects are specific to individual plant species (more on this later), but interactions between plants appear not to be specific: an entire complement of plant species does not need to be present for the ecosystem to function. When the North American chestnut virtually disappeared from the Appalachian forests during the first half of the 20th century (see Chapter 8), this did not result in the disappearance of any other plant species, as far as is known. As the world warmed at the end of the Ice Age, plant species did not migrate at the same rate nor even always in the same direction; so plant communities did not migrate as units but changed (see Chapter 2). This and other evidence indicates substantial redundancy in plant communities.

These examples show that there are keystone species among animals, vertebrate and invertebrate, herbivore and carnivore. This could be one of the criteria in deciding conservation priorities. However, it has not indicated priorities in the conservation of plants.

Species that are specially at risk

Another approach to setting priorities for conservation is to decide which species are in danger of becoming extinct soon and to concentrate our conservation activities on these, arguing that the remaining species can look after themselves. Lists of species at risk have been drawn up, notably the Red Lists and Red Data Books of the International Union for the Conservation of Nature (IUCN), now called the World Conservation Union. Species have been variously categorized as 'critical', 'threatened', 'endangered', 'vulnerable' and 'susceptible'. Often this is based on the species being rare, which can include how restricted its geographical range is, how many populations remain, and how large they are. Decline in abundance is another indicator of a species at risk (Mace 1994). The following section considers in some detail how we can decide whether a species is at risk of extinction because individual populations are too small, or there are too few populations, and what we can do to promote the survival of such rare species.

Conserving wild species in fragmented landscapes

How animals and plants respond to fragmentation of their habitat

Most landscapes contain much natural diversity and so appear patchy. There are many natural causes of vegetation being patchy, including hills and valleys, rivers and lakes, windthrow, fire, and many more, at various scales. This diversity may seem to us a valued feature of the landscape and it may well increase the number of species it can support, but for an individual species it may create problems, by dividing its habitat into fragments. Species have presumably adapted to natural fragmentation, but humans have greatly increased the fragmentation of habitats in some parts of the world. Over much of western Europe during many centuries the formerly extensive forest has been reduced to fragments set among farmland, roads and towns. Deforestation, leaving only fragments, has also occurred in parts of North America, India, New Zealand, West Africa, Malaysia and elsewhere (see Chapter 7). Grassland and other vegetation types have also been reduced to fragments in various parts of the tropical and temperate regions. Two key questions apply to any species inhabiting some of these fragments: are the remaining fragments sufficient to support it long term, and if not, is there anything we can do about it?

Fragmentation of forests

There are also very large areas of near-natural and semi-natural communities that remain unfragmented, for example boreal forest in Russia, tropical forest in Amazonia and Congo (Zaire), savanna woodland and grassland in Africa, semi-desert in Africa and Australia. For these a key question is: if they have to be partially destroyed in the future, what size of patches should we aim to leave, to be sure of conserving desired wild species? The ideal way to answer this question would be by experiment, and in 1979 a group of scientists set out to do just that for Amazonian tropical rainforest, by creating isolated fragments of different sizes and monitoring what happened to the wild species in them. Box 10.3 summarizes the experiment. Although it was not completed quite as originally planned, it has provided much interesting and useful information, some of which will be described later.

Interest in species in habitat fragments evolved from interest in islands. It has been widely observed that there tend to be more species on larger islands than on smaller ones. This applies to various groups of animal, vertebrate and invertebrate, and to plants (MacArthur & Wilson 1967; Connor & McCoy 1979). It also applies to habitat islands on land, in other words any region suitable for some species surrounded by area inhospitable to them. Figure 10.11 shows examples: birds and herbaceous plant species in woods surrounded by farmland. Later in this chapter I shall consider species diversity more directly; here I consider individual species in patches. The clear implication of Fig. 10.11 is that as a fragment is made smaller species are lost. So some species do not survive in small fragments. I first provide some examples of this, and then consider what the cause might be.

Birds in small woods

Moore and Hooper (1975) and Hinsley *et al.* (1996) analysed records, from numerous British woods (i.e. fragments of forest surrounded by arable and grassland), of which bird species were present during the breeding season. They found that some species were found even in small woods, others only in larger ones. Figure 10.2 shows fitted curves summarizing the relationship of five species to wood area. Blackbird and robin were present in many woods of less than 1 hectare; Moore and Hooper (1975) reported blackbird present even in woods less than 0.01 hectare (100 m^2), though robin needed at least 0.01 hectare. In contrast, marsh tit was absent from most woods less than 5 ha in size, an observation confirmed by Moore and Hooper (1975) and by Opdam *et al.* (1985) for woods in the Netherlands. The differences between the five species can be related to their foraging habits. Blackbird and robin often search for food on the ground: they eat earthworms, among other things, and blackbirds also eat slugs and snails. Great tits eat insects and seeds; they forage in hedges as well as woodland, but rarely on the ground. Treecreeper and marsh tit also eat insects and seeds, but confine their foraging much more to woodland. So blackbird, robin and great tit are making positive use of the open country around their nest area to provide alternative sources of food. However, small woods may not be ideal even for blackbirds. Møller (1991) studied nesting by birds in 32 woods, ranging from 0.07 to 3.6 ha, in farmland in Denmark. Blackbird suffered greater loss of eggs to predators the smaller the wood: the number of young successfully fledged ranged from less than one per nest in very small woods to about three in woods over 1 ha. Great tit also suffered some reduction in fledgling numbers in smaller woods. The other three species of Fig. 10.2 were not recorded by Møller.

These European studies are for areas where the landscape has for

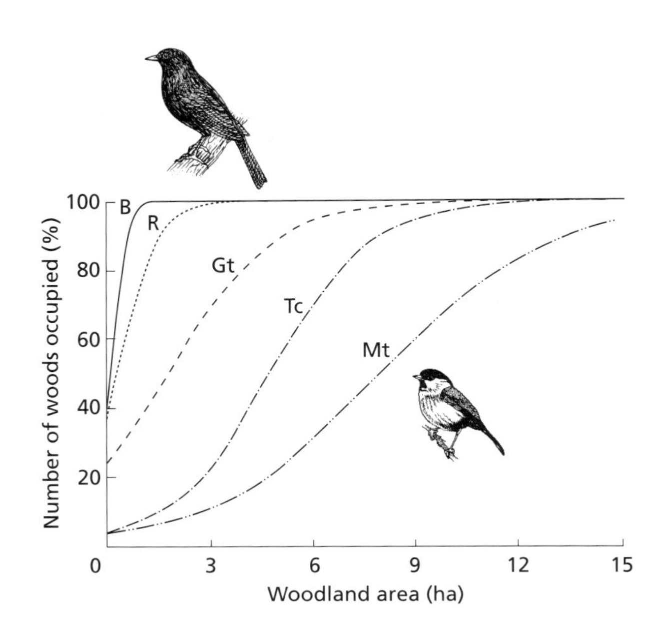

Fig. 10.2 Percentage of English woods in which five bird species occurred during the breeding season, in relation to wood area. B, blackbird (upper picture), R, robin, Gt, great tit, Tc, treecreeper, Mt, marsh tit (lower picture). From Hinsley *et al.* (1996)

centuries been fragments of forest within open land, and the distribution of birds we see today is a response to that. If, instead, we have an enormous extent of forest, broken only by natural disturbances, and then much of it is cut down, how do the birds respond? Bierregaard and Lovejoy (1989) and Stouffer and Bierregaard (1995) described what happened to birds when Amazonian rainforest was cut down and only fragments left, in the fragment experiment (Box 10.3). They studied 1-ha and 10-ha fragments, before, during and after isolation, catching birds in mist nets stretching 2 m above the ground. Birds of the tree canopy were not, therefore, recorded. While tree felling was going on the number of birds caught in the remaining forest increased, presumably because birds were displaced from the felled areas. Soon after isolation was completed overall numbers in 1-ha and 10-ha fragments fell below those in undisturbed forest. However, not all species responded in the same way: some disappeared altogether, some declined in abundance, some were little affected. Before felling there were mixed-species flocks, members of 10–20 insectivorous species foraging together. Each flock foraged over about 8–12 ha. One might therefore expect that they could not survive in a 1-ha fragment, but might survive in 10 ha. In fact, these flocks disintegrated in fragments of both sizes, many of the species disappearing altogether, although a few retained their abundance, foraging alone. In two 10-ha fragments around which forest was allowed to regrow, after 6 years flocks reassembled; by this time the regrowth forest was tall enough for them to make some use of it.

Birds left in rainforest fragments

In this forest before the felling there were army ants. These ants swarm along the ground in a column up to several metres wide, catching and eating other insects. Several species of bird follow them, eating insects that fly up to escape the ants, or dead bits of insect that the ants discard.

Box 10.3. The Amazonia Forest Fragments Experiment, officially called the Biological Dynamics of Forest Fragments Project.

The experiment was set up in central, lowland Amazonia, about 70 km north of Manaus, Brazil, where tropical rainforest was being cleared for cattle ranching. The original aim was to leave replicate patches of forest 1, 10, 100 and 1000 ha in area, surrounded by pastureland. The table gives information on fragments that were actually created.

Area (ha)	Number of fragments	When first isolated	Minimum distance to other forest (m)
1	5	1980–84	70–400
10	4	1980–84	70–650
100	2	1983, 1990	150

7–10 years after isolation, some fragments were surrounded by pasture, others by regrowth forest. Sites within remaining extensive forest were also monitored.

Groups recorded include:
Trees (saplings and mature)
Mammals
Birds
Amphibians
Butterflies
Bees
Beetles

Further information: Laurance & Bierregaard (1997); Lovejoy *et al.* (1986).

Army ant colonies are found about 1 per 30 ha, and when one colony becomes inactive army ant birds have to move between colonies. It is therefore not surprising that after the 1-ha and 10-ha fragments were isolated, the army ants quickly disappeared, together with their bird follower species. After the complete isolation of the first 100-ha fragment the ant-following birds declined and then disappeared (Lovejoy *et al.* 1986), so even this may not be enough to support them. This illustrates that fragmentation can have knock-on effects of from one species to another.

In the Amazonia fragment experiment the abundance of some bee species was measured by the number visiting chemical baits. Figure 10.3 shows that two of the species were much affected by the size of the forest fragment, and were extremely sparse in 1-ha fragments. In contrast, two other species were equally abundant in all fragment sizes.

Why can some species not survive in a small habitat fragment?

One reason why a species may no longer survive in a smaller patch is that its required microhabitat is not there. One possible explanation for there

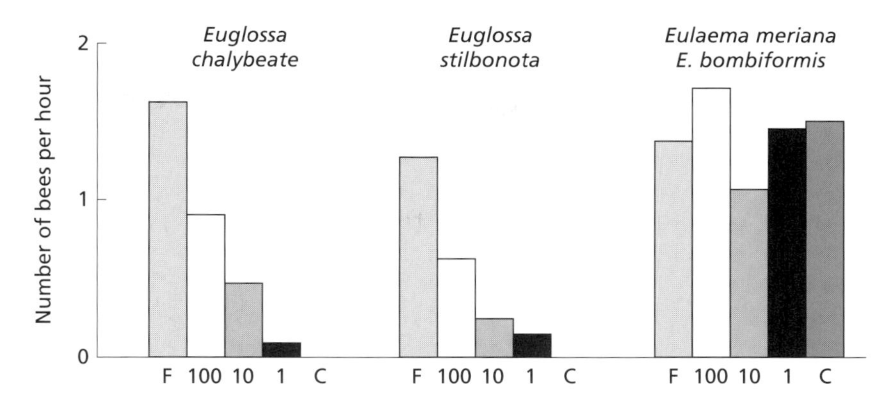

Fig. 10.3 Abundance of bees of named species in tropical rainforest area near Manaus, Brazil, assessed by number visiting baits. F, in large area of continuous forest; 100, 10, 1, areas (ha) of forest fragments; C, cleared area, recently deforested. From Lovejoy *et al.* (1986).

being fewer species in smaller fragments (e.g. Figure 10.11) is that there is probably a smaller range of habitat conditions, and fewer niches. We return to this topic later. Here we consider other possible reasons why a species cannot survive in a fragment smaller than a certain size. Box 10.4 summarizes three sorts of reason.

Edge effects

An example of the first category, edge effects, is provided by the response of tree species to the creation of the fragments in the Amazonia experiment (Box 10.3). Deaths of trees in the large family Myrtaceae were recorded during 7 years after the fragments were isolated. Figure 10.4(a)

Box 10.4. Some reasons why a species may require a certain minimum area of suitable habitat.

1 *Edge effects.* The species cannot survive near the edge of the fragment, e.g. because of different microclimate or because of species invading from other habitats. A small fragment may in effect be all edge.
2 The species *needs more than one site.*
(a) The species needs to migrate between different areas during each year, e.g. because food or water supply are not available in one area throughout the year.
(b) The species needs to inhabit different sites in different years, e.g. to survive different weather conditions.
3 Each individual, pair or family group needs an adequate *home range,* e.g. in which to obtain food. The fragment will need to provide enough area to support a *minimum viable population* of the species. If the population falls below that number, the risks to it may be:
genetic, from inbreeding and genetic drift; and/or
demographic, from fluctuations in population size which could in due course take it to extinction.

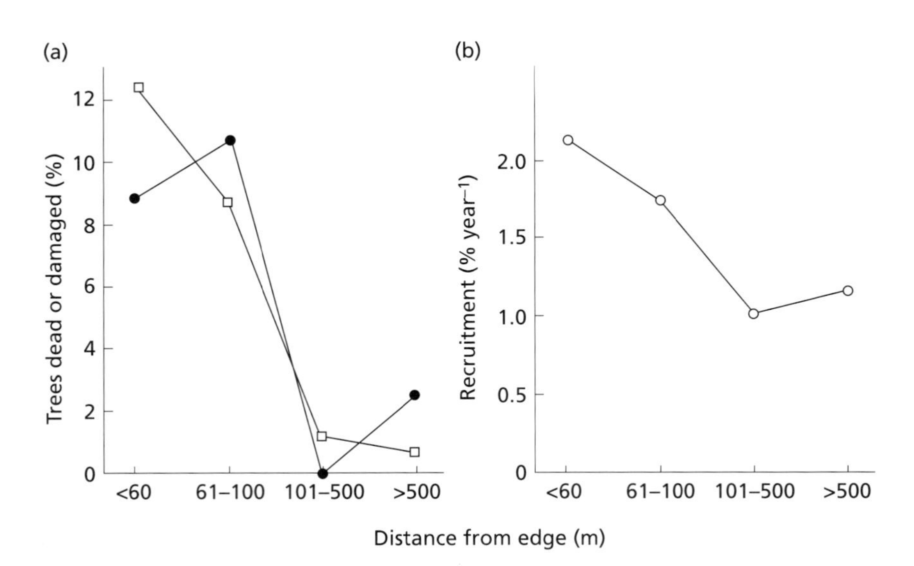

Fig. 10.4 Tree mortality and recruitment at different distances from edge of tropical forest. Results from the Amazonia Forest Fragment Experiment. (a) Deaths and injuries: ● percentage of trees standing dead; □ percentage of trees fallen or damaged; (b) recruitment, i.e. young trees that reach 10 cm diameter during the year. Mortality data for members of one family (Myrtaceae), recruitment for nine families. Data from Ferreira & Laurance (1997), Laurance *et al.* (1998).

shows that deaths were much more frequent within 100 m of an edge than further in. Once this was allowed for, there was little additional effect of fragment size on mortality. The extra deaths were probably related to wind and lower air humidity. Because of the deaths there were more canopy gaps near edges: even 180 m from an edge gaps were more abundant than they were 500 m from an edge (Laurance & Bierregaard 1997, Chapter 3). This is probably the main reason why recruitment of saplings to adult trees was faster near edges (Fig. 10.4(b)). However, a high proportion of the young saplings near edges were pioneer or successional species, characteristic of disturbed sites; there were fewer saplings of primary rainforest species (Laurance *et al.* 1998; Benitez-Malvido 1998). So the tree species composition within 100 m of an edge would change, with species of the primary rainforest being replaced by species more characteristic of disturbed sites. All of a 1-ha fragment, whatever its shape, is within 100 m of an edge; and a square 10-ha fragment will have only a square of 1.35 ha in the middle that is more than 100 m from an edge.

Another sort of edge effect in forests is loss of birds' eggs from nests by predation. In Michigan, Illinois and Sweden the losses were all found to be higher near the edge of a forest than further in (Gates & Gysel 1978; Andren & Angelstam 1988; Marini *et al.* 1995). This is most likely due to predatory mammals or birds coming in from the surrounding open land. The greater loss of blackbird eggs in smaller woods in Denmark (see earlier) may have been an edge effect.

Forest edges provide a sudden contrast in community structure and species composition, where edge effects are likely to be strong. Edge effects in other community types have been less studied, but do occur. An example was mentioned in Chapter 6: heather was more heavily grazed by sheep and deer if it was within a few metres of a grass patch. Hedges are a long, narrow habitat for some birds. Some bird species are more often found in broader hedges (Green *et al.* 1994), which could be a response to distance from the edge.

Some animals need to migrate

A second reason why some species need a large area is that they need access to different sites at different times. An example is provided by Serengeti National Park in Tanzania, and the adjoining Masai Mara reserve in Kenya, which between them form an area of about 200 × 250 km (Sinclair & Norton-Griffiths 1979; Hodgson & Illius 1996, Chapter 9). In the north and west, where the rainfall is higher, there are open savanna woodlands, but in the south and east the lower rainfall supports mainly grassland. There are many herbivorous mammals resident in the moister, wooded region, including giraffe, waterbuck and hippopotamus. However, wildebeest, zebra and Thomson's gazelle migrate from there each year to spend the rainy season in the southeast. In this way they can obtain forage not used by the non-migrants, and also avoid predators such as lion and cheetah for part of the year, as these do not migrate. One reason why wildebeest, zebra and Thomson's gazelle have to migrate back northwestwards at the start of the dry season is that the south-east has no permanent rivers, only temporary pools. If the Serengeti were to be broken up into small fragments, with no opportunity for migration between them, it is unlikely that these three species could survive. Some other large African wildlife parks may already be too small, in the sense that the migration routes of some of their large herbivores formerly extended outside the present park boundaries. Examples are wildebeest in Kruger National Park and elephant in Tsavo (Jordan *et al.* 1987, Chapter 23).

Another example, on a different scale, is the Bay checkerspot butterfly, which occurs only in patches of grassland on outcrops of serpentine rock southwest of San Francisco Bay in California (Harrison *et al.* 1988; Murphy *et al.* 1990). In this Mediterranean climate of mild, wet winters and hot, dry summers, the checkerspot's food plants senesce in late spring. The larvae of the checkerspot feed actively during spring, and then enter diapause. The larvae develop faster on south-facing slopes, and in years when the spring weather is cool and wet larvae from these slopes are the major contributors to the future population. However, in drier years the host plants on these south-facing slopes senesce before the larvae have reached diapause, so they die, and continuation of the butterfly population depends on there being north-facing slopes where the soil remains moist longer and the host plants senesce more slowly. Thus long-term survival of the species requires these contrasting topographical habitats, and this may be the reason why during three consecutive dry years,

1975–77, the species became extinct in small patches of serpentine grassland but survived in a larger one.

Area needed for foraging

Although most land-living animal species do not need to migrate, they all need an adequate area in which to obtain food. An example already mentioned is army ants in Amazonian rainforest; each colony inhabits about 30 ha. Carnivorous mammals and birds can require much more than this. The northern spotted owl has a territory per pair of 5–40 km^2 (see Chapter 7). In the United States the density of cougar in suitable habitat is 1–2 adults per 100 km^2, i.e. 100–200 km^2 for each breeding pair (Beier 1993). Such territory sizes place considerable demands on conservation organizations. Simberloff (1998) reported that, with some difficulty, enough money has been obtained to set aside 120 km^2 in Florida as a Panther National Wildlife Refuge. However, the average home range for the Florida panther is 300 km^2 for females and 550 km^2 for males.

Minimum viable populations

If the remaining habitat for the cougar or panther were reduced to one fragment just large enough to support one pair plus their young, would we feel confident that the species would survive? Today's conservationists would answer no, the species is probably at serious risk of extinction. So how large does a population need to be to have a high chance of surviving for a very long time? In other words, what is the *minimum viable population* for that species? For example, is the giant panda at risk of extinction soon? Fossil remains show that the species was formerly widespread in China, but today it is confined to isolated mountainous areas, where the habitat is further divided, e.g. by rivers and deforested areas. It is thought that some populations have fewer than 20 adults, and the largest is probably less than 200 (O'Brien & Knight 1987). Are panda populations of this size in danger of dying out just because they are too small? There are least 30 species of bird in the world each of which consists of one localized population of 50 individuals or fewer (Remmert 1994). Can they survive?

Direct determination of minimum viable population

The most direct way to determine the minimum viable population of a species is to count the numbers in many separate populations, and to keep records over many years to see whether those populations persist. Such data are available for only a few species. One of them is bighorn sheep in the southwestern USA. This species lives as isolated populations in open vegetation on mountains, so numbers can be counted, at least approximately, from a distance. Berger (1990) collated data extending up to 70 years for 129 populations. Figure 10.5 shows that all populations that had 50 animals or fewer when first recorded had become extinct within 50 years. Populations initially 51–100 survived better, and most populations larger than 100 remained. So the minimum viable population for this species appears to be about 100.

The number of breeding pairs of each bird species were recorded each

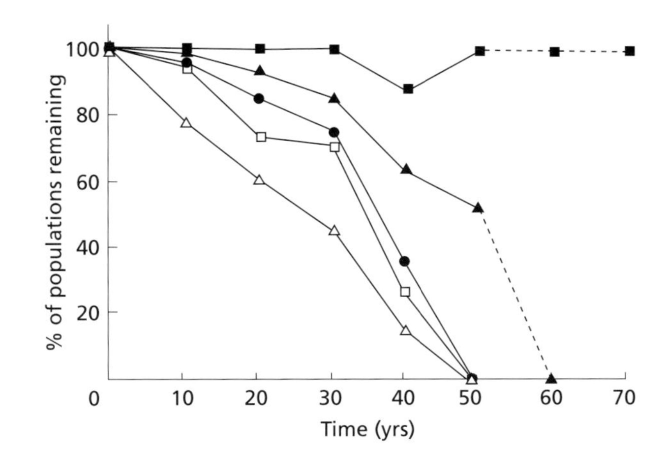

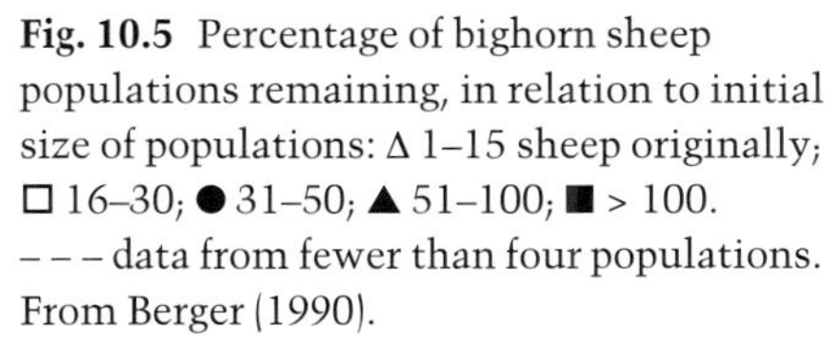
Fig. 10.5 Percentage of bighorn sheep populations remaining, in relation to initial size of populations: Δ 1–15 sheep originally; □ 16–30; ● 31–50; ▲ 51–100; ■ > 100. – – – data from fewer than four populations. From Berger (1990).

year over several decades on 16 islands off the coast of Britain and Ireland (Pimm *et al.* 1988); there were 100 species in all. Many populations smaller than 10 pairs became extinct during the study period, but species with a mean number of nesting pairs per island of 18 or more did not become extinct on any island. These results could be taken to indicate a minimum viable population of about 10–20 pairs for many species. But the recording period was too short to demonstrate continued existence over centuries. Many of the islands were only a few kilometres from other land, and so migration could have helped to maintain populations.

Often a species whose minimum viable population we wish to know has few remaining populations, which may be difficult to count, and we want the answer soon. If we spend decades doing the research the species may go extinct before we start conservation measures. So determining the minimum viable population by direct observation, as used for bighorn sheep, is not appropriate. If we are to estimate minimum viable population size more rapidly we need a basic understanding of what puts small populations at risk. There are two sorts of risk, *genetic* and *demographic* (Box 10.4).

Risks to small populations

Genetic risks Genetic diversity in a population is normally an advantage: a genetically uniform population is at greater risk of being wiped out, for example because it lacks individuals resistant to a disease or tolerant of unusual weather conditions. In all populations there is a tendency for genetic diversity to be lost by *genetic drift*—in one generation the eggs and sperms that happened to meet did not include a particular allele, and so it was lost. Such chance losses are more likely in small populations. The maintenance of genetic diversity depends on genetic drift being balanced by mutation creating new alleles and selection spreading them.

Genetic drift

But in small populations the faster genetic drift will reduce genetic diversity. Among 10 studies of plant species that have isolated populations, in seven of them genetic variation was greater in large populations than in small, although in three species it was not (Ellstrand & Elam 1993). It has been estimated that for quantitative polygenic characters a population of 500 is large enough to maintain high genetic diversity, but for single-locus genes the population would need to be several orders of magnitude larger (Soulé 1987 Chapter 6; Amos & Hoelzel 1992).

Inbreeding depression

Another potential genetic problem for small populations is *inbreeding depression*: if close relatives interbreed their offspring may be less fit. Any population is likely to carry harmful recessive alleles, which are only expressed in individuals that are homozygous for them. In a small population the chance of interbreeding between closely related individuals is increased and hence so is the chance of expression of these harmful characters. There is evidence for the occurrence of inbreeding depression in small populations, of animal and plant species, in the wild (Frankham 1995). It has been estimated that most inbreeding depression can be avoided if the population size is more than a few dozen (Soulé 1987, Chapter 6).

Effective population size

In the two preceding paragraphs 'population size' should in fact refer to *effective population size*. One definition of this is 'the average number of individuals with equal reproductive contribution to the succeeding generation'. It is clear that the effective population size includes only those individuals that are physiologically capable of reproduction: some may be too young or too old. However, in some animal species there are often individuals which are physiologically capable of reproducing but do not do so (in a given year) for social or behavioural reasons (Clemmons & Buchholz 1997). For example, in some mammal species in a particular year some males mate with several females, whereas others mate with none. Examples are elephants and some deer species (Poole 1997; Putman 1988). Monogamy often results in a smaller effective population size than does promiscuity: if the number of males and females is not equal, some adults will have no partner. In a small population the proportion of lone adults can be substantial.

Genetic bottleneck

If the population declines temporarily this will cause a *genetic bottleneck*: the small population will carry little genetic diversity, and after the population has recovered in size its genetic diversity will remain low for a long time. An example is the lions of the Ngorongoro Crater in Tanzania (Packer *et al.* 1991). These are effectively isolated from the nearby Serengeti: migrations in and out of the crater occur rarely. In 1962 the crater population crashed because of an outbreak of biting flies. By the late 1970s the population had recovered to about 50 adults, but they were all descended from seven females and eight males, and the genetic diversity was substantially less than among Serengeti lions.

In very species-rich plant communities such as tropical rainforest, individuals of the same plant species can be far apart, which could limit their gene exchange. When Hubbell and Foster (1986a) identified every woody plant (with stem diameter 1 cm or more) in a 50-ha plot in tropical forest on Barro Colorado Island in Panama, they found that 21 species were present only as one individual. In a 50-ha plot in Malaysian forest, also, some tree species were present as a single individual (Condit 1995). Clearly, if either of these 50-ha areas of forest became isolated by felling of the surrounding forest, there would be a serious genetic risk to these rare species. Even in large areas of forest we may wonder what is the effective population for such widely spaced trees—how often does pollen get from one to another?

We have, then, three different recommendations for minimum effective population size to avoid genetic risks: a few dozen (to avoid inbreeding depression), a few hundred and a few hundred thousand (to avoid loss by genetic drift of polygenic and single-gene characters, respectively). These wide differences may explain why conservation biologists, despite emphasizing the importance of maintaining genetic diversity, have tended to base estimates of the minimum viable population for particular species on demographic risks.

Demographic risks Demographic risk occurs because the numbers fluctuate from year to year. Each year (or short period) some individuals die and some are born. On average births and deaths may be equal, but this is unlikely to be true each year: chance will dictate that in some years deaths will exceed births, and the question is, what is the chance that one year all the remaining individuals (or all those of one sex) will die? So, we are trying to predict the probability that this population will survive 100 years (or 1000 years). This is *population viability analysis*, which is clearly closely related to the idea of a minimum viable population but is answering a more precise question. To answer this question requires a *stochastic* model, one which takes into account chance events and gives an answer in terms of probabilities. This is fundamentally different from the more familiar *deterministic* models, which do not allow chance events and give a definite answer with no uncertainty. Figure 5.4(b), for example, shows predictions by a deterministic model: for a given number of fish this year there is a certain amount of increase next year, with no uncertainty attached.

Population viability analysis

Cougar in southern California

A stochastic model was used to predict how a population of cougar (mountain lion) would respond to loss of part of its habitat area (Beier 1993). An area of 2070 km^2 in the Santa Ana Mountains, southeast of Los Angeles, is inhabited by a population of cougar, thought to number about 20 adults. Some of this area is likely to be developed for housing and other uses that would exclude cougar. However, 1114 km^2 forms a block protected from development. We want to know whether, if the suitable habitat were reduced to 1114 km^2, would cougar survive there? (Cougar is widespread elsewhere in USA, so the species is not at risk.) The model presented by Beier (1993) introduced chance variation by treating each

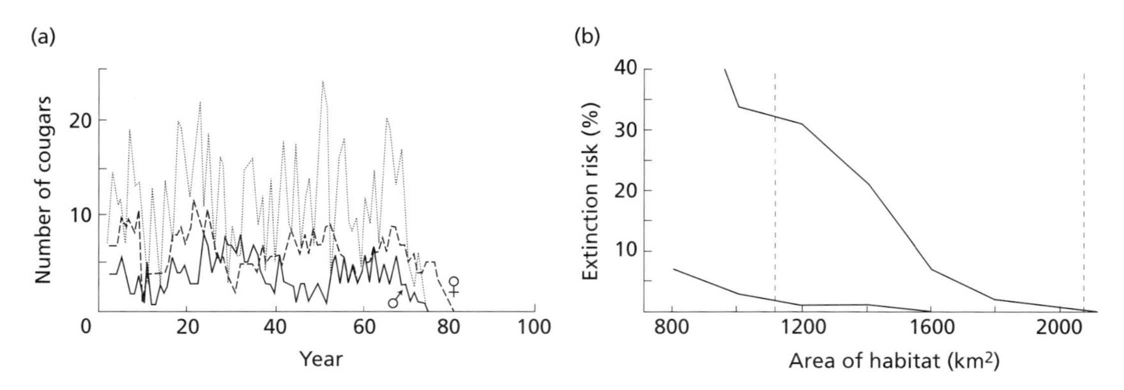

Fig. 10.6 Prediction by a stochastic model of the numbers and survival of a cougar population in an area of southern California. (a) A single run: predicted numbers of male adults (——), female adults (- - - -) and cubs (····). (b) Upper continuous line: extinction risk in habitat of differing total area, if no immigration from outside the area. Lower line: same, except immigration of four adults per decade. Vertical dashed lines: right, present total area; left, area not at risk of development. From Beier (1993).

individual separately. For example, if the mortality risk for adults is 20% per year and the number of female adults is 10, we should expect that on average two would die each year. The model in each year considered each adult in turn and, using a random number generator, decided whether that individual died in that year or not. So although two deaths are the most likely, 0 or 1 or 3 are possible, or even all 10. A similar stochastic method was used to determine how many juveniles died in each year, how many females produced a litter, the size of the litter, and how many of the newborn were female, how many male. Figure 10.6(a) shows an example run. The model population fluctuated widely in numbers from year to year, but survived for 75 years. Then all the males died, so there were no more cubs; and inevitably the remaining females died in due course.

Numerous replicate runs, similar to Fig. 10.6(a), were carried out with a particular set of input values—adult mortality risk, juvenile mortality risk, maximum number of adults that could be supported by 100 km^2, etc. One can then determine extinction risk by the percentage of replicate runs that predicts the population becoming extinct within 100 years. This can then be repeated with alternative input values. The top line of Fig. 10.6(b) shows predictions for different areas of remaining habitat, which determines the maximum size of the population. For the present total area (right-hand dashed line) the risk of extinction within 100 years is low, but not zero. However, as soon as the area begins to fall below 1800 km^2 the extinction risk rises sharply, and the model predicts that if all land except the protected part were developed the cougar population has about one-third chance of dying out within 100 years. Runs using alternative likely values for average mortality and carrying capacity reached broadly similar conclusions. So this model produced a clear prediction about likely effects of loss of habitat on a particular wild species. Whether nearly 1000 km^2 should be barred from human use for the sake of 20 wild cats is a matter for debate at enquiries and decision by planning committees; this model provided relevant evidence.

Close to the eastern corner of this area there is another large habitat area for cougar, separated by a few kilometres of housing and golf course, and crossed by highways. Figure 10.6(b), lower line, shows that if this intervening area could be modified to allow the occasional immigration of cougar the risk of extinction could be greatly reduced. Wildlife corridors will be discussed later.

Risk from a catastrophe

We may need to consider the possibility of some catastrophe (e.g. disease epidemic, exceptional weather) greatly increasing the mortality risk in one year. The cougar model was also run including a 20% or 40% reduction of carrying capacity for 3 years every 25 years, but it had little effect on the results. The difficulty is to know what catastrophe is realistic—how extreme and how often? The population of Capricorn silvereye (a bird) on Heron Island, a 17-ha island in the Great Barrier Reef, Australia, was recorded over 26 years, and varied between 225 and 445. Population viability analysis predicted a 15% chance of the species becoming extinct on the island within 100 years. However, if the birds' mortality increased by only 5%, the model predicted a 97% chance of extinction within 100 years (Brook & Kikkawa 1998). This suggests that the arrival of a new disease or predator on Heron Island could be very serious for this species. Armbruster and Lande (1993) presented a model for African elephant which included mild, medium and severe droughts; these increased elephant mortality to different extents. It predicted that if mild, medium or severe droughts occur every 10, 50 or 250 years, respectively, the chance of an elephant population in a 500 square mile (1294 km^2) area of suitable habitat surviving for 1000 years is 99%, whereas if the droughts occur every 10, 25 or 125 years the survival probability is only 57%. Although there is some information about dry periods in Africa during the last three centuries (see Chapter 3), there are no long-term weather records for tropical Africa that could show how often individual dry years occur, on a timescale of 50 or 100 years. This illustrates how difficult it can be for conservationists to allow adequately for occasional catastrophes. Conservation needs to have a long timescale.

Grizzly bears in Yellowstone National Park

An alternative type of stochastic demographic model involves providing as an input information on how much the existing, real population varies from year to year. An example is the model used by Dennis *et al.* (1991) to predict whether the Yellowstone population of grizzly bears is at risk of extinction. The population of about 100 adult grizzlies inhabits about 20 000 km^2 of Yellowstone National Park and surrounding countryside, separated by more than 200 km from other grizzly populations further north. Figure 10.7 shows the estimated number of adult females each year from 1959 to 1987. During the 1960s and 1970s there seemed to

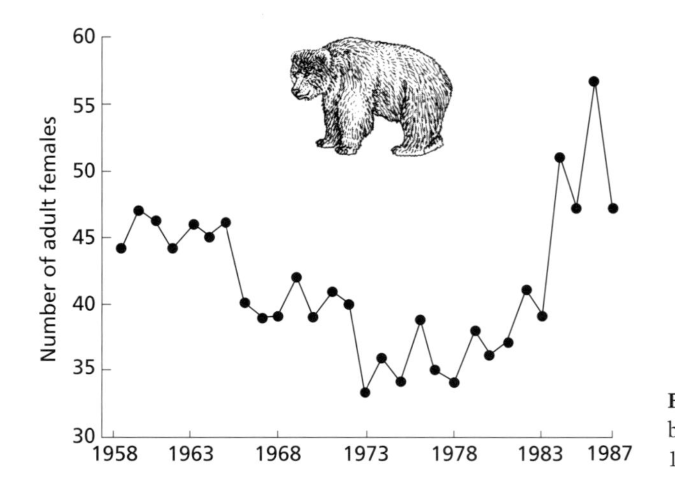

Fig. 10.7 Number of adult female grizzly bears in Yellowstone National Park, 1959–87. From Dennis *et al.* (1991).

be a downward trend in numbers, and people were worried that this might continue. However, during the 1980s numbers rose, which might suggest that there is nothing to worry about. Dennis *et al.* (1991) presented a model which makes use of the fluctuations shown in Fig. 10.7 to predict how likely it is that the number of females will one year in the future reach a very low level. Their model predicted that there is 100% chance of there being only one female at some time during the next 2280 years, with median time to that event being 448 years. So their model predicts that the population is at serious risk of extinction, though not necessarily soon.

A limitation of this type of stochastic model is that the information on how much the population fluctuates may not be representative. The data in Fig. 10.7 are for only 28 years, and may have missed occasional catastrophes. (However, there was no major change in grizzly numbers following the great Yellowstone fire of 1988 (Mattson 1997)). Also, any errors in the determination of numbers can have a serious effect on the year-to-year variability fed into the model. Grizzly bears are not easy to count: they are widespread, live mostly in forests, and tend to avoid humans. Some of the apparent fluctuations in Fig. 10.7 could be due to differences in search efforts, or to chance failure to see some individuals (Mattson 1997).

Species turnover

Models such as those described here often predict that, over a century or more, there is not 100% certainty that the population will survive, i.e. there is some risk of its extinction. This leads to the idea that within a single island or habitat fragment we should expect species *turnover*—some will disappear, others will arrive. Turnover has been reported for various groups of living things, plant and animal; however, it is difficult to be sure that it has really happened. If I say 'Species A used to be on this island, now it has disappeared', you may reply 'Are you *really* sure it is not there?' For example, Diamond (1969) published figures for turnover of bird species on nine islands off the coast of California. These were criticized by Lynch and Johnson (1974) as being based on inadequate recording. Jones and Diamond (1976) replied with further records, from which they concluded that species turnover did occur on all the islands, at rates ranging from 0.5 to 5% per year.

Metapopulations

If we expect a species to go extinct from a habitat fragment sooner or later, then the long-term survival of that species in the world will depend on its being present in several or many fragments. After it goes extinct from one fragment it will later be able to reinvade from another. So the species is a *metapopulation*, meaning a set of local populations which interact via individuals moving between them.

Box 10.5 sets out a very simple model of a metapopulation. It shows that the whole metapopulation can be stable even though populations in individual patches are from time to time becoming extinct, so that at any particular time some of the patches are not inhabited. If each patch is a nature reserve, each with a different manager, we can imagine that

Box 10.5. A simple model of a metapopulation.

The original model of Levins, as summarized by Husband and Barrett (1996).

Patches of habitat suitable for the species are distributed across a landscape; it cannot live in the intervening areas but individuals sometimes migrate across from one patch to another.

p = proportion of patches occupied by the species at a given moment
e = probability that the species becomes extinct in an occupied patch during a year ('extinction factor')
m = probability that the species colonizes a vacant patch ('mobility factor')

e is likely to be related to the size of individual populations, hence to patch area.
m is likely to be related to distances between patches.

The number of vacant patches colonized in a year will depend on how many are vacant, and also on how many are occupied because they provide the colonizers. Hence

$$dp/dt = mp(1-p)-ep \quad (10.1)$$

At equilibrium

$$dp/dt = 0$$

Hence

$$m(1-p) = e \quad (10.2)$$

$$p = 1-e/m \quad (10.3)$$

Extinction of the species throughout the landscape occurs when $p = 0$, i.e. when $e \geq m$.

manager being upset if a cherished species disappears from his or her reserve. Yet this may be part of the normal behaviour of the species, and provided there is adequate migration between reserves the species will in due course recolonize. The model shows that all the patches, even those currently unoccupied, may be important for the species. Suppose a new road is proposed which will destroy some of the patches. The planners may argue that there are other, similar habitat patches nearby, so the species is not at risk. However, removing some patches will presumably increase the distance between the remaining patches, thereby reducing m (the mobility factor). This may merely reduce the proportion of the remaining patches that are occupied at one time, but if m falls below e (the extinction factor) the species is predicted to become extinct altogether. This might take a long time to happen, while one by one populations disappear from patches and too few recolonizations occur.

Metapopulation dynamics has a basic similarity to disease spread in animals (Box 8.3). Each host animal is like a patch of habitat for the disease organism, whose survival depends on how fast it spreads between host individuals and how long it can remain in each. Just as the density of host animals has a critical threshold below which the disease dies out, so the density of habitat patches has a critical threshold below which the inhabiting species dies out.

The model in Box 10.5 assumes that the timing of extinction in each patch is independent; it does not allow for a broad-scale catastrophe. Nor does it consider genetics. The amount of gene flow between populations may be important (Ellstrand & Elam 1993). From the practical point of view, for conservation planners, the model is deficient because it does not relate e to patch size nor m to distance between patches. Models that do this have been developed (e.g. Hanski 1994).

A metapopulation: a butterfly in southeast England

An example of metapopulation dynamics is provided by a butterfly, the silver-spotted skipper, in the North and South Downs, southeast England. The species occurs only on close-grazed grassland on chalk soils. These nowadays occur on the Downs as many separate patches. This butterfly declined following the great decrease in rabbits in Britain in the 1950s, when many grassy areas grew tall, but it has subsequently recovered. Thomas and Jones (1993) recorded, for numerous patches of apparently suitable chalk grassland, whether it was present in 1982 and again in 1991. In some patches it was present in both years, some in neither year, some in 1982 only, and some in 1991 only. So turnover was occurring. Figure 10.8 shows which patches were occupied by the skipper in 1991. As metapopulation theory predicts, both the size of the patch and its distance from other patches are important. The species was rarely found in patches less than 0.05 ha in area or more than 1 km from another populated patch, but there was some trade-off between area and distance.

The plans for conserving the northern spotted owl in old-growth forests of the Pacific Northwest (see Chapter 7) viewed it as having metapopulations, so the fragments of forest left unfelled need to be close

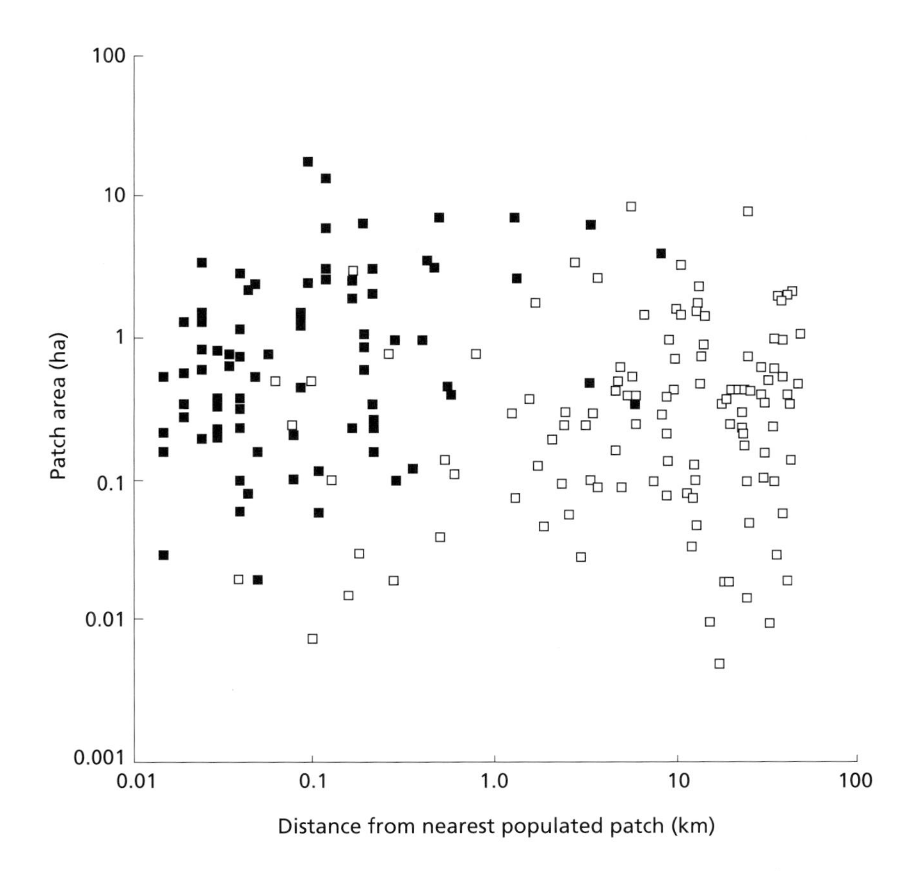

Fig. 10.8 Each point refers to a patch of close-grazed chalk grassland in southeast England, in 1991. Solid symbols: silver-spotted skipper butterfly present. Open symbols: silver-spotted skipper absent. From Thomas & Jones (1993).

enough to allow young birds to spread from one to another. An example of a mammal metapopulation is dormice in English woods, described later.

A Californian butterfly survived only in a large patch

Some species, rather than having a classic metapopulation, have a 'mainland and islands' system, that is, a large habitat patch where the population is large enough to survive indefinitely, and a number of smaller patches where the species becomes extinct from time to time, after which it recolonizes from the 'mainland'. An example is the Bay checkerspot butterfly, which, as described earlier, inhabits patches of serpentine grassland near San Francisco Bay. Following a drought period in 1975–77 the butterfly was still present in Morgan Hill, an area of suitable grassland about 10 km long × 1–2 km wide, but absent from many smaller patches nearby. Harrison *et al.* (1988) identified 59 serpentine outcrops within 21 km of Morgan Hill, all of which carried the chief food plant of the checkerspot and which seemed suitable for it in other respects. In 1987 nine of these sites contained populations of the butterfly. These nine varied in size and other characteristics, but all were

4.4 km or less from Morgan Hill. This strongly suggests that reinvasion from the large Morgan Hill area was the main factor determining how many sites it occupied. If it takes about a decade for the species to migrate that far, there is serious risk of another major drought before it has reached all the sites.

Migration between patches

It is thus clear that for many species the ability to move between habitat fragments is important. So we may need to know how far a species will travel between fragments: what distance would be a complete barrier. There has been research on how wide a gap forest birds and mammals are willing to cross. Desrochers and Hannon (1997) found that the willingness of some small forest birds to cross clear-felled areas or fields between forests in southern Quebec decreased with increasing gap width until scarcely any crossed a 100-m gap. A gap of a few hundred metres can be enough to prevent some British butterfly species from colonizing suitable habitat (Thomas 1991).

Slow spread of a plant species between woods

Figure 10.9 provides information on the ability of a British herbaceous plant species to spread between fragments of deciduous forest (woodland) separated by farmland. The species, dog's mercury, has seeds that are neither fleshy nor windborne, so it would not be expected to spread rapidly. The graph shows its presence in forest fragments that originated about 100–150 years earlier. Evidently during that time dog's mercury can spread a few hundred metres across farmland, but rarely achieves as much as 1 km.

Knowing whether objects such as roads provide a barrier to movement can also be important. Oxley *et al.* (1974) compared eight roads in southern Ontario, varying in width from 11 to 137 m (forest-to-forest); they recorded how often rabbit, hare, squirrel, chipmunk and woodchuck were willing to cross. There were many crossings of gravel roads 11 and 15 m wide, few crossings of roads 27 m wide or more, and none of dual carriageways.

Wildlife corridors

It has frequently been suggested that wildlife corridors connecting habitat patches may increase the ability of animals and plants to move between them, and hence allow metapopulations to survive. This was predicted for the cougar in southern California (Fig. 10.6). If this is so, it would be a reason for leaving corridors in the landscape, and also perhaps for creating new corridors. Potential corridors already exist in many open landscapes, for example hedges, the verges of motorways and railway lines, river banks. Where a large expanse of forest is being felled for the first time it may be possible to leave unfelled strips as corridors. In some parts of Australia strips of forest have been left beside roads. We need to know whether corridors can in fact be effective in promoting migration and the survival of species, and if so what are the features of effective corridors.

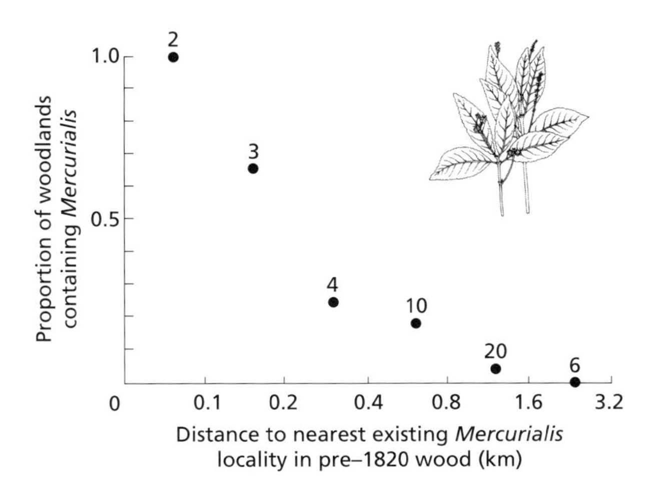

Fig. 10.9 Data on the occurrence of the perennial forb dog's mercury (*Mercurialis perennis*) in small woods in Lincolnshire, England that originated between 1820 and 1887. The woods are grouped according to how far each is from another wood already present in 1820 which contains dog's mercury. The figure above each point shows the number of woods in that class. From Peterken & Game (1981).

There have been many studies of which species occur in potential corridors. For example, Green *et al.* (1994) showed how the occurrence of various species of bird in English hedges was related to characteristics of the hedge, such as height, width, number of trees. However, the presence of a species in a long, narrow habitat does not prove it is moving along it: it may be living there. We need evidence that long features in the landscape are actually promoting movement.

A vole spread along road verges

A convincing example of migration of a small mammal along corridors through farmland involved a vole in rough grassland in road verges. Up to 1970 the southern limit of *Microtus pennsylvanicus* in central Illinois was about 40 km north of Champaign-Urbana. In the late 1960s and early 1970s interstate highways (motorways) were built in the area. Trapping in 1976 (Getz *et al.* 1978) showed that the vole had extended its range about 90 km further south. It was found not only in the verges of the interstate highways, but up to several kilometres away, in other rough grassland, e.g. beside old roads and railways, provided it was connected to an interstate. However, voles were not found in similar grassland if it was unconnected to the interstate system. Interstate highways do not pass through towns, whereas other roads and railways do, which could explain why interstates are more effective corridors.

Corridors could also allow the migration of plants, but I do not know of a clear example of this. Dog's mercury, which migrates slowly between British forest fragments (Fig. 10.9) sometimes occurs in hedges but seems to spread along them only slowly. At one site it had spread only 25 m along a hedge that was about 150 years old; two other woodland floor species that were present at the end of the hedge had not spread along it at all (Pollard 1973).

How wide should a corridor be?

If corridors are to be intentionally left in the landscape, or created, we want to know what characteristics are needed to make them effective, e.g. how wide; what vegetation; would a gap make it ineffective? The answers may differ between species. An experiment to determine the optimum width of corridor for a vole, *Microtus oeconomus*, living in tall grassland in Norway, was carried out by Andreasen *et al.* (1996). They killed meadow vegetation with herbicide to leave corridors of tall vegetation 310 m long and 0.4, 1 or 3 m wide, connected to larger areas of tall vegetation at each end. Radio-collared male voles were released at one end. Since 310 m is much more than the width of their home range, much of the recorded travel was not just foraging but seeking new territory. One might suppose that the wider the corridor the better, but in fact the mean distance achieved was greatest (205 m) in the 1 m corridor, much less in the narrowest and widest corridors. In the widest corridor the voles zigzagged a lot and ended up less distance away from the starting point.

Some studies suggest that a corridor may need to be surprisingly wide to be used by an animal. Lindenmayer and Nix (1993) recorded whether seven arboreal marsupial species occurred in forest corridors 30–264 m wide in Victoria, Australia. Some of the species were absent even from corridors that were wider than their home range. In north Queensland some rainforest insects—some butterfly and dung beetle species—did not occur in corridors 100 m wide (Hill 1995). These animals may be avoiding forests that have been altered by edge effects.

Best corridors for chipmunks

A study of chipmunk's use of fencerows in southern Ontario found, however, that width was not an important feature (Bennett *et al.* 1994). A fencerow is a fence of traditional design which allows plants (herbaceous and woody) to grow on either side. Chipmunks were thought to use fencerows to move across farmland between patches of forest. Mark and recapture was used to distinguish between chipmunks that were resident in fencerows and 'transients' moving along them. The frequency of transients was most strongly correlated (negatively) with the amount of gap in the fencerow path from one wood to another. Also significantly correlated (positively) were the abundance of trees and of tall shrubs. Although the fencerows plus vegetation varied from 1 to 10 m in width, there was no significant correlation between number of transients and width.

Corridors are important for dormice

To show that a species moves along a corridor does not automatically prove that corridors promote the survival of that species in the habitat fragments the corridors connect: maybe it could migrate often enough across open land. One case where hedges as corridors do seem to be important is dormice in English woods. Dormice live most of the time above ground, eating seeds and flowers from trees, and travelling via trees and shrubs. They will travel between woods if there is connecting hedgerow, but are reluctant to cross open ground (Bright & Morris 1991). In a study in Herefordshire, England, Bright *et al.* (1994) examined numerous ancient woods to find out if dormice were present. Figure 10.10 shows how dormouse presence was related to the area of each wood and to whether it was connected to many field boundaries; most of the field boundaries were hedges. Most woods larger than 20 ha contained

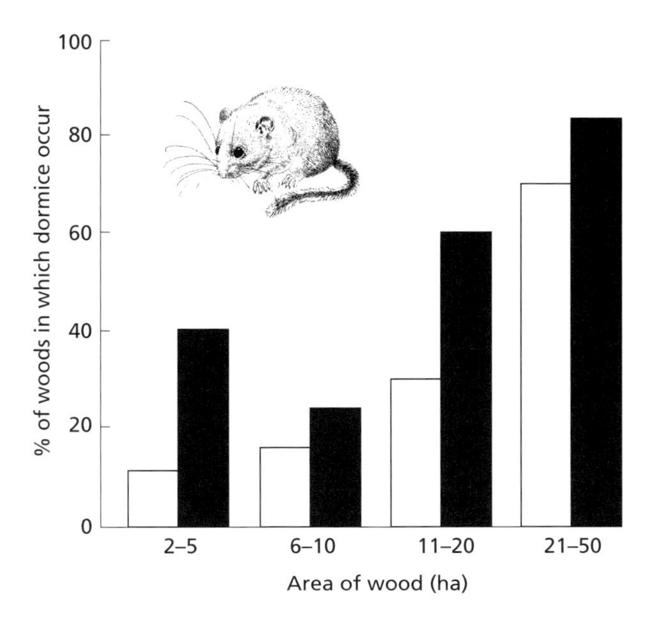

Fig. 10.10 Occurrence of dormice in ancient woods in Herefordshire, England. Open bars: wood attached to few field boundaries; black bars: wood attached to more field boundaries. From Bright *et al.* (1994).

dormice, but in smaller woods dormice were less frequent and connection to hedges became important. This suggests that woods more than 20 ha can support a minimum viable population, in other words extinctions are rare so migration between woods is not important. In terms of Equation 10.3 in Box 10.5, e is low so p is near 1. Below 20 ha extinction in individual woods is more common, so recolonization from other woods becomes important; the species is behaving as a metapopulation. In Equation 10.3 e is larger and the relative sizes of e and m are important in determining p. Anything that increases migration between woods will raise m and hence raise p, the proportion of woods occupied at one time. Even the lowest frequency in Fig. 10.10, about 10% in woods of 2–5 ha with few hedges, may be an equilibrium situation, with $e/m = 0.9$ in Equation 10.3. But clearly, the fewer woods that are occupied at one time the greater the stochastic risk that the species will die out from all of them before other woods have been recolonized.

Alternatives to natural migration

We still do not know the essential characteristics of an effective corridor for most species, but it is likely that a corridor across land used intensively by people, especially a corridor for a large animal, would demand a lot of land and be very expensive. Some corridors are clearly not realistically possible, e.g. to allow Yellowstone grizzlies to exchange with other populations more than 200 km away. So we need to consider alternatives that can reduce demographic and genetic risks to small populations. One is to transport individuals (either alive or as eggs or seeds) from one site to another. Models predict that transferring a few individuals per decade

Transferring individuals between patches

into a small population can greatly reduce the demographic risk of extinction. Figure 10.6(b) shows this for the southern Californian cougar population. Lubow (1996) reached a similar conclusion with a more general model. A rule of thumb concerning genetic risk is that, whatever the population size, one immigrant every second generation will greatly reduce diversity loss due to genetic drift (Ellstrand & Elam 1993). So the amount of work involved in carrying out such transfers need not be enormous. However, trying to increase the genetic diversity of a population can be harmful ('outbreeding depression'). An extreme example was the transport of some individuals of mountain ibex from Turkey and Sinai to add to a population in Slovakia. The imports interbred with the local population, but the young were born in February when it was too cold for them to survive. As a result the whole population died out (Amos & Hoelzel 1992).

Maintaining species in captivity

An insurance against possible extinction of a species is to maintain a population in captivity. Plants may be kept as seeds, or, if seed viability is short, grown in a garden. Maintaining populations of animals can be more difficult (Snyder *et al.* 1996). Among the problems are:

1 Some species produce few young in captivity. Examples include giant panda and northern white rhino.

2 The total population of the species in zoos may preserve little genetic diversity, so if it is used to re-establish the species in the wild there would be a genetic bottleneck.

3 Individuals bred in zoos may have difficulty surviving in the wild because of behaviour they have not learnt. For example, they may be poor at avoiding predators.

So maintaining animals in zoos, although better than allowing the species to become extinct, is not an ideal solution.

Managing for high biodiversity

An alternative aim for conservation is to maintain and promote the maximum amount of species diversity. High diversity can be pleasing for people, whether it is an area of grassland with many species of wild flowers or a varied assemblage of mammals and birds in an African wildlife park. This section considers whether ecological research has provided any helpful suggestions about how to manage natural communities for high biodiversity. Box 10.6 provides some definitions that will be useful.

How is biodiversity possible?

If we are to promote diversity we first need to understand why it occurs at all. Darwin (1859) argued persuasively that species compete and only the fittest survive. Thus natural selection provides a strong pressure towards loss of species, as the less fit are eliminated. From this has

Box 10.6. Biodiversity: some definitions.

Biodiversity includes:
- genetic diversity within species;
- species diversity;
- diversity within landscapes (patchiness).

Species diversity can be at different scales:
- α-diversity: number of species coexisting at a site;
- β-diversity: difference in species complement between patches;
- γ-diversity: number of species in a large area, e.g. a country.

For example, considering plants in Swiss uplands, the number of species in 1 m^2 of hay-meadow would be α-diversity; the contrast between hay-meadows and forest provides β-diversity; and the total number of species in the flora of Switzerland is its γ-diversity.

Diversity is often assessed by species richness, i.e. the number of species present. However, this does not take into account evenness: 10 species of equal abundance may be regarded as a community of greater diversity than one very abundant species plus nine rare ones. Two indices that take into account evenness as well as species richness are the Shannon Index and Simpson's Index.

Shannon Index, $$H = -\sum_{i=1}^{S} P_i \ln P_i \qquad (10.4)$$

where P_i is the abundance of species i, and S is the total number of species.

Simpson's Index, $$\gamma = \Sigma P_i^2 \qquad (10.5)$$

This is an index of dominance. Diversity, D, is measured by

$$D = 1 - \gamma \qquad (10.6)$$

Further information: Krebs (1994); Huston (1994)

arisen the *competitive exclusion principle* (Hardin 1960), which is based on the assumption that no two species can be exactly equally fit. It states that if two or more species exist in the same habitat, ultimately all but one of them will be excluded. We thus have the *paradox of diversity*: we expect few species but we see many. Why are there so many species?

The paradox of diversity

This question has been discussed at great length. The answers proposed by scientists can be grouped into three categories, which are summarized in Box 10.7. It is likely that mechanisms (1) and (2) both operate, though to different extents in different ecosystems. It can be argued that category (3) is not a separate cause of diversity but may contribute to causes (1) and (2). The four 'competition preventers'—disturbance, stress, predation and disease—can slow down the exclusion of a less fit species, and so contribute to cause (2), or alternatively they can provide niche separation. For example, if two plant species are eaten by different insect species, the space near individuals of plant species A may be more

Box 10.7. What allows species to coexist?

Mechanisms that may be responsible for preventing loss of species by competitive exclusion, and hence for allowing diversity to be maintained.

1 Each species has a different ecological niche, a set of conditions where it is fitter than its competitors.

2 Balance of species loss and gain. The species are nearly evenly balanced in fitness. The slightly less fit species are in the process of being eliminated by competitive exclusion, but so slowly that there will be time for other species to arise by evolution or to invade from other regions.

3 Competition is reduced or prevented, because the main controls on abundance are physical disturbance, stresses (e.g. low temperature, toxic substances), predation and/or disease. Hence competitive exclusion does not occur.

suitable for species B than A, because insects that eat A are more likely to invade. If species C is more mobile than D, and so quicker to invade habitat patches made vacant by disturbance, but D, arriving later, is the stronger competitor, this provides a sort of niche separation in time which can allow both the species to continue to exist (Nee & May 1992; Tilman *et al.* 1994).

More species the larger the area

Why are there more species on larger islands?

It was noticed several decades ago that there tend to be more species on larger islands than on smaller ones. This was brought to the attention of many ecologists by MacArthur and Wilson (1967). This also applies to habitat fragments: Fig. 10.11 shows examples. Such correlations to area have been reported for other types of habitat and for other groups of animal and plant. Possible causes of this relationship were discussed at length by MacArthur and Wilson (1967) and by others. There are basically two causes:

1 The larger the area, the greater, on average, will be the variation in environment within it, hence the more opportunity for niche separation among species.

2 There will be turnover of species in each island or fragment; the number of species in a fragment will depend on a balance between the rate at which species become extinct there and the rate at which species colonize from elsewhere. For reasons explained earlier, species are more likely to go extinct in smaller fragments. It may also be that species are more likely to colonize larger fragments, e.g. because a windblown seed has more chance of landing, or a migrating bird more chance of seeing it. In other words, this explanation says that if each species is behaving as a metapopulation the outcome will be more species in larger fragments.

These explanations are essentially the same as the first two in Box 10.7: each habitat fragment is a microcosm of the world.

There is good evidence that both these explanations can apply. For example, Kohn and Walsh (1994) showed that both of them contribute to there being more dicotyledonous plant species on larger islands: among 47 islands off the north of Scotland larger islands had more habitat types, which was one cause of their having more species. But even when islands with the same number of habitat types were compared, larger ones tended to have more species.

Should we preserve few large patches or many small ones?

Relationships such as those in Fig. 10.11 clearly indicate that if our aim is preserve as many species as possible, then the habitat fragments we preserve should be as large as possible. But suppose that only a certain total area can be spared for nature conservation, is it better to save a few large fragments or many small ones? The answer to this question hinges on the cause for there being more species in larger areas. If the main reason is that larger areas have more habitat variation, then survival of the maximum number of species would probably be promoted by having a large number of small fragments well spread out, chosen to provide examples of as wide a range as possible of environmental conditions. Figure 10.12 illustrates this by imagining that in the part of England where the fragments of ancient forest shown in Fig. 10.11(b) occurred, it was possible to save only a certain total area of them. The straight line in Fig. 10.12 is the regression line from Fig. 10.11(b) and shows the expected number of species in a single wood. So if, for example, only 10 ha could be preserved, and the decision was to use it to save a single wood, the number of herbaceous plant species would be about 60. The individual points show a different strategy, aiming for as many separate woods as possible by saving the smallest preferentially; 10 ha would allow the seven smallest to be saved, and they contain about 100 species. Whatever total area is allowed, the strategy of saving many small woods would include more species than saving one large one. Part of the explanation for this is likely to be that the area of this study includes several major soil types, each of which supports certain plant species.

Figures 10.11(b) and 10.12 are based on data from ancient woods, meaning in this case that the woods were already there in AD 1600, and most of them probably much earlier. So the species in them have survived several centuries of fragmentation. Suppose, instead, that we are concerned with an area where clearance and fragmentation are occurring now—for example Amazonian rainforest—and we want to decide whether it is better to leave few large fragments or more, smaller ones. We still need to consider habitat variation: there may be sites or areas within the forest that support particular species. But we also need to consider how species will react to the isolation of small fragments. As described earlier, some species of the Amazonian forest disappeared when confined to a fragment of 1 or 10 ha. This chapter has explained

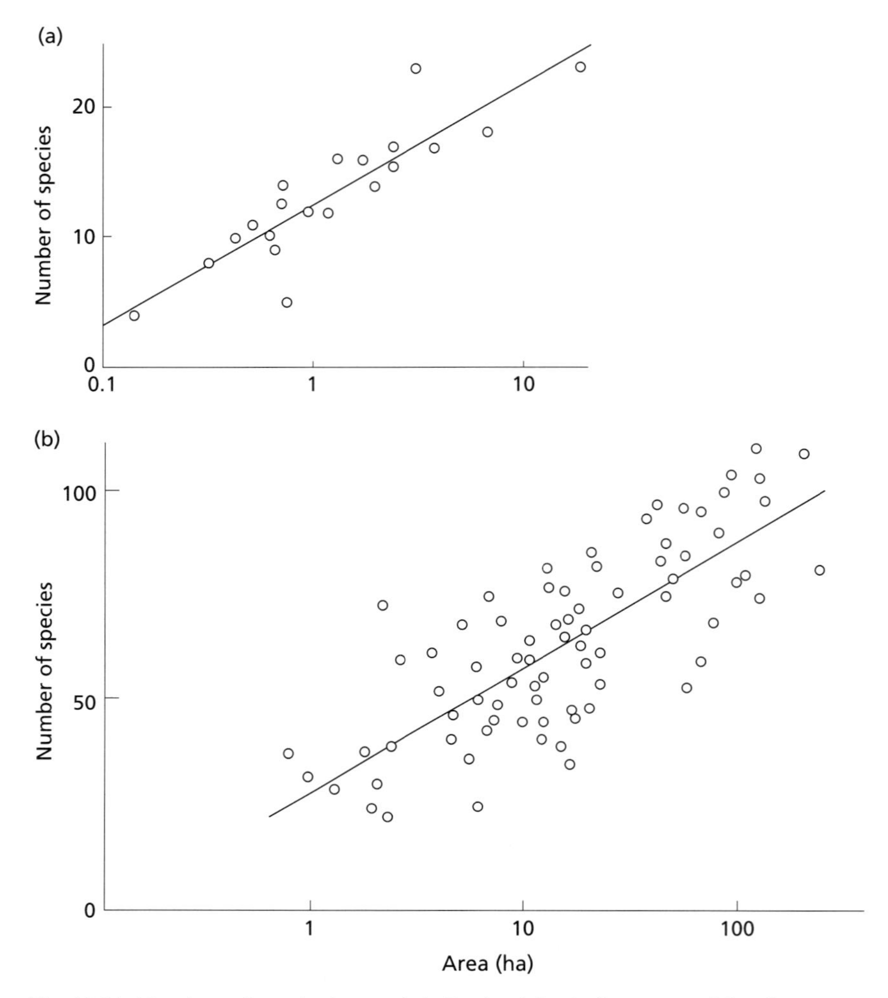

Fig. 10.11 Numbers of species in woods in England, i.e. in fragments of deciduous forest surrounded by farmland, in relation to fragment area. (a) Bird species breeding in woods in Oxfordshire. From Ford (1987). (b) Herbaceous plant species in ancient woods in Lincolnshire. From Game & Peterken (1984); reprinted with permission from Elsevier Science.

in some detail why a particular species may need a habitat patch of a certain minimum size to have a good chance of survival, and that the distance between patches may be important too. So, planning to leave numerous small patches of Amazonian forest (or African savanna, or US semi-desert), chosen to give maximum variety of habitat, would not necessarily preserve the most species. Although these patches might, taken together, have the most species at the time of isolation, some of the species will subsequently disappear if the patches are too small for them.

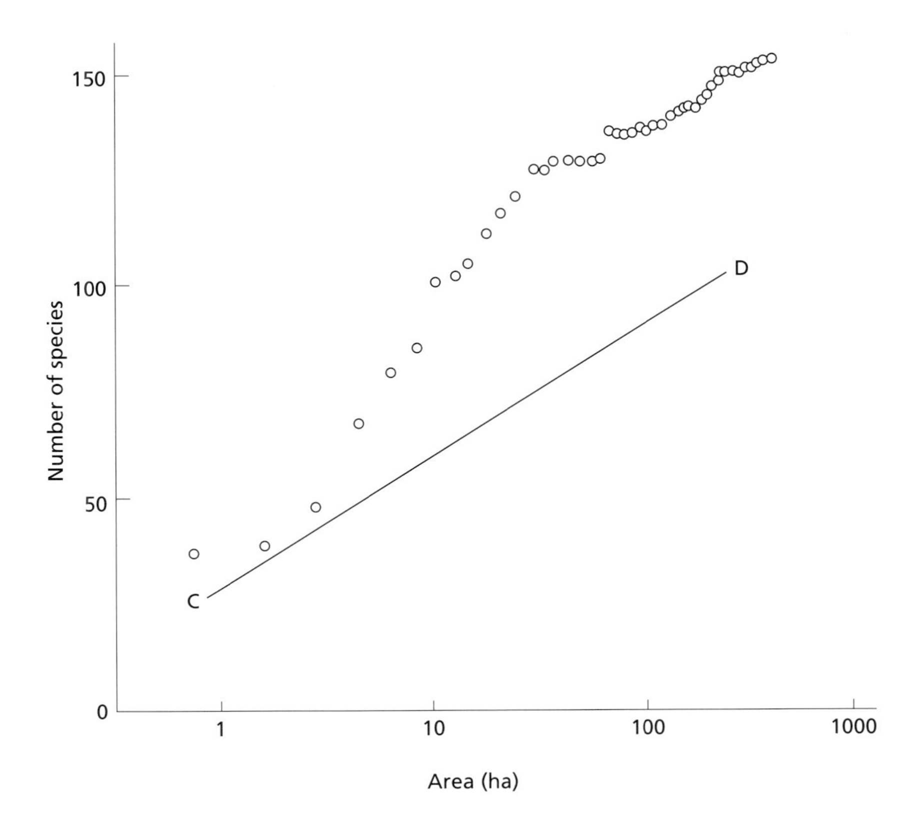

Fig. 10.12 Further analysis of species richness in Lincolnshire woods. CD is the regression line from Fig. 10.11(b). Points show total number of herbaceous plant species in one or more forest fragments, in relation to their total area, combining the smallest first. So the left-hand point refers to the smallest fragment alone, the next point to the two smallest combined, the next to the three smallest, and so on. From Game & Peterken (1984).

Conditions that promote high diversity

Are there more species where conditions are more favourable?

Biodiversity varies greatly from site to site, over large distances and small (Huston 1994, Chapter 2). Figure 10.13 shows an example of variation within a continent, within a field and within a 4-m transect. One might perhaps think that species diversity would be greater the more favourable the conditions for plants and animals; this is not always the case. Currie (1991) collated information on how the numbers of species of tree, mammal, bird, reptile and amphibian vary across the USA and Canada, in relation to environmental conditions, using data for each 'quadrat' 2.5 × 2.5° longitude × latitude (5 × 2.5° when north of 50°N). The species richness of each of the groups showed a strong positive correlation with temperature and incoming solar radiation. Figure 10.13(a) shows the relationship to potential evapotranspiration of the species richness of non-flying terrestrial vertebrates, i.e. amphibians + reptiles + mammals, excluding bats. Potential evapotranspiration is the water vapour formed by

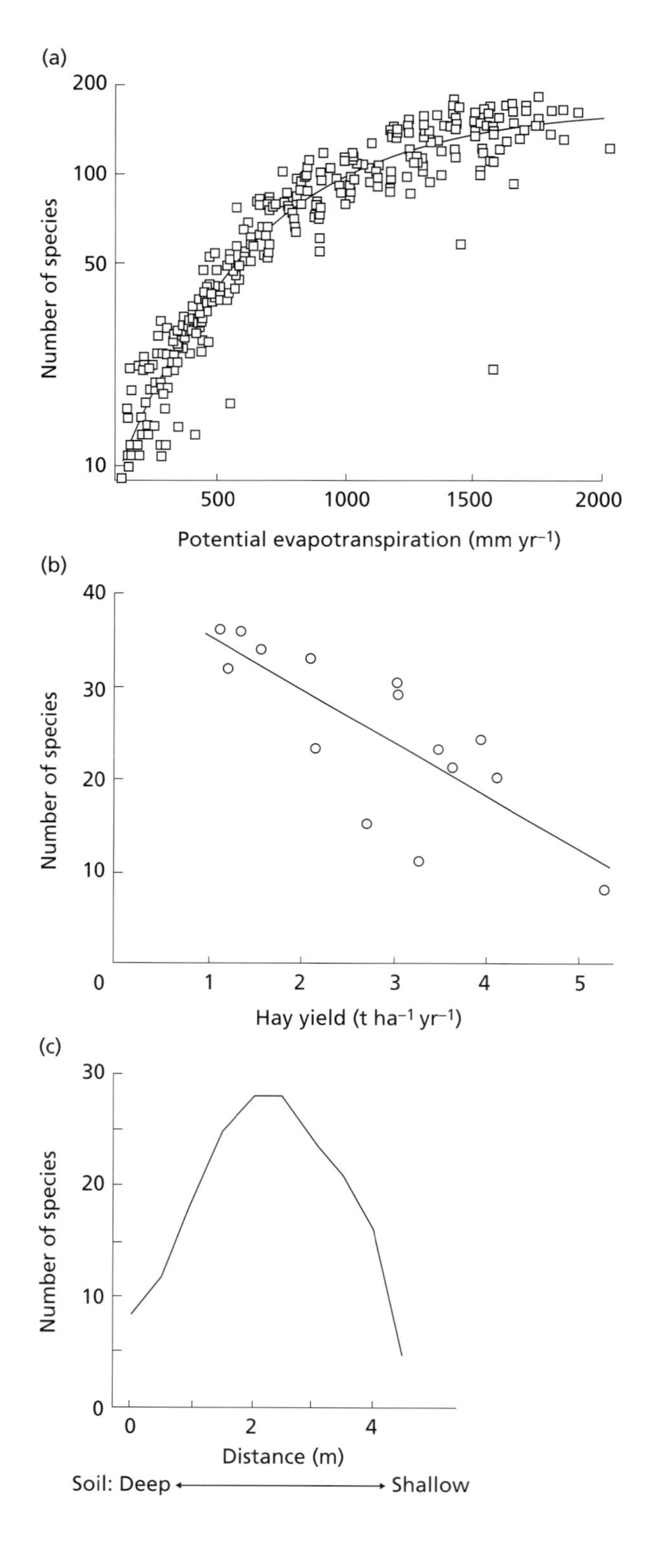

Fig. 10.13 Number of species in relation to favourableness of site. (a) Terrestrial, non-flying vertebrates in North America. Each point refers to an area 2.5° × 2.5° or 5° × 2.5° latitude × longitude. From Currie (1991). (b) Number of plant species and hay yield (air dry) in plots of the long-term Park Grass Experiment, Rothamsted, England. Each plot had received yearly farmyard manure, inorganic fertilizer or no addition. Soil pH was within the range 4.5–5.7. Number of species recorded in 1948 or 1949; hay yield is mean for 1936–49. The correlation coefficient is significant at $P < 0.001$. Data from Brenchley & Warington (1958). (c) Number of plant species in 0.5 × 0.5 m quadrats along a transect in Derbyshire, England, from rough pasture on deep, fairly nutrient-rich soil to a limestone outcrop where the soil is shallow and drought-prone. The most favourable conditions for plant growth are to the left. Simplified from Grime (1973). Reprinted with permission; copyright Macmillan Magazines Limited.

evaporation and transpiration if water supply is never limiting; it is strongly affected by incoming solar radiation and temperature. In relation to these two factors diversity clearly increases with increasing favourableness.

Figure 10.13(b) shows a contrasting result, where favourableness for plant growth varied because of soil nutrient status. These results come from the long-term Park Grass Experiment at Rothamsted, England, where different fertilizer treatments have been applied to a hay-meadow since 1856. No species have been sown or intentionally removed, but the species composition has changed in response to the treatments. The graph shows results only from plots whose pH stayed within the range 4.5–5.7, so the effects of altered pH on species composition are largely avoided. There was a strong negative correlation between hay yield and number of species: the higher the yield—i.e. the more favourable the nutrient regime for plant growth—the lower was the diversity.

Figure 10.13(c) shows a third type of relationship: highest diversity at intermediate favourableness. There are also examples where species richness appears to have no relationship to favourableness, e.g. plant species in forests of southern New Zealand (Wilson *et al.* 1996). However, these are open to the criticism that the relevant measure of favourableness may not have been used.

Grime (1973) proposed that the 'humpback model', as in Fig. 10.13(c), with the highest diversity at intermediate stress or favourableness, is the norm. Although it is possible to argue that parts (a) and (b) of Fig. 10.13 are subsets of part (c), with half of the curve missing, this is not helpful in practice: we need to consider how diversity varies over the range of conditions that actually occurs. It seems therefore that there is no universal relationship between species diversity and favourableness of the environment.

Reducing soil fertility to promote plant diversity

As conservation managers can rarely alter the local climate, relationships between diversity and climate (e.g. Figure 10.13(a)) may be of little relevance. However, it is often possible to alter soil conditions. A negative relationship between diversity and soil fertility, e.g. as shown in Fig. 10.13(b), often holds in grasslands. Janssens *et al.* (1998) analysed data from 281 grassland sites in western Europe, looking for relationships between plant species diversity and various soil factors. The number of species ranged from 3 to 60, but all the sites whose soil had more than 50 mg kg^{-1} of extractable P had 20 species or fewer; all the more species-rich sites had low-P soil. This can lead to conflicts in management where both high productivity and species richness are desired, for example in Swiss alpine meadows. When arable land that has been heavily fertilized is being converted back into grassland for conservation, it may be necessary to reduce soil fertility before high species diversity can be achieved. This is discussed further in the next chapter. The low diversity in high-productivity grassland is basically because a few species, mostly grasses, grow tall, there is intense competition for light, and lower-growing species are eliminated. If the grassland is closely grazed for at least part of the year, it may be possible to apply fertilizer without losing many species (Marrs 1993).

Disturbance

Does disturbance promote biodiversity?

One suggestion arising from Box 10.7 paragraph 3 is that disturbance may help to maintain diversity. Chapter 7 has much to say about how the disturbance of forests by people—for example felling, fires—affects the subsequent structure and species composition. Grazing animals in grassland can be considered as causing disturbance: they can increase diversity (p. 167). Managers have sometimes tried to reduce disturbance, for example trying to prevent fires in boreal forests (see Chapter 7). It can be difficult to decide whether it will favour wildlife if we allow disturbance, or apply artificial disturbance, and if so what disturbance and how much. For example, how should we promote continued species richness in tropical rainforest? According to one school of thought (e.g. Hubbell & Foster 1986b), this diversity occurs because there are many species that are extremely close in their fitness, and the very stable environment results in very slow loss of species. According to this view, we should aim to leave the forest as undisturbed as possible. On the other hand, much tropical rainforest has been used for shifting cultivation in the past. The diversity could then be in part a response to disturbance in the past, the forest of today representing a mosaic of small patches at different stages of succession following disturbance. Connell (1979) argued this point of view strongly. According to this view some disturbance is essential, as it provides niches and contributes to diversity.

In upland Britain heathland is often burnt in strips or patches—different patches in different years—to provide a mosaic of patches dominated by heather of different ages. It has been shown that patches of different ages since fire have different species of invertebrates, e.g. beetles, spiders (Gimingham 1985; Usher & Thompson 1993). If heather is left unburnt until it becomes mixed-age, the invertebrate diversity can be higher than in any one of the even-aged stands, but less than all the different ages taken together. Thus in deciding on the use of disturbance as a management tool we may need to choose between α-diversity and β-diversity. We may also need to consider what timescale is most important. The 1988 fires in Yellowstone National Park (see Chapter 7) left much of the park very unsightly in the first few years, but may be beneficial in the long term.

The basic message is that disturbance can affect diversity, and managers need to consider carefully what disturbance to allow or introduce. But this needs to be decided for each area or community type: a general formula is not possible.

Heterogeneity of the environment

Does environmental heterogeneity increase species diversity? The answer for plants is, β-diversity yes, α-diversity not much. On a landscape scale patches and mosaics of varying vegetation can often be related to differences in exposure, steepness, soil depth, drainage and wetness, rock type affecting soil properties and other factors of microclimate and soil. Each species responds differently to the environmental factors and so the proportions of species change. A clear example of this was provided by Whittaker (1956), who showed how each woody species in the Great Smoky Mountains (on the borders of Tennessee and North Carolina) had a different distribution in relation to altitude and exposure (ridge-top to valley-bottom). There are many other published examples. So if we want to promote β-diversity we should pay attention to heterogeneity in the physical environment. Perhaps there is a slight hillock that should not be levelled off, a pond or waterlogged area that can be left undrained, or a river margin where erosion should be allowed to continue. Or there may be new features that can be exploited, for example a new road embankment or a disused quarry.

Are there small-scale niches for plants?

However, if we are interested in promoting α-diversity of plants, heterogeneity of the physical environment can play only a small part. A few species may be able to coexist because their uptake roots are at different depths (Yeaton *et al.* 1977; Fitter 1986) or because their seeds germinate best in different microsites in the soil (Harper *et al.* 1965). But more often opportunities for niche separation are provided by plants themselves. For example, in some forests the seedlings of some species establish mainly on fallen logs, others mainly in litter on the forest floor (Lusk 1995). Different species may establish more readily in gaps of different sizes. This has been found in grassland and heathland (Bullock *et al.* 1995; Miles 1974). Figure 7.8 shows an example in temperate forest. These are examples of *pattern and process*, a term coined by Watt (1947) to mean that patterns in plant communities are often caused by processes going on within the community. The message to managers is that structurally complex communities will often be species rich, i.e. have high α-diversity for plants.

Plant species diversity may promote insect diversity

If management aims at high plant species diversity, will that automatically also achieve high animal diversity? For insects the answer is yes, probably. This arises primarily because of coevolution between land plants and insects, involving *secondary chemicals* (Harborne 1993). Secondary chemicals are so called because they are not involved in primary metabolism (respiration, nitrogen metabolism and so on). They belong to chemical groups such as alkaloids, terpenoids and flavonoids. Many of the secondary chemicals in plants are poisonous and function as deterrents to herbivorous animals and to fungal pathogens (see Chapter 8). There is a vast array of known secondary chemicals in plants; many are known only from one species. Although most of them are poisonous to most animals, there are many instances of one insect species being tolerant to one secondary chemical: it may have the ability to convert it to a non-toxic compound, to convert it to a form that can be secreted, or to store it in a part of its body where it is not harmful. This gives the insect an ability to eat something that most other insects cannot eat, and it may then specialize in eating that one plant species, using its secondary chemical as an attractant.

Thus many herbivorous insects eat only one or a few plant species, although in contrast some are generalists (Coley & Barone 1996). Barone (1998) collected leaf-eating insects from 10 tree species in tropical forest in Panama. Of the 151 insect species, 46 were experimentally offered leaves of many plant species. Twenty-six per cent of them would eat only one plant species, another 22% ate only within one plant genus. If this applies across the full 151 insect species it suggests that each tree species supports several insect species that will eat nothing else. And that is only the leaf chewers: there may well be specificity in diet among root chewers, leaf miners, gall formers, sap suckers and other groups of plant attackers. This is a tropical example; it is still uncertain whether specificity is equally high among insects in temperate regions (Coley & Barone 1996).

Herbivores in other animal groups tend to show less specificity in their diet. Food selection by herbivorous mammals is considered at length in Chapter 6. They do show preferences between plant species, and may avoid some species altogether, but they rarely confine their feeding to one plant species (see Figs 6.6, 6.12, Table 6.2).

Structural complexity of vegetation can also be important

Thus plant species diversity is likely to promote diversity of insects, but not necessarily of other animals. For vertebrates the structural complexity of the vegetation may be more important than plant species diversity. Figure 7.6 shows how bird diversity in forests can be related to vegetation structural complexity. This classic work by MacArthur and MacArthur (1961) has been supported by later research encompassing a wider range of vegetation types (Recher 1969; Karr & Roth 1971), though Ralph (1985) found in southern Argentina that the greatest bird diversity was associated with intermediate vegetation complexity. Invertebrate groups are also sometimes more species rich in structurally more complex vegetation. Figure 6.13 shows an example for grassland. However, taller grassland often has fewer plant species, so there can be a conflict between managing for plant diversity and managing for animal diversity. Southwood *et al.* (1979) recorded the vegetation and insects during 8 years of succession on abandoned bare soil and in a nearby wood. Plant diversity was highest about 2 years into the succession, but insect species richness was highest in the wood. The authors attributed this to the greater structural complexity of the wood. So structural complexity as well as plant species diversity needs to be considered in management, if a major aim is animal diversity.

Conclusions

- Science can help in deciding priorities in conservation, e.g. by identifying keystone species. But methods for finding out what conservation actions people will support are also important.

- Species differ regarding the size of the smallest habitat patch that can maintain them.
- Habitat fragmentation can affect species through edge effects, blocked migration routes, and inadequate area to support a minimum viable population.
- Demographic risk to small populations—the chance that random fluctuation in population size will lead to extinction—can be assessed by stochastic models (population viability analysis). The biological information available to feed into such models is often inadequate.
- A species that inhabits several or many habitat patches may form a metapopulation: it goes extinct in some patches, but survives by recolonizing from others.
- Corridors can help migration between patches. Characteristics that make corridors more effective are known for some species, but differ between species.
- Larger habitat patches and islands tend to support more species; this is (1) because they have more habitat variation, and/or (2) to do with the balance between extinction and reinvasion. Which of these predominates can determine whether more species will be supported by few large habitat patches or more smaller ones.
- Management to promote biodiversity can include alterations to (1) soil fertility, (2) frequency of disturbance and (3) heterogeneity of the environment.

Further reading

Biodiversity and threats to species:
Wilson (1992)

Conservation biology:
Meffe & Carroll (1994)
Caughley & Gunn (1996)
Primack (1998)

Animal behaviour and conservation:
Clemmons & Buchholz (1997)

Metapopulations:
Harrison (1994)

Population viability analysis:
Boyce (1992)

Prioritizing species for conservation:
Simberloff (1998)

Chapter 11: Restoration of Communities

Questions

- If we want to re-establish a semi-natural ecosystem on abandoned land, can we leave this to natural processes or is active intervention necessary?
- If we do leave it to natural processes, how long will it take for the community to become similar to its original state?
- Is inadequate dispersal of species to the site a limitation to establishment of the restored community?
- If we decide to actively promote the development of a community, what should we be aiming for? The natural state?
- Are there difficulties in reintroducing animal and plant species to areas where they were formerly native?
- Is farm soil sometimes unsuitable for re-establishing wild species? If so, what can we do about it?
- How can we establish wild species on mine waste?

Background science

- Long-term records of forest succession after arable or grassland.
- Seed dispersal to restoration sites.
- Reintroductions of plants, invertebrates, vertebrates: what happened.
- Properties of mine waste, how they affect plants.
- Responses of plants to heavy metals. Evolution of tolerance.

The previous chapter was about *preservation*—how to prevent species from going extinct, how to maintain existing communities and ecosystems that are under threat. In this chapter we consider how to *restore* semi-natural or near-natural communities on former farmland or on industrial wasteland. The emphasis is on the restoration of forests and grasslands, though of course other types of community can be the aim of restoration.

Farmland has been abandoned in the past

The re-establishment of forest and grassland on abandoned farmland is not new: it has happened in the past in various parts of the world. Box 11.1 lists some of the periods when arable farming has been abandoned in Europe, North America and tropical regions. Much of the farmland

Box 11.1. Farmland that has been abandoned in the past.

Time, AD (approximate)	Region	Reason farmland abandoned
300–500	Western Europe	Population decline at end of Roman Empire.
500–1000	Western Europe	Light soils no longer cultivated, as stronger ploughs allowed heavier soils to be cultivated.
1348 onwards	Western Europe	Sudden population decrease caused by Black Death (bubonic plague).
800–1600	Central America	Communities died out, for several possible reasons.
1200–1400	Cambodia + southern Thailand	Invasion by other people
1300–1400	South-western USA	Farming communities left or died out
1850–1920	Eastern USA	Prairieland of Midwest opened up for farming.

Further information: Fussell (1966); Fowler (1981); Scarre *et al.* (1988); Peterken (1991, 1996); Bush & Colinvaux (1994).

became forest, although following the switch in Europe to cultivating heavier soils much of the land on lighter soils became grassland, maintained by grazing. In some of these forests and grasslands the remains of field boundaries and buildings provide visual evidence today that it was formerly cultivated. Other evidence comes from preserved pollen of crop and weed species in what now appears to be pristine forest (e.g. in Panama: Bush & Colinvaux 1994).

Today there is farmland that is no longer required for food production. This is primarily in temperate, developed countries, because of increased yields per hectare; but there is also pastureland created after the felling of tropical forest, which is later abandoned. There is also, in many countries, other unwanted land covered with quarry waste, mine spoil or other industrial waste. If we want one of these areas to bear a natural, semi-natural or species-rich community of plants and animals, what should we do? There are two basic alternatives:

1 Leave it alone, prevent human interference, allow species to colonize naturally and succession to proceed towards a climax community.

2 Take active steps to promote the establishment of a desired community.

Both of these approaches have been tried, at various times in various places, and both will be discussed here.

Natural succession on abandoned farmland

If we leave abandoned farmland alone, what sort of community will develop on it? Will it become something like the natural community that would have occurred on the site? If so, how long will that take? Various sorts of vegetation can develop, including forest, open woodland, grassland and heathland. Here I consider only forest and grassland.

Forest

In this chapter we are concerned with the establishment of forest on farmland, where trees have been absent for some years. If, on the contrary, after forest is felled and the trees have been removed the area is immediately abandoned and regeneration can start, the processes are different and have already been considered in Chapter 7. Sites in the eastern United States provide some of the best evidence for answering the questions posed above. Shenandoah National Park, in Virginia, is an example of an area where the aim was to recreate near-natural forest. The park, which is about 100 km from end to end and several kilometres wide, consists of a ridge which was presumably once almost entirely covered by mixed deciduous (hardwood) forest. Today it is, again, covered by hardwood forest, but none of it is truly natural: when the park was opened in 1935 there was no undisturbed forest left, with the possible exception of a few small patches. At that time some of the park area was being cultivated, mostly as small family farms, and some was used for summer grazing of cattle and sheep belonging to lowland farms. The remainder was regrowth forest on sites which had been clear-felled at least once and left to regenerate naturally. Lambert (1989) describes the history of the national park in a non-scientific way, and the politics and negotiations behind its formation. The inhabitants were offered financial inducements to move out, and those who refused were compulsorily evicted. The Great Smoky Mountains National Park, further southwest in the Appalachians, was formed about the same time (Campbell 1969). Much of its area had also been farmed or cut over, but unlike Shenandoah it had substantial areas of forest that had never been felled, probably totalling about one-quarter of the park area (Ambrose & Bratton 1990).

Shenandoah National Park

Great Smoky Mountains

The policy in both of these national parks was to let natural succession take place, with very little help from planting or other active management. Great Smoky Mountains is the better area to assess the success of this method, because there is near-natural forest to compare it with. Figure 11.1 shows the abundance of predominant tree species in two agricultural fields following the abandonment of farming in about 1920. The two fields were adjacent to an area of old forest that had never been logged, and all were in a sheltered cove, with similar aspect and slope. The two principal early colonizers were black locust and yellow poplar, which are also early colonizers of large gaps in forest (see Fig. 7.8) and

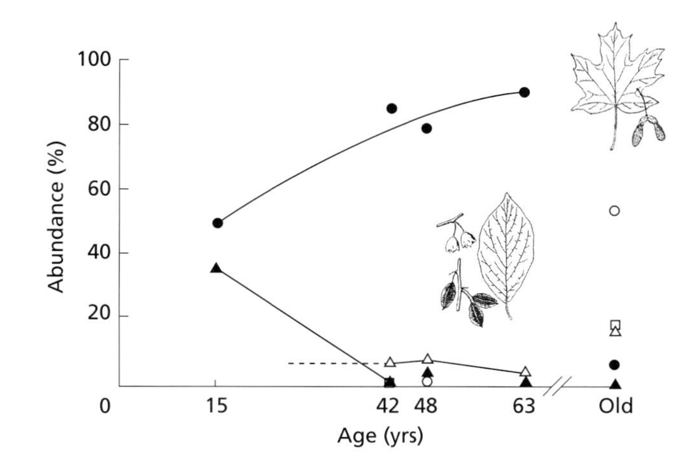

Fig. 11.1 Abundance (percentage of total) of the five most abundant tree species in forest in the Great Smoky Mountains, Tennessee. Measurements in two former arable fields, at various times after abandonment, and in old, unlogged forest nearby. Abundance was assessed after 15 years by foliage cover, and at other times by stem basal area. ● Yellow poplar (*Liriodendron tulipifera*), ▲ black locust (*Robinia pseudoacacia*), ○ sugar maple (*Acer saccharum*, upper picture), □ hemlock (*Tsuga canadensis*) (abundance < 1 except in old forest), Δ silver-bell (*Halesia carolina*, lower picture). Data of Clebsch & Busing (1989).

after clear-felling (see Fig. 7.9) in this part of the USA. On the abandoned farmland of Fig. 11.1 black locust had disappeared by 42 years after abandonment, but yellow poplar was still dominant at 63 years. None of the three principal trees of the old forest—hemlock, silver-bell and sugar maple—had achieved much biomass by 63 years, although silver-bell and sugar maple were by then abundant as small saplings. A computer model predicted that after 150 years sugar maple would overtake yellow poplar in having the largest biomass. Whether the regrowth forests would by then be similar in other respects to the old forest is not known.

Duke Forest

A few hundred kilometres to the east of the Great Smoky Mountains a detailed study of forest development on abandoned farmland was made by Oosting (1942) in Duke Forest, central North Carolina. This 2000-ha area was acquired by Duke University in 1931 and had previously been used for farming by shifting cultivation. In 1931 it included fields still in cultivation and other fields abandoned at various times up to 100 years previously. Unlike the Great Smoky Mountains no unfelled forest remained for comparison, but it was known that the original forest had been mainly hardwoods. During the first few years after the abandonment of a cultivated field herbaceous species dominated, but within 5 years pines began to colonize. As the pine stands grew up dicotyledonous woody species established beneath them. But the oldest stands, 90–110 years from the abandonment of farming, were still dominated by pines; the dominant trees of old undisturbed forest in this area, oaks and

Box 11.2. Summary of features of patches of deciduous forest over limestone in Derbyshire, England.

None of them was planted. Secondary forest resulted from natural colonization of grassland.

	Ancient forest	Secondary forest
Principal trees	Lime	(No lime)
	Oak in some of the patches	(No oak)
	Ash	Ash
	Sycamore	Sycamore
	Elm*	Elm*
Tree ages	Mixed-age	Mostly even-aged; None > 160-year-old
Shrubs	More species than in secondary forest	Much hawthorn and hazel

Ash = *Fraxinus excelsior*
Elm = *Ulmus glabra*
*Since this study elms have been killed by Dutch elm disease.
Hawthorn = *Crataegus monogyna*
Hazel = *Corylus avellana*
Lime = *Tilia cordata* + *T. platyphyllos*
Oak = *Quercus robur*
Sycamore = *Acer pseudoplatanus*

Based on Pigott (1969), Merton (1970).

hickories, were beginning to reach the canopy but were still sparse. So the forest was developing towards a structure and composition similar to what was there previously, but the succession would take well over a century to be completed.

Harvard Forest in Massachusetts is another site of classic studies of succession on abandoned farmland (Spurr 1956). On fields not cultivated for 100 years early successional pines had largely disappeared, hemlock and hardwoods dominated, but the species composition was still substantially different from old forests described by early inhabitants.

Forest in Derbyshire

Natural establishment of forest has occurred in various parts of Europe at various times. Derbyshire in central England provides one example (Box 11.2). There patches of deciduous forest occur on limestone on the steep slopes of valley sides; most of the more level ground is now farmland. Pigott (1969) and Merton (1970) deduced the history of the forests from study, in the 1960s, of the present vegetation and of old maps. There were some patches of lime woodland which were evidently very ancient; the rest of the forest resulted from natural colonization of grassland after grazing was stopped in the early 19th century or later. Although this secondary forest had large, well-established trees, as well as shrubs and

Table 11.1 Tree establishment in fields near Paragominas, eastern Amazonia, 8 years after pasture was abandoned, and in unfelled forest nearby

Intensity of use of pasture*	Light	Moderate	Heavy	Old forest
Number of replicate sites	2	2	1	2
Canopy height (m)	11–14	7–8	3–4	25–35
Trees > 2 m tall (number per 100 m^2)	48, 71	69, 73	0.6	49, 67
Number of tree species (per 100 m^2 plot)	21, 25	16, 19	1	23, 29

* Light: poor establishment of grass cover; pasture abandoned within 4 years after forest felled. Moderate: grass established well; grazed for 6–12 years; some burning. Heavy: similar to 'moderate', except bulldozer used to clear vegetation at start and to remove some topsoil. From Uhl *et al.* (1988).

herbaceous species, there were still substantial differences in the species complement (Box 11.2). Even after 160 years lime, the most characteristic tree of the ancient forests, was not invading the secondary forests.

The message from these examples of natural forest recolonization in temperate regions is that substantial establishment of trees can occur within a few decades, but after 100–160 years the structure and species composition is still markedly different from that of ancient forests in the same area.

Regrowth forest in Amazonia

Table 11.1 summarizes information on the early regeneration of tropical forest in Amazonia, where forest had been felled, grassland sown and used for pasture, then later abandoned. Where a bulldozer had been used to clear the vegetation, regeneration was still very poor 8 years after abandonment of the pasture forest. However, where the use of the pasture had been light or moderate in intensity, by 8 years after its abandonment the number of trees per 100 m^2 was similar to that in unfelled forest, though they were not nearly as tall. The number of tree species was also approaching that in the old forest, though the species composition was not the same: pioneer species were the most abundant. Whether the forest will ever return to near its original state is not known.

Grassland

In many parts of the world abandoned farmland is quickly colonized by herbaceous species, and if woody species are excluded, e.g. by grazing, a herbaceous cover can be maintained. However, the species composition may take a long time to stabilize and conform to 'old grassland'. This is illustrated by work by Wells *et al.* (1976) on the Porton Ranges in Wiltshire, southern England. This 28 km^2 chalkland area is uninhabited and reserved for military use; much of it was cultivated in the past but is now grassland. When Wells *et al.* made their survey, the time since cultivation was abandoned, determined from old maps, ranged from less than 50 years to more than 130. Wells *et al.* were able to list species that were characteristic of grasslands less than 50 years old and others characteristic of grasslands more than 130 years old. The '50-year grassland group' includes species commonly found in roadside verges and others that occur in dune grassland. In other words, they occur in somewhat disturbed habitats, although they are not arable weeds. The '130-year group' are species characteristic of old, long-undisturbed chalk grassland. So the recreation of fairly natural chalk grassland vegetation by unassisted succession can evidently take more than a century.

Seed sources

After felling of forest for timber, and after shifting cultivation, often live roots and stumps of trees remain and re-establishment can start by sprouting (see Chapter 7). However, land that has been under arable for some time contains no live tree remains, and therefore recolonization by forest species must be from seed. This is also true for grassland species. So is seed supply a limiting factor in the regeneration process?

Limited dispersal of seeds

Hutchings and Booth (1996) studied vegetation development on a field 60–70 m wide, in southern England, that had been cultivated for more than 200 years and then left. Surrounding it was old chalk grassland. Ten years after cultivation ceased some chalk grassland species had become established on the field within 20 m of the grassland, but beyond that the vegetation that established had scarcely any species in common with chalk grassland. This was evidently because viable seeds of these chalk grassland species were lacking in the long-cultivated soil, and they had not spread more than 20 m from the remaining chalk grassland. This research suggests that a field further from a seed source (say several hundred metres) would suffer severe restriction in plant colonization.

Table 11.1 showed very poor tree colonization of a tropical site which had been subject to heavy disturbance and use as pasture. Later research nearby (Nepstad *et al.* 1996) showed that shortage of suitable seeds was a problem. There were many viable seeds in the soil of the abandoned pasture, but they were almost all of herbaceous species. Seeds of three woody species were found but they were pioneer trees and shrubs: no species of old forest were represented in the seed bank. However, some tree seeds were brought in by birds, which ate them in the forest and then deposited them (still viable) in their faeces in the field, provided there were shrubs or small trees to perch on. Wind dispersal played little part in transporting seeds into the field.

As described earlier, in central North Carolina the first trees to colonize abandoned fields are pines. Oosting and Humphreys (1940) found that in the Duke Forest area during the herb stage of succession, i.e. a few years after cessation of farming, there were no viable seeds of any tree species in the soil. So all the tree seeds for the subsequent succession must have come from surrounding forest stands. McQuilkin (1940) measured the abundance of pine seedlings colonizing old fields in Virginia,

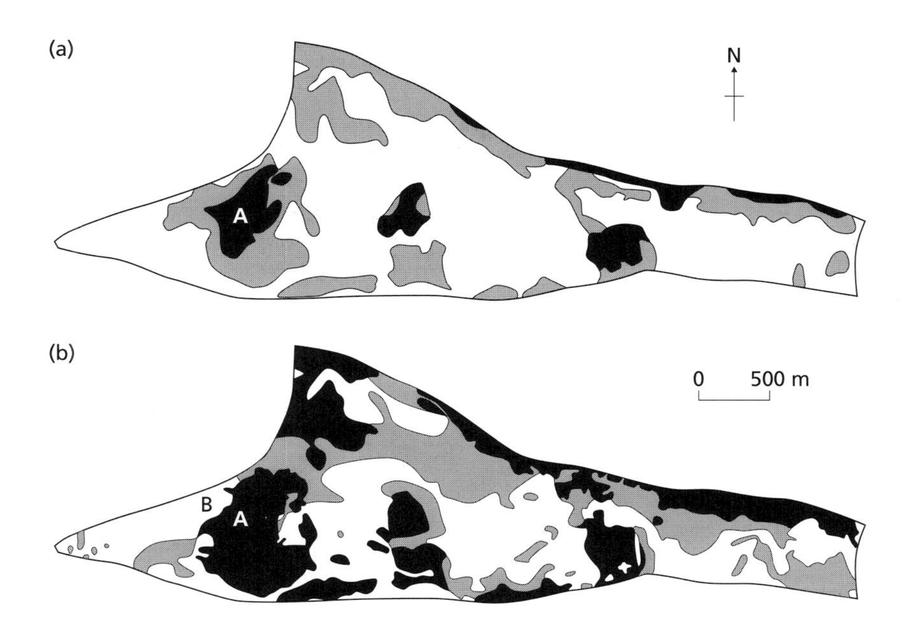

Fig. 11.2 Distribution of Scots pine in Lakenheath Warren, Suffolk, England, (a) in 1971 and (b) in1984. Black: canopy cover of pines more than 10%. Grey: pine present but cover less than 10%, or pine plus deciduous scrub. White: grassland or heath. A, position of house. B, area where pines were cleared in 1983. From Marrs *et al.* (1986). Reprinted with permission from Elsevier Science.

North Carolina and South Carolina, along transects from the edge of the nearest pine stand that could provide seed. The results show that beyond about 100–200 m from the seed source, colonization by pine seedlings would be very slow.

Trees colonizing English heathland

Colonization by trees has occurred during recent decades in the Breckland area of eastern England. This area of freely drained sandy and calcareous soils is known, from pollen records, to have borne mixed deciduous forest until clearance for farming began about 4500 years ago (Bennett 1983, 1986). From mediaeval times parts of Breckland were used for grazing by sheep and rabbits (Crompton & Sheail 1975). The rabbits were killed off in 1954 by myxomatosis, and the sheep had by then been withdrawn, so the grasslands and heaths were left without any grazing. Since then many of these areas have been colonized by trees, which in some parts now form dense forests (Marrs & Hicks 1986; Marrs *et al.* 1986). The species composition varies, however. For example, on Knettishall Heath birch is abundant, with some oak and Scots pine, whereas about 20 km away, at Lakenheath Warren, the developing forest is almost entirely Scots pine. There can be little doubt that the difference is due to the seed sources that were available. Figure 11.2 shows the distribution of pines on Lakenheath Warren 17 and 30 years after the rabbits disappeared. The sources of pine seed were a large plantation abutting the north edge of the Warren, plus a few trees around the house (marked A). After 30 years without grazing pines had established up to 1 km from these sources, but some parts were still uncolonized by trees.

The message from these examples is that seed supply can be a serious limitation to colonization by native plant species. Chapter 2 also gives information on the limited distance that some trees seeds can travel (see Table 2.13). Even where there is suitable vegetation as seed source immediately adjacent, the size of the area left for revegetation is important: part of it may be too far from the seed source. And many areas that become available for restoration are separated by many kilometres of farmland or grazing land from suitable vegetation that could act as a seed source.

Will other species arrive?

So far we have been concerned mainly with colonization by the dominant plant species—the trees in forest, herbaceous vascular plants in grassland. If these become established, can we rely on the normal complement of animals and minor plant species to arrive and establish? To answer this question fully would require a complete census of all the species in the restored forest or grassland and in the natural community it is supposed to emulate. To do this for all species, including all invertebrates and all microorganisms, would at present be technically impossible: at any site many of the species would prove to be previously unknown.

Recolonization by ants

There have been some studies of colonization by particular groups of invertebrates. Majer and Nichols (1998) recorded ant species present in four neighbouring sites in Western Australia, one in eucalypt forest, the others being restored following open-cast bauxite mining. On the mine sites, after the replacement of topsoil, either (1) 23 native forest plant species were sown or planted, (2) one species of eucalypt tree was planted, or (3) nothing further was done. Ants were sampled at intervals over the following 14 years. By the end of that time trees had established well where they had been planted, although even in the 23-species site the vegetation was not as species-rich as in the forest site. The unplanted site still had few trees, though it had much low vegetation. At the three former mine sites the number of ant species increased throughout the 14 years. Figure 11.3 shows that their species composition gradually became more similar to that of the forest site, but after 14 years was still substantially different from it, especially in the unplanted site. In that last year of recording the forest site had 52 species of ant, of which 13 had never been found at any of the mine restoration sites at any time during the 14 years.

Holl (1996) recorded what species of day-flying butterflies and moths occurred on former open-cast coal mines in Virginia that had been restored at various times up to 30 years previously. Sites 25–30 years old were approaching nearby forest sites in their species of butterfly and

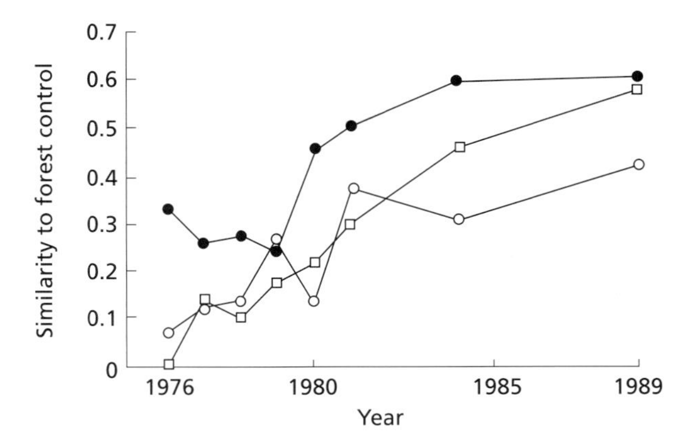

Fig. 11.3 Comparison of ant species present at three restored mine sites in Western Australia with those at a nearby forest site. Similarity to forest expressed by Sorensen's index: 0 indicates no species in common, 1 indicates identical species list. ● 23 plant species sown or planted, □ one tree species planted, ○ nothing planted. Sorensen's Index, I = $2c/(a + b)$, where a and b are number of species present at sites A and B, c is the number of species in common. From Majer & Nichols (1998).

moth, though still lacking a substantial proportion of the rarer forest species.

These two studies covered longer periods than most published research on recolonization by invertebrates, but we would like to know what will happen over a much longer period, giving time for the vegetation to become more like the little-disturbed forest. We know that some species take a very long time to colonize naturally. Figure 10.9 provides evidence that the characteristic British woodland floor plant dog's mercury can in 100–150 years spread a few hundred metres across farmland to a new wood, but is unlikely to spread as much as 1 km. However, once established in a wood it can spread to become the predominant ground floor herb, as it has done in some woodlands 50–70 years old (Pigott 1969; Merton 1970). Figure 10.8 gives information about a butterfly that inhabits short grassland; it rarely migrates more than 1 km between separate patches of grassland. Thomas (1991) gives other examples of butterfly species, of heathland and forest, that fail to migrate between habitat patches if the distance is more than a few hundred metres. Chapter 10 gives examples of the distances that act as a deterrent to certain mammal species, and how habitat corridors can promote their movement.

Woodland floor species that are slow to colonize

Peterken and Game (1984) studied the ground flora of 362 woods set in farmland in Lincolnshire, eastern England, and were able to categorize each as 'ancient'—i.e. there had been forest on the site continuously since before AD 1600—or 'recent', i.e. on land that has been clear of forest for some period since 1600. Some of the woodland floor species were found equally often in ancient and recent woods. However, there were 62 species that were found much more frequently in ancient than in

Table 11.2 Frequency of understorey species, grouped by dispersal mode, in old and regrowth forests in southeastern Pennsylvania and northern Delaware. Regrowth stands were separated from old forest by 10–480 m of non-forested land. Frequency means percentage of sites in which the species occurred

Dispersal mode	Old forest	Regrowth forest	Statistical significance of difference (P =)
Wind	25.4	24.6	not sig
Adhere to animal	26.5	25.4	not sig
Ingested by animal	39.4	30.4	not sig
Spores (ferns)	18.2	6.5	0.051
Ants	34.7	7.3	0.012
None	18.8	5.4	0.016

From Matlack (1994).

recent woods. They have failed to recolonize most of the recent woods, even though some of them are more than 300 years old. The list includes species that most British people, whether trained ecologists or not, would recognize as typical woodland flowers, for example bluebell and wood anemone, as well as dog's mercury (Fig. 10.9). Whitney and Foster (1988) made a similar comparison in New England of old deciduous forest and patches of regrowth forest, mostly 50–90 years old, on former farmland. Again, there were some understorey species that were rarely or never found in the regrowth forests. Long lists have also been produced of British species of lichen and beetle that are characteristic of old forest and rarely found in forest that is less than several centuries old (Rose 1976; Harding & Rose 1986).

These species evidently are very slow to colonize new habitat patches. This may be because they are poor migrators. Matlack (1994) studied the undergrowth vegetation in patches of regrowth forest on former farmland in Delaware and southeastern Pennsylvania, USA, and compared it with old forest, which was mixed hardwoods. The regrowth forests had fewer species in the understorey. This was because of a lower frequency of species whose seeds are dispersed by ants or have no apparent dispersal mechanism, and of species that have spores (Table 11.2). In contrast, species with seeds that are dispersed by attaching to animals' fur, by being ingested or by wind were about equally frequent in old and regrowth forests. This strongly suggests that dispersal was a limiting factor in some undergrowth species reaching these regrowth forests, whose age ranged from 3 years to about 100 years.

Another possible reason for the slow establishment of species in regrowth forests is that forest might take a long time to become a suitable habitat for them. Perhaps gaps do not yet occur in the frequency and size that some understorey plants require: the development of a new forest from even-aged to fully mixed-age can take as long as one or several centuries (see Chapter 7). Chapter 7 also gives examples of animals that require old trees. Many animals have special requirements. Some invertebrates require nectar from particular plants; some inhabit ants' nests; some solitary bees of dead wood nest in the burrows of wood-feeding beetles (Warren & Key 1991). What we are considering here is the re-establishment of a whole ecosystem. There is a vast number of interactions between species, which have to be reconstituted in the right order.

The main message from this section is that if we rely on natural colonization and succession, it will take at least a century—probably several—for near-natural grassland or forest to develop on former farmland, and there is no guarantee that the community will ever reach a state that we could truly consider natural.

Actively promoting restoration

What should we aim to create?

Instead of leaving the restoration process to natural colonization and succession, we can take active steps to promote it. The term 'restoration' may need further definition here. We are aiming to promote the establishment of a semi-natural community or ecosystem on the site, and we need to have a clear idea of what we are aiming for. Box 11.3 sets out the main alternatives. It avoids using the word 'natural', because that term itself needs to be defined. The early development of the science of ecology in the United States was much influenced by the assumption that the landscapes and communities seen by the first European explorers were natural, providing the basis for the key concepts of succession and climax. It is now clear that what the first Europeans saw in North America was already much influenced by people. Many species of large mammal had been made extinct about 11 000 years earlier by hunters (Stuart 1991). Fire had been used, in forest and prairie, to alter the vegetation and so make hunting easier (Spurr 1956). Farming had been carried out in the east and the southwest; much of the apparently primeval forestland of the east was in fact regrowth after shifting cultivation (Williams 1989). In Australia there was no farming before Europeans arrived, but people had

Box 11.3. Possible aims for restoration of an area.

1 Create what would have been here today if people had never interfered.
2 Return it to how it was at some specified time in the past. For example:
In Europe: before the first farming about 7–5000 years ago; or as it was managed in the Middle Ages, about 1000 years ago.
In N. and S. America, Australia, New Zealand: before the first European settlers arrived.
3 Create a suitable habitat for a particular species, or a group of species, e.g. butterflies.
4 Promote recreational enjoyment by people.
5 Improve the appearance of areas of industrial waste.

presumably been responsible for much of the widespread burning of vegetation and for the extinction of many large mammals. Almost everywhere in the world has been influenced by people in the past. So it is not helpful to make the aim 'to return the area to its natural state': we need to be more precise, and in Box 11.3 alternatives (1) and (2) attempt to do that. We can get some information about what the area was like in the past from written records or, further back, from remains of pollen and bones (see Box 2.5). Neither of these will give us much information about invertebrates or microorganisms. Alternative (1) is bound to involve some guessing about what the area would have been like today without people: since people started to be a major influence the climate has changed, and there might have been natural invasions and extinctions of species. Alternatives (3) and (4) do not aim precisely for a natural state: indeed, (4) implies that we can improve on nature. This was the aim of the 18th-century landscape gardeners and park creators of Europe: open landscapes with spaced trees, grassy areas, vistas, lakes, perhaps the occasional statue or temple, were considered more attractive than the dense forest that would naturally cover the land. In the European Alps the characteristic but artificial mosaic of forest and hay-meadows that we see today is considered by many people more attractive than continuous forests, and provides a major recreational area for walking and admiring the scenery, as well as for skiing. This chapter will not discuss further the relative merits of the alternatives in Box 11.3. I will use the term 'near-natural' to encompass them and to describe what we are aiming for.

Restoration of US prairie

The restoration of prairie on farmland in the US mid-West illustrates some of the problems and successes of active restoration. The first attempt to recreate North American prairie on farmland was the Curtis Prairie, a 24-ha site at the University of Wisconsin at Madison, where restoration started in 1935; the 16-ha Greene Prairie was started nearby a few years later. Other prairie restorations, some much larger, have been started since then in various parts of the mid-West.

When an accelerator ring for the study of subatomic particles ('Fermilab') was built in Illinois in 1969–71 it enclosed an area of 314 ha which had originally been prairie but had been cultivated for more than a century. Seeds of prairie species were sown on much of this area to try to restore the original prairie. Figure 11.4 shows the abundance of species, classified into six ecological groups, all recorded in 1985 but in areas where prairie restoration had started at different times. Initially the colonizing vegetation was mostly weeds and other species that are uncommon in prairie. However, after 11 years many of these non-prairie species had become sparse, and prairie graminoids and 'aggressive' prairie forbs had become as abundant as in an old prairie remnant nearby. But the forbs classed as 'less aggressive' had established poorly.

Problems in prairie restoration

Not all attempts at prairie restoration have been so successful. Jordan *et al.* (1987) and Howell and Jordan (1991) summarized the problems that have been encountered.

Fig. 11.4 Above-ground dry weight of groups of plant species in a prairie restoration in Illinois, and in a nearby old prairie. ● Prairie graminoids, ■ 'aggressive' prairie forbs, ▲ 'less aggressive' prairie forbs, Δ weedy perennial forbs, □ weedy annual and biennial forbs, O non-prairie graminoids. Data from Jastrow (1987).

1 Some prairie species that were originally native to the area have proved consistently difficult to re-establish.
2 Some arable weed species and some non-native species (e.g. introductions from Europe) can be very persistent.
3 Woody species often invade.
4 The soil may have changed during the years of cultivation—for example the available nutrient status may have increased but the crumb structure been degraded (see Chapter 4)—and it may be necessary to reverse these changes.

Unwanted species

So one message is that restoration can involve not only introducing desired species but trying to get rid of unwanted ones, or to prevent them invading. A problem tree in Britain is sycamore. If we are trying to recreate forest as it was before the first farmers or in the Middle Ages, sycamore should not be there, as it was only introduced to Britain a few hundred years ago. It is now very widespread in British deciduous forests and regenerates well from seed. For example, in Derbyshire (Box 11.2) it has invaded ancient forest as well as contributing to secondary forest. An example of introduced plant species that have become abundant in North America are European annuals that are now widespread in the west (see Chapter 6). To remove these from a prairie restoration, or sycamore from a British forest, would be technically possible, but if the area is large it would be extremely time-consuming. And the species could reinvade from outside the reserve. Introduced animals can also have a major effect on the survival of plants and on the composition of the vegetation, for example the effects of grazing by deer in New Zealand, and by rabbits in Britain (see Chapter 10). It will not be possible practically to eliminate either of these from the country, though it is sometimes possible to fence them out of a reserve or to eliminate them from an island. Introduced species whose spread cannot be controlled include the causal fungi of chestnut blight in North America and Dutch elm disease in Britain.

Today they prevent chestnut and elm from being major components of forest in those countries, as they once were.

Introducing species

If we are really going to recreate an ecosystem exactly as it was in the past, it needs to have all its original species, but there are some species we can never reintroduce because they are now extinct. There were formerly about 12 species of moa in New Zealand, flightless birds 1–3 m tall (Wardle 1991). These became extinct after the Maori people arrived about 1000 years ago, mainly because of hunting. Moas were herbivores, living mostly in forests. They were probably abundant enough to have a major effect on tree regeneration and on the composition of the understorey, and hence on the structure of forests and on the animals that could live there. This is one reason why New Zealand forests can never return to being exactly how they were before people arrived.

Ways of introducing plants

Sowing seeds into areas where the species was presumably formerly native, and where the site appears suitable now, does not always lead to successful establishment of the species. Such failures have occurred at Curtis Prairie, in British grassland restoration and in eastern US woodland (Jordan *et al.* 1987; Smith *et al.* 1997; Primack & Miao 1992). Alternative methods of introduction, using whole plants or whole turves rather than seed, have been tried: they sometimes help but have not always solved the problem. Sometimes a 'nurse' species can help. For example, when attempting to establish heathland on sandy waste in southern England, it was found that the heathers established better if a grass, *Agrostis capillaris*, was also sown (Jordan *et al.* 1987, Fig. 5.4).

If a large area is to be restored, requirements of time, effort and seed supply may well set a limit on how much direct introduction of plants can be achieved. Simplified methods of collecting and spreading seed may be used. For example, hay with the seed heads can be spread, or it can be fed to horses which are allowed to roam and deposit seeds in their faeces.

Introducing invertebrates

Introductions of invertebrates have usually been only of conspicuous, attractive species, most often butterflies and moths. In Britain there have been successful introductions of several butterfly species, to grassland, heathland and woodland sites (Thomas 1991). A dramatic success was the silver-studded blue: 90 adults were released in 1942 at a site in Wales, and 41 years later there were estimated to be 60 000–90 000, although they had extended their area by only 2 km.

Introductions of other invertebrates have sometimes been suggested. For example, transferring dead logs from old to young forest would introduce many species of invertebrate and fungus, but as far as I know the effectiveness of such methods has not been tested. Some topsoil that has been spread on mine sites is evidently inadequate in mycorrhizal inoculum for the development of forest or shrub vegetation: inoculation with arbuscular mycorrhiza has increased the growth of native species at some sites, but not at others (Jasper *et al.* 1989a, b; Miller, in Jordan *et al.* 1987). As far as I know, inoculation with other soil microbial species has not been used in restoration.

Introducing birds and mammals

There have been many attempts to reintroduce mammals and birds. Wolf *et al.* (1996) used a questionnaire to obtain information about the success of introductions of individual species, either taken from other wild populations or bred in captivity, into sites in North America, Australia and New Zealand. The percentage of introductions resulting in successful establishment was 63% for birds and 73% for mammals. Comparison of the successful and unsuccessful attempts gave some indication of factors favouring establishment. The most consistent was that establishment was more likely in a site near the centre of the species' natural range than near its periphery. Percentage success was in the order omnivores > carnivores > herbivores. Success was more likely if the area of suitable habitat was larger, and if more individuals were introduced. Chapter 10 explained the special risks to small populations. Some early introductions involved few individuals. For example, during the second half of the 19th century red deer were released in the South Island of New Zealand on at least 10 occasions, but probably fewer than 10 individuals on each occasion and sometimes only one male and one female. Some of the populations died out, although several multiplied and formed the basis of the present very abundant and widespread deer population (Clarke 1971).

Introducing captive-bred animals has been markedly less successful in some species than transferring individuals from another site. Attempts have been made to reintroduce the swift fox to prairie sites in Canada (Bowles & Whelan 1994). Only 11% of the introductions involving captive-reared animals led to establishment, as against 47% using wild-caught animals. Predation by coyotes was the greatest cause of death, and captive-bred animals may be poorer at escaping them.

The area of suitable habitat available may be the critical determinant of whether the introduction of a large mammal is successful. Chapter 10 explained methods of determining the area needed to support a minimum viable population; alternatively, a mosaic of habitat patches between which animals can migrate may support a metapopulation. Howells and Edwards-Jones (1997) attempted to predict whether there is adequate habitat in Scotland to support wild boar if they were reintroduced. Wild boar occur today in several countries in Europe, including France, Germany, Poland and Russia, but they died out in Britain in the 16th century. Acorns and beech mast provide a major component of their diet, so they need oak or beech forest. Howells and Edwards-Jones concluded that there is at present no suitable habitat in Scotland large enough to support a minimum viable population (MVP) of boar, which they estimated to be 300. However, this MVP estimate involved major uncertainties, including the effects of inbreeding depression and year-to-year environmental variation.

Wolves

Wolves were formerly widespread in North America and Europe but have disappeared from most of their former range, presumably partly because of hunting and partly because of loss of habitat. As explained in Chapter 10, the loss of wolves and other large carnivores is likely to lead to an increased abundance of herbivores such as deer, which were their prey; this in turn will affect the regeneration of trees. So wolves were keystone species. Wolves are expanding their range again in some parts of Europe and North America (Boitani 1992; Forbes & Boyd 1997), but there is also interest in reintroducing them to suitable habitat areas. In 1995–96 wolves caught in Canada were released at two sites in the USA, 31 in Yellowstone National Park and 35 in central Idaho (Forbes & Boyd 1997). An indication of the minimum area needed is provided by the much-studied wolf population of Isle Royale in Lake Superior, which is probably near to an MVP: in the mid-1980s it suddenly dropped from about 50 to 12 (Vucetich *et al.* 1997). The island is 544 km^2. In Italy wolf pack territories were found to be 200–400 km^2 (Boitani 1992). These areas indicate that Yellowstone National Park, at 20 000 km^2, should be easily large enough to support a viable population of wolves, but Shenandoah National Park's few hundred square kilometres seem marginal. Wolves are more willing than most large carnivores to migrate across farmland (Forbes & Boyd 1997), but while doing so they might attack lambs and other domestic animals. If the introduction of wolves is to be widely accepted by humans the wolves will need an adequate amount of suitable semi-natural habitat.

Management of restored areas

Should we create disturbances?

There may be other things we can do, besides introducing species, to promote the development of the desired community. We need to decide what is our attitude to disturbance. Should we create artificial gaps in a developing forest? Gaps of different sizes allow the regeneration of a variety of tree species (see Fig. 7.8). In a new forest, which is likely to be approximately even-aged, gaps will eventually form naturally, but it may well take more than a century for something like a natural mixed-age structure to develop (e.g. Fig. 7.3). So we may want to speed up the process. Disturbance was probably important in the former prairies. The presence of some weedy perennials in the old prairie remnant of Fig. 11.4 probably depended on disturbance, either recent or past. Bison wallows were open patches that weedy species could colonize. If our prairie restoration is too small to support bison, should we make simulated wallows? Fire was normal in the prairies and must have influenced the vegetation composition, including killing trees. The restored Fermilab prairie in Illinois is burnt regularly, but effective burning has proved difficult at the smaller Curtis and Greene Prairies in Wisconsin.

Invasion by unwanted species from outside the restoration area is often a problem, especially if the area is small. Common problem invaders are trees in prairie and deer in forests. Their presence is due to the combination of opportunity to invade and lack of the former control (by fire or carnivores). There may be no alternative but to kill them individually. Culling deer upsets some wildlife managers, but if the former large carnivores cannot be reintroduced culling may be the nearest simulation of what their effect would be.

Sites with special problems

Fertile farmland

This section is mainly about problems with soil. If the site is covered with mine waste it may not be soil at all, in any normal sense. However, let us first consider a common problem with former arable land: if fertilizers have been applied over many years the soil may be too fertile to allow the desired community to develop. More fertile sites tend to support fewer plant species, e.g. in tall grassland (Fig. 10.13(b)): many species are unable to establish in the dense, tall vegetation that grows on fertile soil. Simply stopping the addition of fertilizer is unlikely to cure the problem. At some sites nutrient inputs will exceed outputs: this was true for nitrogen at a site in the Netherlands, if nothing was removed from the site by people (Table 11.3, left-hand column). For more information on nitrogen input in rain and as gases see Chapter 4, including Box 4.4 and Table 4.4.

Reducing soil fertility

One possible way to increase nutrient loss from grassland is to cut and remove vegetation as hay. Berendse *et al.* (1992) tried this as part of an investigation into the best way to restore species-rich hay-meadows in the Netherlands. Cutting and removing hay twice a year resulted in nitrogen losses considerably exceeding inputs (Table 11.3). However, during the 4 years of the experiment little difference in plant species composition developed between the hay-removal and non-removal treatments. Such treatments may need to continue much longer to allow

Table 11.3 Nitrogen balance of hay-meadows on a wet site in Netherlands. Hay cut twice per year, and either left on the field or removed

Inputs		$kg\ ha^{-1}\ yr^{-1}$
Atmospheric deposition (rain + gases)		60
N_2 fixation		<1
Total		**60–61**
Outputs	No hay removal	Hay removed
Denitrification	16	16
Leaching	3	3
In hay removed	—	136
Total	**19**	**155**

Data of Berendse *et al.* (1992).

species characteristic of low-nutrient soils to establish. If the cut hay is left to dry some of the nutrients (especially potassium) leach back into the soil during the first few weeks, so the cut vegetation needs to be removed promptly to achieve maximum nutrient loss (Schaffers *et al.* 1998). Chapter 4 gives more information on long-term nutrient balances.

It may be possible to make the soil more suitable by adding organic or mineral material. Dunsford *et al.* (1998) investigated ways of re-establishing heathland vegetation at a site in Suffolk, England, that had formerly been heathland but had been cultivated for about 50 years. Heather plants with many ripe seeds were spread over the soil, but after 5 years the heather had established only about 10% cover. However, if peat was mixed into the top 50 cm of soil, forming a 50:50 mixture, before the seeds were spread, after 5 years the heather cover was about 50%. Whether this technique could be applied over a large area, such as several square kilometres, is doubtful: to obtain enough peat would probably involve destroying other vegetation.

Industrial waste

At other sites revegetation may be difficult because the soil is too infertile. Mining and industrial processes can result in large dumps of material which is very unfavourable for plant growth. Coal, after mining, is separated from shale and mudstone, much of it in large fragments. China clay is separated from almost pure sand. Pulverized fly ash is a waste product from coal-burning power stations. These materials are unsuitable for plant establishment because of their coarse particle size and lack of organic matter, and hence low water-holding capacity and low ability to supply or retain mineral nutrients. At some sites there are additional problems from toxic heavy metals. At such sites the initial aim may be the last one in Box 11.3, to improve the appearance when viewed from a distance. If we can establish a complete vegetation cover this will help a conspicuous eyesore to blend in with the surrounding countryside.

Recolonization of coal-mine waste

In spite of the problems just listed, at some of these sites a substantial vegetation cover can develop, given sufficient time. Hall (1957) made a survey of more than 200 heaps of coal-mine waste in England that had been standing for various lengths of time up to 100 years. No attempt had been made to improve the soil or to revegetate any of them. A few of the oldest had developed a closed-canopy forest of oak, but on most heaps woody plants were sparse. On many heaps over 40 years old there was a dense grass–forb sward. At some sites the rock fragments in the upper few centimetres had within 20–30 years weathered to finer particles, though at other sites there was no sign of soil structure developing even after 100 years. Forest can also establish naturally on pure sand waste from china clay mining: a site in Cornwall, England, where such waste had been tipped 116 years earlier, had well-developed oak–birch forest (Roberts *et al.* 1981).

So vegetation can establish naturally on industrial waste; but even on the more favourable sites it takes a long time. One way to promote revegetation is to spread topsoil on the waste. This is often done on mine waste nowadays. If opencast mining is continuing there should be a supply of topsoil, but that may not be true for other types of site, such as deep mine waste, fly ash or old building sites. Substitutes for topsoil have been tested. In the treatment of domestic refuse, if paper and plastic are removed and the remainder passed through a fine screen, the resulting pulverized refuse fines have a higher concentration of organic matter and of available N, P and K than many topsoils. In a field test at two sites, Chu and Bradshaw (1996) spread a 20-cm layer of pulverized refuse fines over colliery spoil and brick waste and then sowed perennial ryegrass. The grass grew well, better than on the uncovered waste even with fertilizer added, and better than on 20 cm of topsoil. On an abandoned sandpit in Quebec, grass establishment was promoted by adding paper de-inking sludge, which is mainly wood fibres containing inks and other chemicals (Fierro *et al.* 1999).

Adding organic matter

Adding fertilizer . . .

On some waste material, sowing plants with an application of inorganic fertilizer but without adding any organic material can promote early vegetation establishment. Marrs *et al.* (1980) surveyed 68 sites of china clay waste (sand) on which grass and clover seed had been sown up to 8 years earlier, followed by NPK fertilizer and lime. All sites a year or more old had a vegetation cover of at least 40%; on the older sites it varied from 40 to 100%. Up to 11 native species had invaded per site. An increase in the nitrogen content of the vegetation plus substrate showed that N accumulation was faster than the fertilizer input. Presumably there was some input from N fixation by the sown clovers or legume invaders. Because fresh waste is commonly coarse and has little organic matter, if fertilizer is applied early nutrient losses by leaching can be substantial (Marrs & Bradshaw 1980). Legumes may be a more effective means of nitrogen input. Finding out which legume species are best suited to particular waste materials could help in their revegetation. There are large differences between legume species in their ability to establish on colliery waste and sand waste from china clay, and hence in the N input they provide (Jefferies *et al.* 1981). One legume which has been successfully sown on waste tips is the tree lupin *Lupinus arboreus*. This has a lifespan of about 6 years, and during that time can provide substantial N input (Palaniappan *et al.* 1979).

. . . or legumes

Toxicity

Waste material from mining for a heavy metal such as copper or lead is likely to contain toxic concentrations of that heavy metal. Heavy metals also accumulate where sewage sludge is dumped. Plants growing on such media will take up the heavy metals, which may reach toxic concentrations in their tissues (see Chapter 9, especially Fig. 9.9). Some plants do colonize such sites, but usually only a few species, and the same species recur at different polluted sites. For example, in Britain the grass *Agrostis*

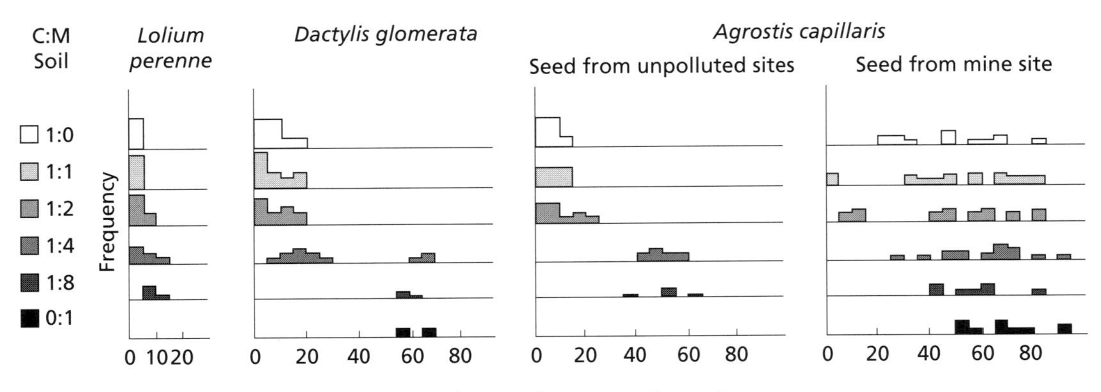

Fig. 11.5 Frequency of plants of different copper tolerances in three British grass species. All plants were grown from seed collected at unpolluted sites, except for the mine population of *Agrostis capillaris*. The seed was sown on mixtures of potting compost and Cu-polluted mine soil; the figures under C:M show the ratio of compost:mine soil. The surviving plants were assessed for Cu tolerance, by their root extension in Cu solution. From Gartside & McNeilly (1974).

capillaris is often seen on old waste dumps from heavy metal mining. These observations raise the question of whether the species found at these sites are always tolerant of heavy metals, or whether they have races that are specially tolerant.

Plants that are tolerant of heavy metals

The genetics of heavy metal tolerance has been extensively investigated (see review by Macnair 1993). It has been consistently found that plants growing on heavily polluted sites are genetically different from members of the same species growing close by on unpolluted soil: they are ecotypes with genetically based heavy metal tolerance characters. Figure 11.5 shows the results of tests for copper tolerance on three grass species. Tolerance was assessed by the length of roots that grew in a solution containing copper (Cu) at 0.25 mg l^{-1} as a percentage of the length in Cu-free solution. Two populations of *Agrostis capillaris* were compared, one from an old copper mine, the other from uncontaminated soil. When numerous plants were grown on uncontaminated soil and then tested (see top line of Fig. 11.5), the mine population showed a wide range of copper tolerance, some plants being extremely tolerant; in contrast, no members of the 'normal' population showed more than slight tolerance. If the ability to grow on strongly polluted soil depends on such genetic adaptation, why do not more species occur on these sites? Are only a few species able to undergo the necessary genetic change? To investigate this, Gartside and McNeilly (1974) looked for Cu-tolerant individuals within nine British herbaceous species (seven grasses and two forbs). They made up mixtures of potting compost and Cu-polluted mine soil in various proportions, and sowed 10 000 seeds of each species on each mixture. The seeds all came from populations on unpolluted sites. Some of the seeds from each of the species germinated, but three of them had no survivors on the mixtures with a high proportion of mine soil. Of the remaining species, some (e.g. *Lolium perenne*, Fig. 11.5) had no individuals that showed high Cu tolerance in the root extension test. In contrast, two grass species, *Agrostis capillaris* (unpolluted site ecotype) and *Dactylis glomerata*, had a few individuals, among the survivors on the most polluted soils, that had high Cu tolerance (see Fig. 11.5). It seems then, that in these populations from unpolluted sites, two out of the nine species had the ability to produce mutants that were Cu-tolerant, but the other seven did not.

Several other projects have also searched for the existence of Cu-tolerant or zinc (Zn)-tolerant individuals among large populations in non-contaminated sites. This was done either by testing many individuals for tolerance of the heavy metal; or by studying sites where Cu or Zn contamination was local, recording which species occurred on uncontaminated soil nearby and how many of these could survive in the contaminated area (Al-Hiyali *et al.* 1990; Bradshaw 1991). All the research was performed in Britain; I am not aware of similar work elsewhere. Taken together with the results of Gartside and McNeilly (1974), this indicates 11 species, all grasses, which in non-contaminated sites can have occasional individuals tolerant of Cu or Zn. Six of the species appeared in more than one of the studies. In contrast, there are 10 other grass species and 14 non-grass herbaceous species in which individuals tolerant to Cu or Zn have never been found. It appears that they do not have the ability to evolve Cu or Zn tolerance, or can do so only extremely rarely. Bradshaw (1991) discusses why this might be.

This inability of many species to produce heavy metal-tolerant ecotypes places a severe restriction on the restoration of vegetation on the sites: unless the toxic waste can be detoxified or covered with uncontaminated material, the vegetation that can be established will be restricted to a few species, as far as we know all of them grasses. If we can identify and isolate a gene for heavy metal tolerance it might be possible to transfer it to other species. This would only be possible if one or a few genes confer the tolerance. Macnair (1993) listed examples where this has been found to be the case. However, in other cases inheritance of heavy metal tolerance appears to be polygenic: the wide range of tolerance among *Agrostis capillaris* individuals in Fig. 11.5 (right-hand side) strongly suggests polygenic control there. An ecotype tolerant of one heavy metal is sometimes found to be tolerant of another, but often tolerance of each heavy metal is inherited independently. There are several possible types of tolerance mechanism. Reduced uptake of the toxic ion by the tolerant ecotype has occasionally been found, e.g. arsenic (Meharg & Macnair 1990, 1991). Other possibilities are chemicals in the plant that complex with the heavy metal and so take it out of solution within the cell (Steffens 1990), and enzymes within the plant that are less affected by the heavy metal (Mathys 1975). The mechanisms of heavy metal tolerance in plants are still little understood, which makes it difficult to devise ways of inserting tolerance into other species.

If plants do grow on polluted sites animals will eat them and ingest the heavy metal from them, which they will in turn pass on to carnivores.

Chapter 9 gives information about this movement of heavy metals along food chains, and the tissue concentrations that develop (see Tables 9.6 and 9.7). Heavy metals can have a substantial effect on the biomass and metabolic activity of microorganisms in soil; this includes reduced abundance of *Rhizobium*, leading to reduced nodulation and nitrogen fixation by legumes (Giller *et al.* 1998). These effects on microbes could interfere with attempts to restore fertile soil on waste sites.

Conclusions

- Forest and grassland have in many parts of the world established on abandoned farmland without active intervention by people. However, after a century or even several centuries the structure and species composition may still be markedly different from that of old forest or grassland in the same area.
- Seeds of some species fail to arrive, even over a long time, at apparently suitable sites, sometimes when seed sources are only 100 m or less away. Some invertebrates also fail to colonize.
- It may not be possible to return communities to the state before the first people arrived, for example because some species are now extinct.
- There are various possible aims for restoration. We need to be clear for each site what our aims are.
- Restoration can be greatly speeded up by planting required species. But some native plant species have proved very difficult to establish.
- Reintroducing animals also poses problems. For invertebrates these relate especially to the large number of species. A large mammal may require a large area of suitable habitat.
- Fresh mine waste is often unfavourable for plant growth because of coarse particle size and low organic matter content. Unassisted, vegetation tends to require decades to establish. It can be speeded up by adding waste organic matter, adding mineral nutrients, planting legumes.
- A substantial proportion of plant species appear to be unable to evolve heavy metal tolerance, and this places a restriction on what vegetation can be established on mine sites polluted by heavy metals.

Further reading

Restoration, general:
Meffe & Carroll (1994) Chapter 14
Jordan, Gilpin & Aber (1987)
Bowles & Whelan (1994)

Temperate forests:
Peterken (1996), especially Chapter 15

Heavy metal tolerance:
Macnair (1993)

Glossary

This glossary has several functions.

1 It gives the meaning of specialist terms and abbreviations used in the book. If a term is not defined here, the index may lead you to a longer explanation within the book.

2 If a species has been called by its English name (only) in the text, the Latin (scientific) name is given here. However, the Latin name is not given for common domestic species, such as cow and wheat.

3 If a species was called only by its Latin name in the text an English name is given here, or else some indication of which major group it belongs to.

Acer = maple. *Acer campestre* = field maple; *A. saccharum* = sugar maple.

Aerosol. A solid particle or liquid droplet so small that it remains suspended in air almost indefinitely.

Aggregate = soil crumb. Formed by numerous soil mineral particles (e.g. individual clay particles) bound together, though with pores between them able to hold water. A water-stable aggregate retains its structure even when sprayed with water or shaken in water.

Agrostis = bentgrass. *Agrostis capillaris* was formerly called *Agrostis tenuis*; *Agrostis vinealis* was formerly called *Agrostis canina*.

Al = aluminium.

Albedo = reflection coefficient. The proportion of the incoming short-wave radiation that an object reflects.

Alder = *Alnus*.

Alewife = *Alosa pseudoharengus*.

Alfalfa = lucerne = *Medicago sativa*.

Amaranthus hybridus = amaranth, Prince's feather.

Amazonia. The region of South America drained by the Amazon River and its tributaries; or the lowland part of that region.

Andropogon scoparius = *Schizachyrium scoparium* = little bluestem. A bunchgrass.

Anemone, wood = *Anemone nemorosa*.

Anchovy = anchoveta = *Engraulis*. Peruvian anchoveta = *Engraulis ringens*; anchovy off southern Africa = *E. capensis*.

Antelope, pronghorn = *Antilocapra americana*.

Aphid = Aphidoidea.

Aquifer. Where water is held underground in porous rock.

Artemia = brine shrimp. In the Anostraca (fairy shrimp) group of Crustacea.

Ash = *Fraxinus*.

Aspen. Several species of *Populus*.

Autotroph. An organism that does not obtain its energy from organic matter but by photosynthesis (photoautotroph), or by oxidizing inorganic materials.

Avocado = *Persea americana*.

Baboon = *Papio anubis*.

Badger = *Meles meles.*
Barramundi = *Lates calcarifer.*
Bear = *Ursus.* Grizzly bear = *Ursus arctos.*
Beaver = *Castor canadensis.*
Beech = *Fagus.*
Benthic. Species that live within the bottom deposit of a lake or sea, or on the surface of the deposit.
Bilberry = *Vaccinium myrtillus.*
Biocontrol. Biological control of pests.
Biodiversity. Diversity among living things. Sometimes broadened to mean 'nature' or natural communities, but not in this book.
Biomagnification. Increase in concentration of a substance in the tissues of living things as it passes up a food chain.
Biopesticide. Biological pesticide. See Box 8.1
Bioremediation. Using living things to remove or break down a toxic substance.
Birch = *Betula.*
Bison = *Bison bison.*
Blackbird = *Turdus merula.*
Black locust = *Robinia pseudoacacia* (a tree)
Bluebell = *Hyacinthoides nonscripta.*
Bluebird, mountain = *Sialia currucoides.*
Bluegill fish = *Lepomis macrochirus.*
Bluejay = *Cyanocitta cristata.*
BMWP. Biological Monitoring Working Party, who devised a scheme for assessing the degree of aquatic pollution by which invertebrates are present.
Boar, wild = *Sus scrofa.*
Boreal. Cold temperate region; cold winters, but summers warm enough for all soil to thaw. Boreal forest: native forest of boreal regions, in which conifers usually predominate.
Bouteloua curtipendula = side-oats grama. A bunchgrass.
BP. Years before present.
Bq. Bequerel. Amount of radioactive material that produces 1 disintegration per second.
Browser. Mammal that eats predominantly leaves and stem material of trees and bushes.
Brussels sprouts = *Brassica oleracea* var. *gemmifera.*
Buffalo, African = *Syncer caffer.* North American buffalo = bison.
Bushbuck = *Tragelaphus scriptus.*
Business-as-usual. Scenario for the future in which no attempt is made to restrain the use of fossil fuels and the production of greenhouse gases.
C3, C4, CAM photosynthesis. Three alternative carbon pathways involved in photosynthetic CO_2 fixation. In the C4 and CAM pathways there is an additional initial CO_2 capture and concentration step. In CAM photosynthesis the CO_2 is taken in mostly at night, the stomata can remain closed by day, and hence transpirational water loss per unit C fixed is reduced.
Ca = Calcium.
Cabbage white butterfly = *Pieris rapae.*
CAI. Current annual increment. New timber growth per year in a forest, usually expressed as $m^3\,ha^{-1}\,year^{-1}$.
Calluna vulgaris = heather (a low shrub).
CAM. Crassulacean acid metabolism. See under C3.
Camel = *Camelus bactrianus* + *C. dromedarius.*
Capricorn silvereye (an Australian bird) = *Zosterops lateralis chlorocephala.*
Carp. Common carp = *Cyprinus carpio*; grass carp = *Ctenopharyngodon idellus.*
Carya = hickory.
Cassava = *Manihot esculenta.*
Casuarina. Tropical and subtropical trees, sometimes known as she-oak.
Cd = Cadmium

Cedar = *Cedrus*. Red cedar (in eastern USA) = *Juniperus virginiana*.
CFC. Chlorofluorocarbon.
Cheatgrass = *Bromus tectorum*.
Checkerspot. Bay checkerspot butterfly = *Euphydryas editha bayensis*.
Cheetah = *Acinonyx jubatus*.
Cherry, black = *Prunus serotina*.
Chestnut blight (fungus) = *Cryphonectria parasitica*, formerly called *Endothia parasitica*.
Chimpanzee = *Pan troglodytes*.
Chipmunk = *Tamias striatus*.
Chironomids. Non-biting midges. In the group Nematocera, in the Diptera.
Cholinesterase. Enzyme which breaks down acetylcholine and is essential for the functioning of nervous systems.
Cl = Chlorine.
Cladoceran. Water flea, a group in the Crustacea.
Classical biological control. See Box 8.1.
Clear-felling. Felling all trees in a large area in one episode.
Clover = *Trifolium*. White clover = *Trifolium repens*.
Cod. Atlantic cod = *Gadus morhua*; Greenland cod = *G. ogac*.
Colorado beetle = *Leptinotarsa decemlineata*.
Copepod. Copepoda are a major group of Crustacea, mainly marine.
Coppice. System of tree harvesting in which several or many shoots grow from the cut stump; these are later harvested and the stump left for further cycles of growth and harvesting. The term also applies to the stems produced by this system.
Corn = maize = *Zea mays*.
Corncockle = *Agrostemma githago*.
Cougar = mountain lion = *Felis concolor*.
Cowpea = *Vigna sinensis* = *V. unguiculata*.
CPUE. Catch per unit effort. See Box 5.3.
Cross-resistance. Individuals that evolve resistance to one pesticide also simultaneously become resistant to another.
Cryphonectria parasitica. Parasitic fungus that causes chestnut blight.
Cs = Caesium.
Cu = Copper.
Dactylis glomerata = cock's-foot grass.
Daphnia. In the Cladocera (water flea) group of the Crustacea.
Deer. Red deer = *Cervus elaphus*; white-tailed deer = *Odocoileus virginianus*.
Demersal. Living in water, near the bottom.
Denitrification. Conversion by microorganisms of nitrate to the gases nitrous oxide and nitrogen.
Desertification. Decrease of vegetation and degradation of soil in semiarid areas. A word not precisely defined and therefore not often used in this book.
Deterministic model. A mathematical model that allows for no random variation. So a particular set of input values will always lead to exactly the same predicted outcome.
Diapause. A period during the life of an insect, e.g. during the larval stage, when the metabolic rate is greatly reduced and the insect becomes better able to survive unfavourable conditions. Equivalent to dormancy, but applies only to insects.
Dogwood = *Cornus*.
Dog's mercury = *Mercurialis perennis*.
Dolphin. Aquatic mammals, mostly in family Delphinidae.
Dormouse = *Muscardinus avellanarius*.
Douglas fir = *Pseudotsuga menziesii*. A conifer.
Eagle. Bald eagle = *Haliaeetus leucocephalus*; golden eagle = *Aquila chrysaeetos*.
EC_{50}. (EC = effective concentration). Concentration of chemical that causes 50% reduction in measured activity of a species, e.g. in growth rate or respiration rate.
Ecotoxicology. Study of the effects and fate of potentially toxic chemicals in the environment.

Ecotypes. Genetically different populations within a species that are adapted to local environmental conditions.
Elephant. African elephant = *Loxodonta africana*; Indian elephant = *Elephas maximus*.
Elk = moose = *Alces alces*.
Epizootic. Sudden and temporary increase of a disease in an animal population.
Eucalyptus = eucalypt. Trees of the southern hemisphere, especially in Australia. *Eucalyptus regnans* = mountain ash.
Eutrophic. Applied to water of rivers, lakes and oceans: contains mineral nutrients in sufficient concentrations to support rapid algal growth.
Evapotranspiration. Total loss of water vapour by transpiration from plant plus evaporation from any surface (e.g. soil water, rain on plant surfaces).
Falcon, peregrine = *Falco peregrinus*.
FAO. Food and Agriculture Organization of the United Nations.
Farmyard manure. Mixture of plant material (often straw) with faeces and urine from farm animals.
Fe = Iron.
Festuca ovina = sheep's fescue (a grass).
Fir = *Abies*. Balsam fir = *Abies balsamea*.
Fixation of nitrogen. Incorporation of N from gaseous N_2 into an organic compound.
Flea. Siphonaptera. Rabbit flea = *Spilopsyllus cuniculi*.
Foraminifera. Protozoa which produce a complex shell.
Forb. A herbaceous plant other than a grass.
Forest. Vegetation in which trees form a continuous canopy.
Fox = *Vulpes*. European fox = *Vulpes vulpes*; swift fox = *V. velox*.
FYM. Farmyard manure.
G = giga = $\times 10^9$.
Gazelle. Grant's gazelle = *Gazella granti*; Thomson's gazelle = *G. thomsonii*.
gbh. Girth at breast height, i.e. circumference of tree trunk at about 1.5 m above the ground.
Genetic drift. Change in the genetic composition of a population due to chance processes, not selection.
Giraffe = *Giraffa camelopardalis*.
Gorse = *Ulex*.
Graminoids. Grasses, sedges and rushes.
Gull, herring = *Larus argentatus*.
Guppy = *Poecilia reticulata*.
Gypsy moth = *Lymantria dispar*.
h. Hour.
ha. Hectare = 10^4 m^2.
Haddock = *Melanogrammus aeglefinus*.
Halesia carolina = silver-bell tree.
Half-life. The time taken for a radioactive isotope to lose half its radioactivity. Can be applied to some populations of living things: time taken for half of population to die.
Halocarbon. An organic chemical containing fluorine, chlorine, bromine or iodine.
Halophyte. A plant species that occurs naturally on saline soil.
Hardwood. In temperate regions: dicotyledonous tree species, and forests in which these dominate (in contrast to conifers which are called softwoods). In tropics: tree species that have dense, strong wood; they are usually slow-growing.
Hare, Canadian or snowshoe = *Lepus americanus*.
Hawthorn = *Crataegus*.
Hazel = *Corylus avellana*.
Heather. In this book means *Calluna vulgaris*; but *Erica* spp. are also heathers. Low shrubs.
Heavy metal. Metal with relative density greater than 5, e.g. cadmium, copper, lead, nickel, zinc.
Hemlock = *Tsuga*. A North American tree. (A poisonous herb of Europe, called hemlock in Britain, is not mentioned in this book.)

Herring, Atlantic = *Clupea harengus*.
Heterotroph. An organism that obtains its energy from organic material.
Hg = Mercury.
Hickory = *Carya*.
Hippopotamus = *Hippopotamus amphibius*.
Holocene. Postglacial period = last 10 000 years.
Humus. Dark-coloured organic material in soil. It is amorphous, i.e. so far decomposed that no detectable cell structures remain.
Ibex, mountain = *Capra ibex ibex*.
Impala = *Aepyceros melampus*.
Isohyet. Line on map joining points which have the same rainfall.
ITCZ. Intertropical Convergence Zone. See p. 60.
J. Joule.
Jay, blue = *Cyanocitta cristata*.
K = Potassium
k, K Kilo = $\times 10^3$.
Kaoliang = *Sorghum nervosum*, a cereal
Kinglet, golden-crowned = *Regulus satrapa*.
Klamath-weed = St John's wort = *Hypericum perforatum*.
Kudu, greater = *Tragelaphus strepsiceros*.
Ladybird = vedalia beetle = *Rodolia cardinalis*.
LAI. Leaf area index = Total area of leaves/Area of ground
Larch = *Larix*.
Lark, horned = *Eremophila alpestris*.
Lark bunting = *Calamospiza melanocorys*.
Latent heat of evaporation. Energy required to convert liquid water to vapour (at the same temperature).
LC_{50}. (LC = lethal concentration.) Concentration of a chemical in water surrounding a population that is sufficient to kill 50% of them.
LD_{50}. (LD = lethal dose.) When this amount of a chemical is fed to each animal it kills 50% of the individuals.
LEA protein. Late embryogenesis abundant protein. These proteins increase in plant tissue after water stress, and may confer drought or salt tolerance.
Leaf-hopper = Auchenorhyncha. In Hemiptera (bugs).
Legume. Member of flowering plant family Fabaceae (formerly Leguminosae). Many of them form a symbiotic association with *Rhizobium* bacteria, which live in nodules on their roots and fix nitrogen from the air.
Lemming, Hudson Bay collared = *Dichrostonyx hudsonius*.
Lepidoptera. Butterflies and moths.
Lettuce = *Lactuca sativa*.
Lion, African = *Panthera leo*.
Liriodendron tulipifera = yellow poplar, tulip tree.
Locust, black (North American tree) = *Robinia pseudoacacia*.
Lolium perenne = perennial ryegrass.
Long-wave radiation. Has wavelength more than 3 μm. See Box 2.1.
Lucerne = alfalfa = *Medicago sativa*.
Lynx, Canadian = *Felis lynx* = *Lynx canadensis*.
m Milli = $\times 10^{-3}$.
M Mega = $\times 10^6$.
Mahogany = *Swietenia macrophylla*.
MAI. Mean annual increment = (volume of timber in a forest stand)/(age of stand).
Maize = corn = *Zea mays*.
Maple = *Acer*. Red maple = *Acer rubrum*; sugar maple = *A. saccharum*.
Mark and recapture. See Box 5.3.
Mayfly = Ephemeroptera.
Melon = *Cucumis melo*.
Merlin = *Falco columbarius*.

Mg = Magnesium.
Milkfish = *Chanos chanos.*
Millet. Cereals; several species including *Setaria italica* and *Pennisetum glaucum.*
Minnow, fathead = *Pimephales promelas.*
Moa. Dinorthiformes. Flightless birds, all now extinct.
Mongoose, Indian = *Herpestes auropunctatus.*
Monkey. Blue monkey = *Cercopithecus mitis*; redtail monkey = *C. ascanius*; colobus monkey = *Colobus guereza.*
Monophagous. Animal that eats only one species of plant or animal.
Moose = elk = *Alces alces.*
Mosquito = *Culicidae.*
MPa. Megapascal, a unit of pressure. 1 MPa = 10 bars.
Mullet = *Mugil* and *Mullus* spp.
Musk ox = *Ovibos moschatus.*
MVP. Minimum viable population.
Mycoherbicide. Fungus used to kill weeds, by regular application, usually as spores.
Mycorrhiza. Symbiotic association of plant root and fungus. Arbuscular (= vesicular-arbuscular) mycorrhiza: the fungi form characteristic vesicles and arbuscules inside the root, but have little effect on the root's external appearance. This is the principal type of mycorrhiza in herbaceous plants and also occurs in some woody species.
n. Nano = $\times 10^{-9}$.
N = Nitrogen.
Na = Sodium.
Nardus stricta = mat-grass.
NDVI. Normalized difference vegetation index, a 'greenness index' used for assessing vegetation from satellites (see Box 3.2).
Nekton. Actively swimming aquatic animals, including fish.
Nettle = *Urtica dioica.*
Ni = Nickel.
Nutcracker = *Nucifraga.* Clark's nutcracker = *Nucifraga columbiana*; Eurasian nutcracker = *N. caryocatactes.*
Oak = *Quercus.* Chestnut oak = *Q. prinus*; red oak = *Q. rubra* = *Q. borealis.*
Octanol. Octyl alcohol. $CH_3(CH_2)_7OH$.
OECD. Organization for Economic Cooperation and Development.
Opuntia = prickly pear cactus.
Orange = *Citrus aurantium.*
Organochlorine. Any organic compound that includes chlorine.
Organophosphorus. Any organic compound that includes phosphorus.
Osmoticum. Any dissolved substance which creates an osmotic effect.
Osprey = *Pandion haliaetus.*
Owl. Northern spotted owl = *Strix occidentalis caurina.*
p Pico = $\times 10^{-12}$.
P = Phosphorus.
P = probability.
Panda, giant = *Ailuropoda melanoleuca.*
Panther, Florida = *Felis concolor coryi.*
Parasitoid. An animal that is a parasite at one stage of its lifecycle but free-living at another. Especially applied to insects whose larvae are parasitic on another insect species but whose adult stage is free-living.
Pb = Lead.
PCB. Polychlorinated biphenyl. See Fig. 9.10.
Pea = *Pisum sativum.*
Pelagic. Lives suspended in water.
Penicillium. Fungi, saprophytes, in the Ascomycetes. Some produce the antibiotic penicillin.
Peregrine falcon = *Falco peregrinus.*

Phenolic. Any compound containing a benzene ring with one or more OH groups directly attached to it, e.g. Figure 9.10(a) and (b).

Pheromone. A chemical emitted by an animal as a signal to another individual, e.g. to attract a potential mate.

Photic. Upper layer of water in an ocean or lake, in which the light intensity is great enough to allow photosynthesis.

Photochemical reaction. A chemical reaction which is promoted by light.

Photorespiration. Oxidation of ribulose-1,5-bisphosphate, which happens in light and is catalysed by Rubisco, the same enzyme that catalyses the reaction of this substance with CO_2. Photorespiration therefore competes against photosynthesis for substrate and reduces photosynthetic carbon fixation.

Photovoltaic cell. A physical device which when illuminated generates an electric current.

Phytophthora cinnamomi. Plant-parasitic fungus, in Phycomycetes. Causes root rot and dieback in many woody species.

Phytoplankton. Planktonic algae and cyanobacteria; can include photosynthetic protozoa.

Pine = *Pinus*. Lodgepole pine = *P. contorta*; Scots pine = *P. sylvestris*; white pine = *P. strobus*.

Pioneer. A tree species whose seedlings can establish and grow only in a large area free of trees, not in undisturbed forest or small gaps.

Plaice, European = *Pleuronectes platessa*.

Plankton. Small organisms that live suspended in water. Some are unicellular, some multicellular, but all are small enough to be much affected by currents. See Box 5.1.

Pleistocene. Period starting 2 million years ago and ending 10 000 years ago. Comprised a series of glacials (cold periods) and interglacials (warmer periods).

Plover, mountain = *Charadrius montanus*.

Polygenic. A character controlled by several or many genes.

Polypeptide. Several or many amino acids linked together.

Polysaccharide. Several or many monosaccharide (simple sugar) molecules linked together.

Poplar = *Populus*. Yellow poplar = *Liriodendron tulipifera*.

Postglacial period. Holocene = last 10 000 years.

Productivity. The rate at which an organism captures energy or increases in dry matter. Primary productivity: productivity by photosynthetic organisms. Net primary productivity: increase in dry matter content or energy content of plants, after allowing for loss by their own respiration. Secondary productivity: productivity by heterotrophs.

Quercus = oak. *Quercus prinus* = chestnut oak; *Q. rubra* = red oak.

Rabbit, European = *Oryctolagus cuniculus*.

Radiation. Long-wave, short-wave, photosynthetically active: see Box 2.1.

Radiocarbon dating. See Box 2.5.

Rape = *Brassica napus*.

Recruitment. Applied to fish: the number of fish which, in a particular year have joined the catchable stock, i.e. have during the year grown large enough to be caught by the nets used.

Red scale = *Aonidiella aurantii*.

Reindeer = *Rangifer tarandus*.

Rhinoceros, black = *Diceros bicornis*; white = *Ceratotherium simum*.

Rhizobium. Bacteria which form symbiotic N-fixing association in nodules on roots of leguminous plants.

Rhizosphere. The region of soil very close to an individual root. The abundance of microorganisms is higher than elsewhere in the soil, because of organic materials from the root and dying root cells.

Robin, American = *Turdus migratorius*; European = *Erithacus rubecula*.

Robinia pseudoacacia = black locust.

Rubisco. Ribulose-1,5-bisphosphate carboxylase/oxygenase, a key enzyme in CO_2 capture in photosynthesis.

Ruminant. A herbivorous mammal whose stomach includes a rumen, a chamber containing microorganisms that break down cellulose. For examples see Box 6.1.
Rust fungus = Uredinales, in the Basidiomycetes. Obligate plant parasites.
Ryegrass = *Lolium*. Perennial ryegrass = *Lolium perenne*.
S = Sulphur.
Sahel. Region in Africa south of the Sahara Desert, where the climate is semiarid.
Salmon, Atlantic = *Salmo salar*.
Sapsucker, red-naped = *Sphyrapicus nuchalis*.
Sardine, Peruvian = southern African = *Sardinops sagax*.
Savanna. Tropical vegetation in which perennial grasses are prominent. Trees or shrubs may be present, but they do not form a continuous canopy.
Secondary chemical. Compound produced within a plant which is not involved in primary metabolism, such as respiration or photosynthesis.
Senecio vulgaris = groundsel.
Serpentine. A basic rock with high magnesium content.
Shag = *Phalacrocorax aristotelis*.
Shannon Index = Shannon–Wiener Index. A measure of species diversity. See Box 10.6.
Sheep = *Ovis*. Bighorn sheep = *Ovis canadensis*.
Shelterwood system. See Box 7.3.
Shrew = *Sorex*. Common shrew = *Sorex araneus*; long-tailed shrew = *S. dispar*.
Silt. Soil particles 2–20 μm or 2–50 μm across (definitions vary).
Silver-bell tree = *Halesia carolina*.
Silver-spotted skipper butterfly = *Hesperia comma*.
Silver-studded blue butterfly = *Plebejus argus*.
Silviculture. Care and management of a forest; often refers to a plantation, after the trees have been planted.
Simpson Index of dominance. See Box 10.6.
Smelt. Several genera, in family Osmeridae.
Sole, common = *Solea vulgaris*.
Sorghum = *Sorghum bicolor* = *S. vulgare*. A cereal.
Sparrowhawk = *Accipiter nisus*.
Springbok = *Antidorcas marsupialis*.
Spruce = *Picea*. Engelmann's spruce = *P. engelmannii*; red spruce = *P. rubens*; sitka spruce = *P. sitchensis*.
Squirrel, North American = *Sciurus* (grey), *Tamiasciurus* (red); 13-lined ground squirrel = *Spermophilus tridecemlineatus*.
Stoat = *Mustela erminea*.
Stochastic model. Mathematical model which includes the possibility of chance events or random variation.
Stock. Applied to fish: the number or biomass of fish (or of one species of fish) in the area that are large enough to be caught by the nets used.
Stocking density. Applied to grazing mammals: the number of animals per unit ground area (e.g. per hectare).
Sugar-cane = *Saccharum officinarum*.
Sunflower = *Helianthus annuus*.
Swallow. Several genera, in family Hirundinidae. Cliff swallow = *Hirundo pyrrhonota*.
Sycamore = *Acer pseudoplatanus*.
t = tonne = metric ton = 10^6g.
T Tera = $\times 10^{12}$.
Takahe = *Notornis mantellii*. A flightless bird of New Zealand.
Tannin. Secondary chemicals in plants which combine with proteins; hence they often inactivate enzymes. Many tannins are polyphenolics.
Teleosts. The major group of fish, containing most of the bony fishes.
Thrip. Thysanoptera.
Thylakoid membrane. Within each chloroplast stacks of thylakoid membranes, containing chlorophyll, are a key structure in light capture for photosynthesis.
Tit, great = *Parus major*; marsh = *P. palustris*.

Treecreeper = *Certhia familiaris.*
Tsuga canadensis = eastern hemlock.
Trout, lake (= char) = *Salvelinus namaycush*; rainbow trout = *Oncorhynchus mykiss.*
Tuna. Several species, mostly in genus *Thunnus.*
Turbot = *Psetta maxima.*
UN. United Nations.
UNEP. United Nations Environment Programme.
Ungulate. A hooved herbivorous mammal.
Vetch = *Vicia.*
Virulence. How much a disease-causing organism affects the attacked species; often assessed by what percentage of infected individuals die and how quickly they die.
Warbler, yellow-rumped = *Dendroica coronata.*
Warthog = *Phacochoerus africanus.*
Wasp. Gall wasp: Cynipidae.
Water use efficiency. Amount of plant growth per amount of water used. There are several variants of this definition: see p. 64–5.
Waterbuck = *Kobus ellipsiprymnus.*
Waxwing, cedar = *Bombycilla cedrorum.*
Weevil = Cuculionidae, in the Coleoptera.
Whale, humpback = *Megaptera novaeanglie.*
Wildebeest = *Connochaetes taurinus.*
Wolf = *Canis.* Grey wolf of Europe and N. America = *Canis lupus.*
Woodchuck = *Marmota monax.*
Woodland. Vegetation in which trees are present but are far enough apart to form an open canopy.
Woodlouse = *Isopoda.* In the Crustacea.
Woodpecker, black-backed = *Picoides arcticus*; three-toed = *P. tridactylus.*
Wren, house = *Troglodytes aedon.*
Zebra = *Equus burchelli.*
Zn = Zinc.
Zooplankton Small animals that are a component of plankton. See Boxes 5.1, 5.2.
α, β-, γ-diversity. See Box 10.6.
μ Micro = $\times 10^{-6}$.

References

If there are more than four authors, the first only is named, followed by *et al.* Latin names have been omitted from some titles. [] indicates that the title has been shortened in some other way.

Abaye, A.O., Allen, V.G. & Fontenot, J.P. (1994). Influence of grazing cattle and sheep together and separately on animal performance and forage quality. *Journal of Animal Science* **72,** 1013–1022.

Abbott, I. & Loneragan, O. (1986). [*Ecology of Jarrah* (Eucalyptus marginata) *in the Northern Jarrah Forest of Western Australia*]. Department of Conservation and Land Management, Perth.

Ahrens, W.H. & Stoller, E.W. (1983). Competition, growth rate, and CO_2 fixation in triazine-susceptible and resistant smooth pigweed (*Amaranthus hybridus*). *Weed Science* **31,** 438–444.

Al-Hiyaly, S.A.K., McNeilly, T. & Bradshaw, A.D. (1990). The effects of zinc contamination from electricity pylons. Contrasting patterns of evolution in five species. *New Phytologist* **114,** 183–190.

Alabaster, J.S., Garland, J.H.N., Hart, I.C., de Solbe, J.F.L.G. (1972). An approach to the problem of pollution and fisheries. *Symposia of the Zoological Society of London* **29,** 87–114.

Alexander, M. (1994). *Biodegradation and Bioremediation*. Academic Press, London.

Allaby, M. (1998). *A Dictionary of Ecology*, 2nd edn. Oxford University Press, Oxford.

Allden, W.G. & Whittaker, I.A.M. (1970). [The determinants of herbage intake by grazing sheep]. *Australian Journal of Agricutural Research* **21,** 755–766.

Altman, P.L. & Dittmer, D.S. (1968). *Metabolism*. Federation of American Societies for Experimental Biology, Bethesda, MD.

Alverson, W.S., Waller, D.M. & Solheim, S.L. (1988). Forests too deer: edge effects in northern Wisconsin. *Conservation Biology* **2,** 348–358.

Ambrose, J.P. & Bratton, S.P. (1990). Trends in landscape heterogeneity along the borders of Great Smoky Mountains National Park. *Conservation Biology* **4,** 135–143.

Amos, B. & Hoelzel, A.R. (1992). Applications of molecular genetic techniques to the conservation of small populations. *Biological Conservation* **61,** 133–144.

Ampong-Nyarko, K., Seshu Reddy, K.V. & Saxena, K.N. (1994). *Chilo partellus* oviposition on non-hosts: a mechanism for reducing pest incidence in intercropping. *Acta Oecologica* **15,** 469–475.

Anagnostakis, S.L. (1995). The pathogens and pests of chestnut. *Advances in Botanical Research* **21,** 125–145.

Andersen, M.C. & Mahato, D. (1995). Demographic models and reserve designs for the Californian spotted owl. *Ecological Applications* **5,** 639–647.

Anderson, P. & Radford, E. (1994). [Changes in vegetation following reduction in grazing in Derbyshire, England]. *Biological Conservation* **69,** 55–63.

Anderson, R.M., ed. (1982). *Population Dynamics of Infectious Diseases*. Chapman & Hall, London.

Anderson, R.M., Jackson, H.C., May, R.M. & Smith, A.M. (1981). Population dynamics of fox rabies in Europe. *Nature* **289,** 765–771.

Anderson, R.M. & May, R.M. (1979). Population biology of infectious diseases. *Nature* **280,** 361–367.

Anderson, R.M. & May, R.M. (1986). The invasion, persistence and spread of infectious diseases within animal and plant communities. *Philosophical Transactions of the Royal Society B* **314,** 533–570.

Andreasen, C., Stryhn, H. & Streibig, J.C. (1996). Decline of the flora in Danish arable fields. *Journal of Applied Ecology* **33,** 619–626.

Andren, H. & Angelstam, P. (1988). [Elevated predation rates as an edge effect in habitat islands]. *Ecology* **69,** 544–547.

Annala, J.H. (1996). New Zealand's ITQ system: have the first eight years been a success or a failure? *Reviews in Fish Biology and Fisheries* **6,** 43–62.

Armbruster, P. & Lande, R. (1993). A population viability analysis for African elephant: how big should reserves be? *Conservation Biology* **7,** 602–610.

Arnason, R. (1996). On the ITQ fisheries management system in Iceland. *Reviews in Fish Biology and Fisheries* **6,** 63–90.

Ashenden, T.W. (1993). Sulphur dioxide pollution. In: *Methods in Comparative Plant Ecology* (eds G.A.F. Hendry & J.P. Grime), pp. 65–68. Chapman & Hall, London.

Ashenden, T.W. *et al.* (1996). Responses to SO_2 pollution in 41 British herbaceous species. *Functional Ecology* **10,** 483–490.

Ashworth, A.C. (1997). The response of beetles to Quaternary climate changes. In: *Past and Future Rapid Environmental Changes* (eds B. Huntley *et al.*), pp. 119–127. Springer, Berlin.

Asman, W.A.H., Slanina, J. & Baard, J.H. (1981). Meteorological interpretation of the chemical composition of rain-water at one measuring site. *Water, Air and Soil Pollution* **16,** 159–175.

Bailey, K.M. & Houde, E.D. (1989). Predation on eggs and larvae of marine fishes and the recruitment problem. *Advances in Marine Biology* **25,** 1–83.

Baker, K.F. & Cook, R.J. (1974). *Biological Control of Plant Pathogens.* Freeman, San Francisco, CA.

Bañuls, J., Legaz, F. & Primo-Millo, E. (1991). Salinity–calcium interactions on growth and ionic concentrations of Citrus plants. *Plant and Soil* **133,** 39–46.

Barlow, N.D. (1996). The ecology of wildlife disease control: simple models revisited. *Journal of Applied Ecology* **33,** 303–314.

Barnabé, G., ed. (1994). *Aquaculture.* Ellis Horwood, New York.

Barnes, R.S.K. & Hughes, R.N. (1999). *An Introduction to Marine Ecology,* 3rd edn. Blackwell Science, Oxford.

Barnes, R.S.K. & Mann, K.H. (1991). *Fundamentals of Aquatic Ecology.* Blackwell Science, Oxford.

Barone, J.A. (1998). Host-specificity of folivorous insects in a moist tropical forest. *Journal of Animal Ecology* **67,** 400–409.

Barrow, N.J. (1961). Mineralization of nitrogen and sulphur from sheep faeces. *Australian Journal of Agricultural Research* **12,** 644–650.

Battiston, G.A. *et al.* (1991). Transfer of Chernobyl fallout radionuclides from feed to growing rabbits: cesium-137 balance. *Science of the Total Environment* **105,** 1–12.

Bayliss-Smith, T.P. (1982). *The Ecology of Agricultural Systems.* Cambridge University Press, Cambridge.

Bazzaz, F.A. & Miao, S.L. (1993). Successional status, seed size, and responses of tree seedlings to CO_2, light, and nutrients. *Ecology* **74,** 104–112.

Beare, M.H., Hendrix, P.F. & Coleman, D.C. (1994). Water-stable aggregates and organic matter fractions in conventional and no-tillage soils. *Soil Science Society of America Journal* **58,** 777–786.

Beckett, P.H.T. & Davis, R.D. (1977). Upper critical levels of toxic elements in plants. *New Phytologist* **79,** 95–106.

Beddington, J.R., Free, C.A. & Lawton, J.H. (1978). Characteristics of successful natural enemies in models of biological control of insect pests. *Nature* **273,** 513–519.

Begon, M., Harper, J.L. & Townsend, C.R. (1996). *Ecology,* 3rd edn. Blackwell Science, Oxford.

Behrenfeld, M.J. & Falkowski, P.G. (1997). Photosynthetic rates derived from satellite-based chlorophyll concentration. *Limnology and Oceanography* **42,** 1–20.
Beier, P. (1993). Determining minimum habitat areas and habitat corridors for cougar. *Conservation Biology* **7,** 94–108.
Belsky, A.J. (1992). Effects of grazing, competition, disturbance and fire on species composition and diversity in grassland communities. *Journal of Vegetation Science* **3,** 187–200.
Benitez-Malvido, J. (1998). Impact of forest fragmentation on seedling abundance in a tropical rain forest. *Conservation Biology* **12,** 380–389.
Bennett, A.F., Henein, K. & Merriam, G. (1994). Corridor use and the elements of corridor quality: chipmunks and fencerows in a farmland mosaic. *Biological Conservation* **68,** 155–165.
Bennett, K.D. (1983). [Devensian Late-glacial and Flandrian vegetational history of Hockham Mere, Norfolk, England]. *New Phytologist* **95,** 457–487.
Bennett, K.D. (1986). Competitive interactions among forest tree populations in Norfolk, England, during the last 10 000 years. *New Phytologist* **103,** 603–620.
Bennett, R.N. & Wallsgrove, R.M. (1994). Secondary metabolites in plant defense mechanisms. *New Phytologist* **127,** 617–633.
Berendse, F., Oomes, M.J.M., Altena, H.J. & Elberse, W.T. (1992). Experiments on the restoration of species-rich meadows in the Netherlands. *Biological Conservation* **62,** 59–65.
Berg, B., Ekbohm, G., Söderström, B. & Staaf, H. (1991). Reduction of decomposition rates of Scots pine needle litter due to heavy-metal pollution. *Water, Air and Soil Pollution* **59,** 165–177.
Berger, J. (1990). Persistence of different sized populations: an empirical assessment of rapid extinctions in bighorn sheep. *Conservation Biology* **4,** 91–98.
Berglund, S., Davis, R.D. & L'Hermite, P., eds. (1984). *Utilisation of Sewage Sludge on Land*. Reidel, Dordrecht.
Berner, E.K. & Berner, R.A. (1996). *Global Environment: Water, Air and Geochemical Cycles*. Prentice Hall, Upper Saddle River, NJ.
Berner, R.A. (1998). The carbon cycle and CO_2 over Phanerozoic time: the role of land plants. *Philosophical Transactions of the Royal Society B* **353,** 75–82.
Betts, W.B. (1991). *Biodegradation*. Springer, London.
Beverton, R.J.H. & Holt, S.J. (1956). The theory of fishing. In: *Sea Fisheries: Their Investigation in the United Kingdom* (ed. M. Graham), pp. 372–441. Arnold, London.
Bewley, R. *et al.* (1989). Microbial clean-up of contaminated soil. *Chemistry and Industry* **1989,** 778–783.
Bezemer, T.M. & Jones, T.H. (1998). [Plant–insect herbivore interactions in elevated atmospheric CO_2]. *Oikos* **82,** 212–222.
Bierregaard, R.O. & Lovejoy, T.E. (1989). Effects of forest fragmentation on Amazonian understorey bird commmunities. *Acta Amazonica* **19,** 215–241.
Billing, E. (1981). Hawthorn as a source of fireblight. In: *Pests, Pathogens and Vegetation* (ed. J.M. Thresh), pp. 121–130. Pitman, Boston.
Birks, H.J.B. (1989). Holocene isochrone maps and patterns of tree-spreading in the British Isles. *Journal of Biogoegraphy* **16,** 503–540.
Black, J.L. & Kenney, P.A. (1984). Factors affecting diet selection by sheep. II. Height and density of pasture. *Australian Journal of Agricultural Research* **35,** 565–578.
Bland, W. & Rolls, D. (1998). *Weathering*. Arnold, London.
Bockman, O.C., Kaarstad, O., Lie, O.H. & Richards, I. (1990). *Agriculture and Fertilizers*. Norsk Hydro, Oslo.
Boddey, R.M. *et al.* (1995). [Biological nitrogen fixation associated with sugar cane and rice]. *Plant and Soil* **174,** 195–209.
Boddey, R.M., Urquiaga, S., Reis, V. & Dobereiner, J. (1991). Biological nitrogen fixation associated with sugar cane. *Plant and Soil* **137,** 111–117.
Boitani, L. (1992). Wolf research and conservation in Italy. *Biological Conservation* **61,** 125–132.

Boonyaratpalin, M. (1997). Nutrient requirements of marine food fish cultured in Southeast Asia. *Aquaculture* **151,** 283–313.

Bormann, F.H. & Likens, G.E. (1979). *Pattern and Process in a Forested Ecosystem.* Springer, New York.

Bowen, G.D. & Nambiar, E.K.S. (1984). *Nutrition of Plantation Forests.* Academic Press, London.

Bowles, M.L. & Whelan, C.J. (1994). *Restoration of Endangered Species.* Cambridge University Press, Cambridge.

Boyce, M.S. (1992). Population viability analysis. *Annual Review of Ecology and Systematics* **23,** 481–506.

Boyle, G., ed. (1996). *Renewable Energy.* Oxford University Press, Oxford.

Bradshaw, A.D. (1991). Genostasis and the limits to evolution. *Philosophical Transactions of the Royal Society B* **333,** 289–305.

Brady, N.C. & Weil, R.R. (1999). *The Nature and Properties of Soils,* 12th edn. Prentice Hall, Upper Saddle River, NJ.

Brafield, A.E. & Llewellyn, M.J. (1982). *Animal Energetics.* Blackie, Glasgow.

Bragg, J.R., Prince, R.C., Harner, E.J. & Atlas, R.M. (1994). Effectiveness of bioremediation for the Exxon Valdez oil spill. *Nature* **368,** 413–418.

Bray, E.A. (1993). Molecular responses to water deficit. *Plant Physiology* **103,** 1035–1040.

Brenchley, W.E. & Warington, K. (1958). *The Park Grass Plots at Rothamsted 1856–1949.* Rothamsted Experimental Station, Harpenden, England.

Brewer, R. (1994). *The Science of Ecology,* 2nd edn. Saunders, Fort Worth, TX.

Briggs, C.J., Hails, R.S., Barlow, N.D. & Godfray, H.C.J. (1995). Dynamics of insect–pathogen interactions. In: *Ecology of Infectious Diseases in Natural Populations* (eds B.T. Grenfell & A.P. Dobson), pp. 295–320. Cambridge University Press, Cambridge.

Bright, P.W. & Morris, P.A. (1991). Ranging and nesting behaviour of the dormouse in diverse low-growing woodland. *Journal of Zoology* **224,** 177–190.

Bright, P.W., Mitchell, P. & Morris, P.A. (1994). Dormouse distribution: survey techniques, insular ecology and selection of sites for conservation. *Journal of Applied Ecology* **31,** 329–339.

Bromfield, S.M. & Jones, O.L. (1970). The effect of sheep on the recycling of phosphorus in hayed-off pastures. *Australian Journal of Agricultural Research* **21,** 699–711.

Brook, B.W. & Kikkawa, J. (1998). [A population viability analysis on the Capricorn silvereye]. *Journal of Applied Ecology* **35,** 491–503.

Brown, C.R. & Brown, M.B. (1986). Ectoparasitism as a cost of coloniality in cliff swallows. *Ecology* **67,** 1206–1218.

Brown, K. (1994). [Approaches to valuing plant medicines]. *Biodiversity and Conservation* **3,** 734–750.

Brown, L.R., Renner, M. & Flavin, C. (1997). *Vital Signs 1997.* W.W. Norton, New York.

Brown, N.D. & Whitmore, T.C. (1992). Do dipterocarp seedlings really partition tropical rain forest gaps?. *Philosophical Transactions of the Royal Society B* **335,** 369–378.

de Bruijn, F.J., Jing, Y. & Dazzo, F.B. (1995). [Potentials and pitfalls of trying to extend symbiotic interactions of N-fixing organisms to presently non-nodulated plants]. *Plant and Soil* **174,** 225–240.

Brunner, W., Sutherland, F.H. & Focht, D.D. (1985). Enhanced biodegradation of polychlorinated biphenyls in soil by analog enrichment and bacterial inoculation. *Journal of Environmental Quality* **14,** 324–328.

Buck, K.W. (1988). Control of plant pathogens with viruses and related agents. *Philosophical Transactions of the Royal Society B* **318,** 295–317.

Bullock, J.M., Clear Hill, B., Silvertown, J. & Sutton, M. (1995). [Gap colonization as a source of grassland community change: effects of gap size and grazing on colonization by different species]. *Oikos* **72,** 273–282.

Buol, S.W. (1995). Sustainability of soil use. *Annual Review of Ecology and Systematics* **26,** 25–44.

Burbank, D.W. *et al.* (1996). Bedrock incision, rock uplift and threshold hillslopes in the northwestern Himalayas. *Nature* **379,** 505–510.

Burdon, J.J. & Chilvers, G.A. (1982). Host density as a factor in plant disease ecology. *Annual Review of Phytopathology* **20,** 143–166.

Bush, M.B. & Colinvaux, P.A. (1994). Tropical forest disturbance: paleoecological records from Darien, Panama. *Ecology* **75,** 1761–1768.

Butcher, S.S., Charlson, R.J., Orians, G.H. & Wolfe, G.V. (1992). *Global Biogeochemical Cycles*. Academic Press, London.

Butler, G.C., ed. (1978). *Principles of Ecotoxicology*. Wiley, Chichester.

Butterfield, J.E.L. & Coulson, J.C. (1997). [Terrestrial invertebrates and climate change]. In: *Past and Future Rapid Environmental Changes* (eds B. Huntley *et al.*), pp. 401–412. Springer, Berlin.

Cain, M.L., Damman, H. & Muir, A. (1998). Seed dispersal and the Holocene migration of woodland herbs. *Ecological Monographs* **68,** 325–347.

Caldwell, M.M. *et al.* (1981). Coping with herbivory: photosynthetic capacity and resource capture in two semiarid *Agropyron* bunchgrasses. *Oecologia* **50,** 14–24.

Calow, P. (1992). The three Rs of ecotoxicology. *Functional Ecology* **6,** 617–619.

Campbell, C.C. (1969). *Birth of a National Park in the Great Smoky Mountains*. University of Tennessee Press, Knoxville, TN.

Campbell, J.B. (1996). *Introduction to Remote Sensing*. Taylor & Francis, London.

Campbell, M.M. (1976). [Colonisation of *Aphytis melinus* in *Aonidiella aurantii* on citrus in South Australia]. *Bulletin of Entomological Research* **65,** 659–668.

Campbell, R. (1989). *Biological Control of Microbial Plant Pathogens*. Cambridge University Press, Cambridge.

Cannell, M.G.R. (1980). Productivity of closely-spaced young poplar on agricultural soils in Britain. *Forestry* **53,** 1–21.

Cape, J.N. *et al.* (1991). Sulphate and ammonium in mist impair the frost hardening of red spruce seedlings. *New Phytologist* **118,** 119–126.

Casella, E., Soussana, J.F. & Loiseau, P. (1996). Long-term effects of CO_2 enrichment and temperature increase on a temperate grass sward. I. Productivity and water use. *Plant and Soil* **182,** 83–99.

Castle, M.E., Foot, A.S. & Halley, R.J. (1950). Some observations on the behaviour of dairy cattle with particular reference to grazing. *Journal of Dairy Research* **17,** 215–230.

Caughley, G. & Gunn, A. (1996). *Conservation Biology in Theory and Practice*. Blackwell Science, Oxford.

Cerdá, X., Retana, J. & Cros, S. (1998). [Critical thermal limits in Mediterranean ant species]. *Functional Ecology* **12,** 45–55.

Ceulemans, R. & Mousseau, M. (1994). Effects of elevated atmospheric CO_2 on woody plants. *New Phytologist* **127,** 425–446.

Chaîneau, C.H. *et al.* (1999). Comparison of the fuel oil biodegradation potential of hydrocarbon-assimilating microorganisms isolated from a temperate agricultural soil. *Science of the Total Environment* **227,** 237–247.

Chatelain, C., Gautier, L. & Spichinger, R. (1996). A recent history of forest fragmentation in southwestern Ivory Coast. *Biodiversity and Conservation* **5,** 37–53.

Chu, L.M. & Bradshaw, A.D. (1996). The value of pulverized refuse fines as a substitute for topsoil in land reclamation. *Journal of Applied Ecology* **33,** 851–857.

Church, D.C., ed. (1988). *The Ruminant Animal*. Prentice-Hall, Englewood Cliffs, NJ.

Clapham, A.R. & Godwin, H. (1948). [Studies of the post-glacial history of British vegetation. The Somerset Levels]. *Philosophical Transactions of the Royal Society B* **233,** 233–273.

Clark, J.S. (1990). Fire and climate change during the last 750 yr in northwestern Minnesota. *Ecological Monographs* **60,** 135–159.

Clark, J.S., Royall, P.D. & Chumbley, C. (1996). The role of fire during climate change in an eastern deciduous forest at Devil's Bathtub, New York. *Ecology* **77,** 2148–2166.

Clark, K.E., Gobas, F.A.P.C. & Mackay, D. (1990). Model of organic chemical uptake and clearance by fish from food and water. *Environmental Science and Technology* **24,** 1203–1213.

Clark, T. *et al.* (1988). Wildlife monitoring, modeling, and fugacity. *Environmental Science and Technology* **22,** 120–127.

Clarke, C.M.H. (1971). Liberation and dispersal of deer in northern South Island districts. *New Zealand Journal of Forest Science* **1,** 194–207.

Clarke, J.L., Welch, D. & Gordon, I.J. (1995). [The influence of vegetation pattern on the grazing of moorland by red deer and sheep. I and II.]. *Journal of Applied Ecology* **32,** 166–176 & 177–186.

Clausen, J., Keck, D.D. & Hiesey, W.M. (1948). [*Experimental Studies on the Nature of Species*, 3]. Carnegie Institution of Washington Publication 581, Washington DC.

Clebsch, E.E.C. & Busing, R.T. (1989). Secondary succession, gap dynamics, and community structure in a southern Appalachian cove forest. *Ecology* **70,** 728–735.

Clemmons, J.R. & Buchholz, R. (1997). *Behavioural Approaches to Conservation in the Wild*. Cambridge Univerity Press, Cambridge.

Coates, K.D. & Burton, P.J. (1997). A gap-based approach for development of silvicultural systems to address ecosystem management objectives. *Forest Ecology and Management* **99,** 337–354.

Cocchio, L.A., Rodgers, D.W. & Beamish, F.W.H. (1995). Effects of water chemistry and temperature on radiocesium dynamics in rainbow trout. *Canadian Journal of Fisheries and Aquatic Sciences* **52,** 607–613.

Cochrane, K.L., Butterworth. D.S., De Oliveira, J.A.A. & Roel, B.A. (1998). Management procedures in a fishery based on highly variable stocks and with conflicting objectives: experiences in the South African pelagic fishery. *Reviews in Fish Biology and Fisheries* **8,** 177–214.

Coley, P.D. & Barone, J.A. (1996). Herbivory and plant defenses in tropical forests. *Annual Review of Ecology and Systematics* **27,** 305–335.

Collier, B.D., Cox, G.W., Johnson, A.W. & Miller, P.C. (1973). *Dynamic Ecology*. Prentice-Hall, Englewood Cliffs, NJ.

Collins, N.M., Sayer, J.A. & Whitmore, T.C., eds. (1991). *The Conservation Atlas of Tropical Forests. Asia and the Pacific*. Macmillan, London.

Condit, R. (1995). Research in large, long-term tropical forest plots. *TREE* **10,** 18–22.

Connell, J.H. (1979). Tropical rain forests and coral reefs as open non-equilibrium systems. In: *Population Dynamics* (eds R.M. Anderson, B.D. Turner & L.R. Taylor), pp. 141–163. Blackwell Science, Oxford.

Connolly, J. & Nolan, T. (1976). Design and analysis of mixed grazing experiments. *Animal Production* **23,** 63–71.

Connor, E.F. & McCoy, E.D. (1979). The statistics and biology of the species–area relationship. *American Naturalist* **113,** 791–833.

Cook, R.J. & Baker, K.F. (1983). *The Nature and Practice of Biological Control of Plant Pathogens*. American Phytopathological Society, St Paul, MN.

Cook, R.M. (1997). Stock trends in six North Sea stocks as revealed by an analysis of research vessel surveys. *ICES Journal of Marine Science* **54,** 924–933.

Cooke, G.W. (1976). A review of the effects of agriculture on the chemical composition and quality of surface and underground waters. In: *Agriculture and Water Quality. Technical Bulletin 32 of the Ministry of Agriculture, Fisheries and Food*, pp. 5–57. Her Majesty's Stationery Office, London.

Coope, G.R. (1987). The response of late Quaternary insect communities to sudden climatic changes. In: *Organization of Communities, Past and Present* (eds J.H.R. Gee & P.S. Giller), pp. 421–438. Blackwell Science, Oxford.

Coppock, D.L., Ellis, J.E. & Swift, D.M. (1986). Livestock feeding ecology and resource utilization in a nomadic pastoral ecosystem. *Journal of Applied Ecology* **23,** 573–583.

Corten, A. (1996). The widening gap between fisheries biology and fisheries management in the European Union. *Fisheries Research* **27,** 1–15.

Coughenour, M.B. *et al.* (1985). Energy extraction and use in a nomadic pastoral system. *Science* **230,** 619–625.

Cousens, R. & Mortimer, M. (1995). *Dynamics of Weed Populations.* Cambridge University Press, Cambridge.

Crawley, M.J. (1983). *Herbivory. The Dynamics of Animal–Plant Interactions.* Blackwell Science, Oxford.

Cresser, M., Killham, K. & Edwards, T. (1993). *Soil Chemistry and its Applications.* Cambridge University Press, Cambridge.

Crews, T.E. *et al.* (1995). Changes in soil phosphorus fractions and ecosystem dynamics across a long chronosequence in Hawaii. *Ecology* **76,** 1407–1424.

Crisp, D.T. (1966). [Input and output of minerals for an area of Pennine moorland]. *Journal of Applied Ecology* **3,** 327–348.

Crompton, G. & Sheail, J. (1975). [The historical ecology of Lakenheath Warren in Suffolk, England]. *Biological Conservation* **8,** 299–313.

Currie, D.J. (1991). Energy and large-scale patterns of animal- and plant-species richness. *American Naturalist* **137,** 27–49.

Curry, J.P. (1994). *Grassland Invertebrates.* Chapman & Hall, London.

Curtis, P.S. & Wang, X. (1998). A meta-analysis of elevated CO_2 effects on woody plant mass, form, and physiology. *Oecologia* **113,** 299–313.

Curtis, R.O., Glendenen, G.W., Reukema, D.L. & DeMars, D.K. (1982). *Yield Tables for Managed Stands of Coast Douglas-Fir.* USDA Forest Service General Technical Report PNW-135. Olympia, WA.

Cushing, D.H. (1981). *Fisheries Biology: a Study in Population Dynamics.* 2nd edn. University of Wisconsin Press. Madison, WI.

Cushing, D.H. (1990). Plankton production and year-class strength in fish populations: an update of the match/mismatch hypothesis. *Advances in Marine Biology* **26,** 249–293.

Cussans, G.W. (1995). Integrated weed management. In: *Ecology and Integrated Farming Systems* (eds D.M. Glen, M.P. Greaves & H.M. Anderson), pp. 17–29. Wiley, Chichester.

da Silva, J.G. *et al.* (1978). Energy balance for ethyl alcohol production from crops. *Science* **201,** 903–906.

Dalling, J.W., Hubbell, S.P. & Silvera, K. (1998). Seed dispersal, seedling establishment and gap partitioning among tropical pioneer trees. *Journal of Ecology* **86,** 674–689.

Dansgaard, W., White, J.W.C. & Johnsen, S.J. (1989). The abrupt termination of the Younger Dryas climate event. *Nature* **339,** 532–534.

Danson, F.M. & Plummer, S.E., eds. (1995). *Advances in Environmental Remote Sensing.* Wiley, Chichester.

Darwin, C. (1859). *The Origin of Species by Means of Natural Selection.* Murray, London.

Daubenmire, R.F. (1940). Plant succession due to overgrazing in the *Agropyron* bunchgrass prairie of southeastern Washington. *Ecology* **21,** 55–64.

Davidson, W.R., Hayes, F.A., Nettles, V.F. & Kellogg, F.E. (1981). *Diseases and Parasites of White-Tailed Deer.* Miscellaneous Publication no. 7 of Tall Timbers Research Station, Tallahassee, FL.

Davis, M.B. (1976). Erosion rates and land-use history in southern Michigan. *Environmental Conservation* **3,** 139–148.

Davis, M.B. (1981). Quaternary history and the stability of forest commmunities. In: *Forest Succession* (eds D.C. West, H.H. Shugart & D.B. Botkin), pp. 132–153. Springer, New York.

Day, R.J. (1972). Stand structure, succession, and use of southern Alberta's Rocky Mountain forest. *Ecology* **53,** 472–478.

Debach, P. & Rosen, D. (1991). *Biological Control by Natural Enemies,* 2nd edn. Cambridge University Press, Cambridge.

Delcourt, H.R. & Delcourt, P.A. (1991). *Quaternary Ecology.* Chapman & Hall, London.

Dempster, J.P. (1969). Some effects of weed control on the numbers of the small cabbage white (*Pieris rapae*) on Brussels sprouts. *Journal of Applied Ecology* **6,** 339–345.

Dennis, B., Munholland, P.A. & Scott, J.M. (1991). Examination of growth and extinction parameters for endangered species. *Ecological Monographs* **61,** 115–143.
Desrochers, A. & Hannon, S.J. (1997). Gap crossing decisions by forest songbirds during the post-fledging period. *Conservation Biology* **11,** 1204–1210.
De Vleeschauwer, D. & Lal, R. (1981). Properties of worm casts under secondary tropical forest regrowth. *Soil Science* **132,** 175–181.
Devonshire, A.L. & Field, L.M. (1991). Gene amplification and insecticide resistance. *Annual Review of Entomology* **36,** 1–23.
DeYoung, B. & Rose, G.A. (1993). On recruitment and distribution of Atlantic cod off Newfoundland. *Canadian Journal of Fisheries and Aquatic Sciences* **50,** 2729–2741.
Diamond, J.M. (1969). Avifaunal equilibria and species turnover rates on the Channel Islands of California. *Proceedings of the National Academy of Sciences* **64,** 57–63.
Dickinson, R.E. & Henderson-Sellers, A. (1988). Modelling tropical deforestation: a study of GCM land–surface parametrizations. *Quarterly Journal of the Royal Meteorological Society* **114,** 439–462.
Dirmeyer, P.A. & Shukla, J. (1996). The effect on regional and global climate of expansion of the world's deserts. *Quarterly Journal of the Royal Meteorological Society* **122,** 451–482.
Dixon, R.K. *et al.* (1994). Carbon pools and flux of global forest ecosystems. *Science* **263,** 185–190.
Drake, B.G., Gonzales-Meler, M.A. & Long. S.P. (1997). More efficient plants: a consequence of rising atmospheric CO_2? *Annual Review of Plant Physiology* **48,** 609–639.
Drury, S.A. (1998). *Images of the Earth: a Guide to Remote Sensing.* Oxford University Press, Oxford.
Dueck, T.H., Van der Eerden, L.J. & Berdowski, J.J.M. (1992). Estimation of SO_2 effect thresholds for heathland species. *Functional Ecology* **6,** 291–296.
Duncan, P. (1992). *Horses and Grasses.* Springer, New York.
Dunsford, S.J., Free, A.J. & Davy, A.J. (1998). Acidifying peat as an aid to the reconstruction of lowland heath on arable soil: a field experiment. *Journal of Applied Ecology* **35,** 660–672.
During, H.J. & Willems, J.H. (1984). Diversity models applied to chalk grassland. *Vegetatio* **57,** 103–114.
Dwyer, G., Levin, S.A. & Buttel, L. (1990). A simulation model of the population dynamics and evolution of myxomatosis. *Ecological Monographs* **60,** 423–447.
Dyke, G.V. *et al.* (1983). [The Broadbalk wheat experiment 1968–78: yields and plant nutrients in crops]. *Rothamsted Experimental Station Report for 1982, Part* **2,** 5–44.
Eamus, D. (1991). The interaction of rising CO_2 and temperatures with water use efficiency. *Plant, Cell and Environment* **14,** 843–852.
Edwards, C.A. & Lofty, J.R. (1982). The effect of direct drilling and minimal cultivation on earthworm populations. *Journal of Applied Ecology* **19,** 723–734.
Elliott, K.J., Boring, L.R., Swank, W.T. & Haines, B.R. (1997). Successional changes in plant species diversity and composition after clearcutting a southern Appalachian watershed. *Forest Ecology and Management* **92,** 67–85.
Ellstrand, N.C. & Elam, D.R. (1993). Population genetic consequences of small population size: implications for plant conservation. *Annual Review of Ecology and Systematics* **24,** 217–242.
Evans, D.G. & Miller, M.H. (1988). Vesicular-arbuscular mycorrhizas and the soil-disturbance-induced reduction of nutrient absorption in maize. *New Phytologist* **110,** 67–74.
Evans, J. (1992). *Plantation Forestry in the Tropics,* 2nd edn. Clarendon Press, Oxford.
Evans, L.T. (1997). Adapting and improving crops: the endless task. *Philosophical Transactions of the Royal Society B* **352,** 901–906.
Everson, I. (1992). Managing Southern Ocean krill and fish stocks in a changing environment. *Philosophical Transactions of the Royal Society B* **338,** 311–317.
Falkenmark, M. (1997). Meeting water requirements of an expanding world population. *Philosophical Transactions of the Royal Society B* **352,** 929–936.

Fan, S. *et al.* (1998). A large terrestrial carbon sink in North America implied by atmospheric and oceanic carbon dioxide data and models. *Science* **282,** 442–446.
FAO (1979). *Eucalypts for Planting*. Food and Agriculture Organization of the United Nations, Rome.
FAO Yearbooks. Published by the Food and Agriculture Organization of the United Nations, Rome.
Fearnside, P.M. (1987). Rethinking continuous cultivation in Amazonia. *Bioscience* **37,** 209–214.
Fenchel, T. (1992). What can ecologists learn from microbes: life beneath a square centimetre of sediment surface. *Functional Ecology* **6,** 499–507.
Fenner, F. (1983). Biological control, as exemplified by smallpox eradication and myxomatosis. *Proceedings of the Royal Society B* **218,** 259–285.
Ferreira, L.V. & Laurance, W.F. (1997). Effects of forest fragmention on mortality and damage of selected trees in central Amazonia. *Conservation Biology* **11,** 797–801.
Fierro, A., Angers, D.A. & Beauchamp, C.J. (1999). Restoration of ecosystem function in an abandoned sandpit: plant and soil responses to paper de-inking sludge. *Journal of Applied Ecology* **36,** 244–253.
Firbank, L.G. & Watkinson, A.R. (1986). Modelling the population dynamics of an arable weed and its effects upon crop yield. *Journal of Applied Ecology* **23,** 147–159.
Fischer, G. & Heilig, G.K. (1997). Population momentum and the demand on land and water resources. *Philosophical Transactions of the Royal Society B* **352,** 869–889.
Fitter, A.H. (1986). Spatial and temporal patterns of root activity in a species-rich alluvial grassland. *Oecologia* **69,** 594–599.
Floate, M.J.S. (1970a). [Mineralization of nitrogen and phosphorus from organic materials of plant and animal origin]. *Journal of the British Grassland Society* **25,** 295–302.
Floate, M.J.S. (1970b). [Decomposition of organic materials from hill soils and pastures. II]. *Soil Biology and Biochemistry* **2,** 173–185.
Flowers, T.J. & Yeo, A.R. (1986). Ion relations of plants under drought and salinity. *Australian Journal of Plant Physiology* **13,** 75–91.
Fogg, G.E. (1991). The planktonic way of life. *New Phytologist* **118,** 191–232.
Forbes, S.H. & Boyd, D.K. (1997). Genetic structure and migration in native and reintroduced Rocky Mountain wolf populations. *Conservation Biology* **11,** 1226–1234.
Forcella, F., Eradat-Oskoui, K. & Wagner, S.W. (1993). Application of weed seedbank ecology to low-input crop management. *Ecological Applications* **3,** 74–83.
Ford, H.A. (1987). Bird communities on habitat islands in England. *Bird Study* **34,** 205–218.
Ford, T.E., ed. (1993). *Aquatic Microbiology: an Ecological Approach*. Blackwell Science, Oxford.
Fordham, M. *et al.* (1997). The impact of elevated CO_2 on growth and photosynthesis in *Agrostis canina* L. ssp. *monteluccii* adapted to contrasting atmospheric CO_2 concentrations. *Oecologia* **110,** 169–178.
Fowler, P.J. (1981). *The Farming of Prehistoric Britain*. Cambridge University Press, Cambridge.
Frank, J.H. (1998). How risky is biological control? Comment. *Ecology* **79,** 1829–1834.
Frankham, R. (1995). Inbreeding and extinction: a threshold effect. *Conservation Biology* **9,** 792–799.
Freedman, B. (1989). *Environmental Ecology*. Academic Press, San Diego, CA.
French, R.J. & Schultz, J.E. (1984). [Water use efficiency of wheat in a Mediterranean-type environment. I & II]. *Australian Journal of Agricultural Research* **35,** 743–764 & 765–775.
Fritz, H. & Duncan, P. (1994). On the carrying capacity for large ungulates of African savanna ecosystems. *Proceedings of the Royal Society B* **256,** 77–82.
Fritz, H., de Garine-Wichatitsky, M. & Letessier, G. (1996). Habitat use by wild and domestic herbivores in an African savanna woodland: the influence of cattle spatial behaviour. *Journal of Applied Ecology* **33,** 589–598.

Fryrear, D.W. (1995). Soil losses by wind erosion. *Soil Science Society of America Journal* **59,** 668–672.

Fussell, G.E. (1966). *Farming Technique from Prehistoric to Modern Times*. Pergamon, Oxford.

Game, M. & Peterken, G.F. (1984). Nature reserve selection strategies in the woodlands of central Lincolnshire, England. *Biological Conservation* **29,** 157–181.

Gartside, D.W. & McNeilly, T. (1974). The potential for evolution of heavy metal tolerance in plants. II. Copper tolerance. *Heredity* **32,** 335–348.

Gaston, K.J. & Hudson, E. (1994). Regional patterns of diversity and estimates of global insect species richness. *Biodiversity and Conservation* **3,** 493–500.

Gates, J.E. & Gysel, L.W. (1978). Avian nest dispersion and fledgling success in field–forest ecotones. *Ecology* **59,** 871–883.

Getz, L.L., Cole, F.R. & Gates, D.L. (1978). Interstate roadsides as dispersal routes for *Microtus pennsylvanicus*. *Journal of Mammalogy* **59,** 208–212.

Giesy, J.P. & Graney, R.L. (1989). Recent developments in and intercomparisons of acute and chronic bioassays and indicators. *Hydrobiologia* **188/189,** 21–60.

Gilbert, D.J. (1997). Towards a new recruitment paradigm for fish stocks. *Canadian Journal of Fisheries and Aquatic Sciences* **54,** 969–977.

Giller, K.E. & Day, J.M. (1985). [Nitrogen fixation in the rhizosphere]. In: *Ecological Interactions in Soil* (eds A.H. Fitter, D. Atkinson, D.J. Read & M.B. Usher), pp. 127–147. Blackwell Science, Oxford.

Giller, K.E., Witter, E. & McGrath, S.P. (1998). Toxicity of heavy metals to microorganisms and microbial processes in agricultural soils: a review. *Soil Biology and Biochemistry* **30,** 1389–1414.

Gimingham, C.H. (1985). Age-related interactions between *Calluna vulgaris* and phytophagous insects. *Oikos* **44,** 12–16.

Gleick, P.H. (1993). *Water in Crisis*. Oxford University Press, Oxford.

Goldenberg, J. (1996). The evolution of ethanol costs in Brazil. *Energy Policy* **24,** 1127–1128.

Golley, F.B., ed. (1983). *Tropical Forest Ecosystems*. Elsevier, Amsterdam.

Gomez-Pompa, A., Whitmore, T.C. & Hadley, M. (1991). *Rain Forest Regeneration and Management*. UNESCO/Parthenon, Paris.

Gorchov, D.L., Cornejo, F., Ascorra, C. & Jaramillo, M. (1993). The role of seed dispersal in the natural regeneration of rain forest after strip-cutting in the Peruvian Amazon. *Vegetatio* **107/108,** 339–349.

Gordon, I.J. & Illius, A.W. (1994). The functional significance of the browser–grazer dichotomy in African ruminants. *Oecologia* **98,** 167–175.

Gorham, J., Wyn Jones, R.G. & McDonnell, E. (1985). Some mechanisms of salt tolerance in crop plants. *Plant and Soil* **89,** 15–40.

Goulding, K.W.T. *et al.* (1998). Nitrogen deposition and its contribution to nitrogen cycling and associated soil processes. *New Phytologist* **139,** 49–58.

Grabherr, G., Gottfried, M. & Pauli, H. (1994). Climate effects on mountain plants. *Nature* **369,** 448.

Grafton, R.Q. (1996). Individual transferable quotas: theory and practice. *Reviews in Fish Biology and Fisheries* **6,** 5–20.

Graham, R.W. (1997). The spatial response of mammals to Quaternary climate changes. In: *Past and Future Rapid Environmental Changes* (eds B. Huntley *et al.*), pp. 153–162. Springer, Berlin.

Grahame, J. (1987). *Plankton and Fisheries*. Arnold, London.

Grainger, A. (1993). Rates of deforestation in the humid tropics: estimates and measurements. *Geographical Journal* **159,** 33–44.

Grant, S.A. *et al.* (1985). Comparative studies of diet selection by sheep and cattle: the hill grasslands. *Journal of Ecology* **73,** 987–1004.

Grant, S.A. *et al.* (1987). Comparative studies of diet selection by sheep and cattle: blanket bog and heather moor. *Journal of Ecology* **75,** 947–960.

Gray, N.F. (1989). *Biology of Wastewater Treatment*. Oxford University Press, Oxford.

Green, R.E., Osborne, P.E. & Sears, E.J. (1994). [The distribution of passerine birds in hedgerows in relation to characteristics of the hedgerow and adjacent farmland]. *Journal of Applied Ecology* **31,** 677–692.

Grenfell, B.T. & Dobson, A.P. (1995). *Ecology of Infectious Diseases in Natural Populations.* Cambridge University Press. Cambridge.

Gressel, J. & Segel, L.A. (1990). Modelling the effectiveness of herbicide rotations and mixtures as strategies to delay or preclude resistance. *Weed Technology* **4,** 186–198.

Grime, J.P. (1973). Competitive exclusion in herbaceous vegetation. *Nature* **242,** 344–347.

Grime, J.P. (1977). Evidence for the existence of three primary strategies in plants and its relevance to ecological and evolutionary theory. *American Naturalist* **111,** 1169–1194.

Grime, J.P. (1998). Benefits of plant diversity to ecosystems: immediate, filter and founder effects. *Journal of Ecology* **86,** 902–910.

GRIP Members (1993). Climate instability during the last interglacial period recorded in the GRIP ice core. *Nature* **364,** 203–207.

Gross, M.R. (1998). One species with two biologies: Atlantic salmon in the wild and in aquaculture. *Canadian Journal of Fisheries and Aquatic Sciences* **55** (Suppl. 1), 131–144.

Grue, C.E., Powell, G.V.N. & McChesney, M.J. (1982). Care of nestlings by wild female starlings exposed to an organophosphate pesticide. *Journal of Applied Ecology* **19,** 327–335.

Guénette, S., Luack, T. & Clark, C. (1998). Marine reserves: from Beverton and Holt to the present. *Reviews in Fish Biology and Fisheries* **8,** 251–272.

Hajibagheri, M.A., Harvey, D.M.R. & Flowers, T.J. (1987). Quantitative ion distribution within root cells of salt-sensitive and salt-tolerant maize varieties. *New Phytologist* **105,** 367–379.

Hall, I.G. (1957). The ecology of disused pit heaps in England. *Journal of Ecology* **45,** 689–720.

Hall, J.L., Harvey, D.M.R. & Flowers, T.J. (1978). Evidence for the cytoplasmic localization of betaine in leaf cells of *Suaeda maritima. Planta* **140,** 59–62.

Hamilton, G.J. & Christie, J.M. (1971). *Forest Management Tables.* Forestry Commission. Her Majesty's Stationery Office, London.

Hanley, T.A. (1997). A nutritional view of understanding and complexity in the problem of diet selection by deer. *Oikos* **79,** 209–218.

Hanski, I. (1994). A practical model of metapopulation dynamics. *Journal of Animal Ecology* **63,** 151–162.

Harborne, J.B. (1993). *Introduction to Ecological Biochemistry,* 4th edn. Academic Press, London.

Hardarson, G. (1993). Methods for enhancing symbiotic nitrogen fixation. *Plant and Soil* **152,** 1–17.

Hardarson, G. *et al.* (1993). Genotypic variation in biological nitrogen fixation by common bean. *Plant and Soil* **152,** 59–70.

Hardin, G. (1960). The competitive exclusion principle. *Science* **131,** 1292–1297.

Hardin, G. (1968). The tragedy of the commons. *Science* **162,** 1243–1248.

Harding, P.T. & Rose, F. (1986). [*Pasture Woodlands in Lowland Britain*]. Institute of Terrestrial Ecology, Abbots Ripton, England.

Hardison, W.A., Fisher, H.L., Graf, G.C. & Thompson, N.R. (1956). Some observations on the behavior of grazing lactating cows. *Journal of Dairy Science* **39,** 1735–1741.

Harper, J.L. (1977). *Population Biology of Plants.* Academic Press, London.

Harper, J.L., Williams, J.T. & Sagar, G.R. (1965). [The heterogeneity of soil surfaces and its role in determining the establishment of plants from seed]. *Journal of Ecology* **53,** 273–286.

Harrison, R.M., ed. (1996). *Pollution: Causes, Effects and Control,* 3rd edn. Royal Society of Chemistry, Cambridge.

Harrison, S. (1994). Metapopulations and conservation. In: *Large-Scale Ecology and Conservation Biology* (eds P.J. Edwards, R.M. May, & N.R. Webb), pp. 111–128. Blackwell Science, Oxford.

Harrison, S., Murphy, D.D. & Ehrlich, P.R. (1988). Distribution of the Bay checkerspot butterfly: evidence for a metapopulation model. *American Naturalist* **132,** 360–382.

Harte, J. (1985). *Consider a Spherical Cow.* Kaufmann, Los Altos, CA.

Hartshorn, G.S. (1989). Application of gap theory to tropical forest management: natural regeneration in strip clear-cuts in the Peruvian Amazon. *Ecology* **70,** 567–569.

Harvey, D.M.R., Hall, J.L., Flowers, T.J. & Kent, B. (1981). Quantitative ion localization within *Suaeda maritima* leaf mesophyll cells. *Planta* **151,** 555–560.

Havel, J. & Reineke, W. (1991). Total degradation of various chlorobiphenyls by cocultures and in vivo constructed hybrid pseudomonads. *FEMS Microbiology Letters* **78,** 163–170.

Haygarth, P.M., Chapman, P.J., Jarvis, S.C. & Smith, R.V. (1998). Phosphorus budgets for two contrasting grassland farming systems in the UK. *Soil Use and Management* **14,** 160–167.

He, D.-Y. *et al.* (1994). The fate of nitrogen from ^{15}N-labeled straw and green manure in soil–crop–domestic animal systems. *Soil Science* **158,** 65–73.

Hein, M. & Sand-Jensen, K. (1997). CO_2 increases oceanic primary production. *Nature* **388,** 526–527.

Heinselman, M.L. (1973). Fire in the virgin forest of the Boundary Waters Canoe Area. *Quaternary Research* **3,** 329–382.

Heitefuss, R. (1989). *Crop and Plant Protection.* Wiley, New York.

Hellawell, J.M. (1986). *Biological Indicators of Fresh Water Pollution and Environmental Management.* Elsevier, London.

Henderson-Sellers, A. *et al.* (1993). Tropical deforestation: modeling local- to regional-scale climate change. *Journal of Geophysical Research* **98,** 7289–7315.

Henter, H.J. & Via, S. (1995). The potential for coevolution in a host–parasitoid system. I. Genetic variation within an aphid population in susceptibility to a parasitic wasp. *Evolution* **49,** 427–438.

Herridge, D.F. & Danso, S.K.A. (1995). Enhancing crop legume N_2 fixation through selection and breeding. *Plant and Soil* **174,** 51–82.

Heske, F. (1938). *German Forestry.* Yale University Press, New Haven, CT.

Hester, A.J. & Bailie, G.J. (1998). Spatial and temporal patterns of heather use by sheep and deer within natural heather/grass mosaics. *Journal of Applied Ecology* **35,** 772–784.

Hickey, J.J. & Anderson, D.W. (1968). Chlorinated hydrocarbons and eggshell changes in raptorial and fish-eating birds. *Science* **162,** 271–273.

Hilborn, R. (1997). Comment: recruitment paradigms for fish stocks. *Canadian Journal of Fisheries and Aquatic Sciences* **54,** 984–985.

Hilborn, R. & Walters, C.J. (1992). *Quantitative Fisheries Stock Assessment.* Chapman & Hall, New York.

Hill, C.J. (1995). Linear strips of rain forest vegetation as potential dispersal corridors for rain forest insects. *Conservation Biology* **9,** 1559–1566.

Hill, M.O., Evans, D.F. & Bell, S.A. (1992). Long-term effects of excluding sheep from hill pastures in North Wales. *Journal of Ecology* **80,** 1–13.

Hill, P. (1982). *Dry Grain Farming Families: Hausaland (Nigeria) and Karnataka (India) Compared.* Cambridge University Press, Cambridge.

Hillel, D. (1994). *Rivers of Eden.* Oxford University Press, Oxford.

Hillier, S.H., Walton, D.W.H. & Wells, D.A. (1990). *Calcareous Grasslands—Ecology and Management.* Bluntisham Books, Huntingdon, England.

Hinsley, S.A., Bellamy, P.E., Newton, I. & Sparks, T.H. (1996). Influences of population size and woodland area on bird species distributions in small woods. *Oecologia* **105,** 100–106.

Hodgson, J. & Illius, A.W. (1996). *The Ecology and Management of Grazing Systems.* CAB International, Wallingford, England.

Hodkinson, I.D., Coulson, S.J., Webb, N.R. & Block, W. (1996). Can high arctic soil microarthropods survive elevated summer temperatures? *Functional Ecology* **10,** 314–321.

Hof, J. & Raphael, M.G. (1997). Optimization of habitat placement: a case study of the northern spotted owl in the Olympic Peninsula. *Ecological Applications* **7,** 1160–1169.

Hoffmann, J.H., Moran, V.C. & Zeller, D.A. (1998). Long-term population studies and the development of an integrated management programme for control of *Opuntia stricta* in Kruger National Park, South Africa. *Journal of Applied Ecology* **35,** 156–160.

Hokkanen, H.M.T. & Lynch, J.M., eds. (1995). *Biological Control: Benefits and Risks.* Cambridge University Press, Cambridge.

Holdgate, M.W. (1991). Conservation in a world context. In: *The Scientific Management of Temperate Communities for Conservation* (eds I.F. Spellerberg, F.B. Goldsmith & M.G. Morris), pp. 1–26. Blackwell Science, Oxford.

Holl, K.D. (1996). The effect of coal surface mine reclamation on diurnal lepidopteran conservation. *Journal of Applied Ecology* **33,** 225–236.

Holland, W.E., Smith, M.H., Gibbons, J.W. & Brown, D.H. (1974). Thermal tolerances of fish from a reservoir receiving heated effluent from a nuclear reactor. *Physiological Zoology* **47,** 110–118.

Holt, J.S. (1988). Reduced growth, competitiveness, and photosynthetic efficiency of triazine-resistant *Senecio vulgaris* from California. *Journal of Applied Ecology* **25,** 307–318.

Holt, J.S., Powles, S.B. & Holtum, J.A.M. (1993). Mechanisms and agronomic aspects of herbicide resistance. *Annual Review of Plant Physiology* **44,** 203–229.

Holt, R.D. & Hochberg, M.E. (1997). When is biological control evolutionarily stable (or is it)? *Ecology* **78,** 1673–1683.

Hopkin, S.P. (1990). Species-specific differences in the net assimilation of zinc, cadmium, lead, copper and iron by the terrestrial isopods *Oniscus asellus* and *Porcellio scaber*. *Journal of Applied Ecology* **27,** 460–474.

Hopkinson, C.S. & Day, J.W. (1980). Net energy analysis of alcohol production from sugarcane. *Science* **207,** 302–304.

Horwood, J.W., Nichols, J.H. & Milligan, S. (1998). Evaluation of closed areas for fish stock conservation. *Journal of Applied Ecology* **35,** 893–903.

Houghton, J. (1997). *Global Warming. The Complete Briefing,* 2nd edn. Cambridge Univerity Press, Cambridge.

Houghton, J.T. *et al.*, eds. (1996). *Climate Change 1995.* Cambridge University Press. Cambridge.

Howard, P.H. *et al.* (1991). [Development of a predictive model for biodegradability]. *Science of the Total Environment* **109/110,** 635–641.

Howell, D.G. & Murray, R.W. (1986). A budget for continental growth and denudation. *Science* **233,** 446–449.

Howell, E.A. & Jordan, W.R. (1991). Tallgrass prairie restoration in the North American midwest. In: *The Scientific Management of Temperate Communities for Conservation* (eds I.F. Spellerberg, F.B. Goldsmith & M.G. Morris), pp. 395–414. Blackwell Science, Oxford.

Howells, O. & Edwards-Jones, G. (1997). A feasibility study of reintroducing wild boar to Scotland: are existing woodlands large enough to support minimum viable populations? *Biological Conservation* **81,** 77–89.

Hsieh, Y.-P. (1992). Pool size and mean age of stable soil organic carbon in cropland. *Soil Science Society of America Journal* **56,** 460–464.

Hubbell, S.P. & Foster, R.B. (1986a). [Commonness and rarity in a neotropical forest]. In: *Conservation Biology* (ed. M.E. Soulé), pp. 205–231. Sinauer, Sunderland, MA.

Hubbell, S.P. & Foster, R.B. (1986b). Biology, chance, and history and the structure of tropical rain forest tree communities. In: *Community Ecology* (eds J. Diamond & T.J. Case), pp. 314–329. Harper & Row, New York.

Hudson, R.J., Drew, K.R. & Baskin, L.M., eds. (1989). *Wildlife Production Systems: Economic Utilization of Wild Ungulates.* Cambridge University Press, Cambridge.

Huey, R.B., Partridge, L. & Fowler, K. (1991). Thermal sensitivity of *Drosophila melanogaster* responds rapidly to laboratory natural selection. *Evolution* **45,** 751–756.

Huffaker, C.B. & Kennett, C.E. (1959). A ten-year study of vegetational changes associated with biological control of Klamath weed. *Journal of Range Management* **12,** 69–82.

Hughes, D.E. & McKenzie, P. (1975). The microbial degradation of oil in the sea. *Proceedings of the Royal Society B* **189,** 375–390.

Hulme, M. (1992). Rainfall changes in Africa: 1931–60 to 1961–90. *International Journal of Climatology* **12,** 685–699.

Hulzebos, E.M. *et al.* (1991). QSARs in phytotoxicity. *Science of the Total Environment* **109/110,** 493–497.

Hunter, B.A., Johnson, M.S. & Thompson, D.J. (1987a,b,c). Ecotoxicology of copper and cadmium in a contaminated grassland ecosystem. *Journal of Applied Ecology* **24.** (a). I. Soil and vegetation contamination: pp. 573–586. (b). II. Invertebrates: pp. 587–599. (c). III. Small mammals: pp. 601–614.

Huntley, B. & Birks, H.J.B. (1983). *An Atlas of Past and Present Pollen Maps for Europe*: 0–13 000 *Years Ago*. Cambridge University Press, Cambridge.

Huntley, B. *et al.* (1997). *Past and Future Rapid Environmental Changes*. Springer, Berlin.

Husband, B.C. & Barrett, S.C.H. (1996). A metapopulation perspective on plant population biology. *Journal of Ecology* **84,** 461–469.

Huston, M.A. (1994). *Biological Diversity: the Coexistence of Species on Changing Landscapes*. Cambridge University Press, Cambridge.

Hutchings, J.A. & Myers, R.A. (1994). What can be learned from the collapse of a renewable resource? Atlantic cod of Newfoundland and Labrador. *Canadian Journal of Fisheries and Aquatic Sciences* **51,** 2216–2246.

Hutchings, M.J. & Booth, K.D. (1996). Studies on the feasibility of recreating chalk grassland vegetation on ex-arable land. I. The potential roles of the seed bank and the seed rain. *Journal of Applied Ecology* **33,** 1171–1181.

Hutchins, H.E. & Lanner, R.M. (1982). The central role of Clark's nutcracker in the dispersal and establishment of whitebark pine. *Oecologia* **55,** 192–201.

Hutchinson, K.J. & King, K.L. (1980). The effects of sheep stocking levels on invertebrate abundance, biomass and energy utilization in a temperate sown grassland. *Journal of Applied Ecology* **17,** 369–387.

Hutto, R.L. (1995). Composition of bird communities following stand-replacement fires in northern Rocky Mountain conifer forests. *Conservation Biology* **9,** 1041–1058.

Illius, A.W. & Gordon, I.J. (1992). Modelling the nutritional ecology of ungulate herbivores: evolution of body size and competitive interactions. *Oecologia* **89,** 428–434.

Illius, A.W., Gordon, I.J., Milne, J.D. & Wright, W. (1995). Costs and benefits of foraging on grasses varying in canopy structure and resistance to defoliation. *Functional Ecology* **9,** 894–903.

Issar, A. (1985). Fossil water under the Sinai–Negev Peninsula. *Scientific American* **253** (1), 82–88.

Ives, J.D. & Messerli, B. (1989). *The Himalayan Dilemma*. Routledge, London.

Jacobsen, T. & Adams, R.M. (1958). Salt and silt in ancient Mesopotamian agriculture. *Science* **128,** 1251–1258.

Jacobson, L.D. & MacCall, A.D. (1995). Stock-recruitmant models for Pacific sardine. *Canadian Journal of Fisheries and Aquatic Sciences* **52,** 566–577.

Janssens, F. *et al.* (1998). Relationship between soil chemical factors and grassland diversity. *Plant and Soil* **202,** 69–78.

Jaramillo, V.J. & Detling, J.K. (1992). [Cattle grazing of simulated urine patches in North American grassland]. *Journal of Applied Ecology* **29,** 9–13.

Jarvis, S.C. (1993). Nitrogen cycling and losses from dairy farms. *Soil Use and Management* **9,** 99–105.

Jasper, D.A., Abbott, L.K. & Robson, A.D. (1989a). Acacias respond to phosphorus and VA mycorrhizal fungi in soils stockpiled during mineral sand mining. *Plant and Soil* **115,** 99–108.

Jasper, D.A., Abbott, L.K. & Robson, A.D. (1989b). The loss of VA mycorrhizal infection during bauxite mining may limit the growth of *Acacia pulchella*. *Australian Journal of Botany* **37,** 33–42.

Jasper, D.A., Robson, A.D. & Abbott, L.K. (1987). The effect of surface mining on the infectivity of vesicular–arbuscular mycorrhizal fungi. *Australian Journal of Botany* **35,** 641–652.

Jastrow, J.D. (1987). Changes in soil aggregation associated with tallgrass prairie restoration. *American Journal of Botany* **74,** 1656–1664.

Jastrow, J.D. (1996). Soil aggregate formation and the accrual of particulate and mineral-associated organic matter. *Soil Biology and Biochemistry* **28,** 665–676.

Jefferies, R.A., Bradshaw, A.D. & Putwain, P.D. (1981). Growth, nitrogen accumulation and nitrogen transfer by legume species established on mine spoils. *Journal of Applied Ecology* **18,** 945–956.

Jenkinson, D.S. (1991). [The Rothamsted long-term experiments]. *Agronomy Journal* **83,** 2–10.

Jenkinson, D.S. & Rayner, J.H. (1977). The turnover of soil organic matter in some of the Rothamsted classical experiments. *Soil Science* **123,** 298–305.

Jennings, S. & Kaiser, M.J. (1998). The effects of fishing on marine ecosystems. *Advances in Marine Biology* **34,** 201–352.

Johns, A.D. (1992). Species conservation in managed tropical forests. In: *Tropical Deforestation and Species Extinction* (eds T.C. Whitmore & J.A. Sayer), pp. 15–53. Chapman & Hall, London.

Johns, A.G. (1996). Bird population persistence in Sabahan logging concessions. *Biological Conservation* **75,** 3–10.

Johnsen, S.J. *et al.* (1992). Irregular glacial interstadials recorded in a new Greenland ice core. *Nature* **359,** 311–313.

Johnson, A.H. (1992). The role of abiotic stresses in the decline of red spruce in high elevation forests of the eastern United States. *Annual Review of Phytopathology* **30,** 349–367.

Johnson, E.A., Miyanishi, K. & Kleb, H. (1994). The hazards of interpretation of static age structures as shown by stand reconstructions in a *Pinus contorta–Picea engelmannii* forest. *Journal of Ecology* **82,** 923–931.

Johnson, W.C. & Webb, T. (1989). [The role of blue jays in the postglacial dispersal of trees in eastern North America]. *Journal of Biogeography* **16,** 561–571.

Johnston, A.E. (1969). The plant nutrients in crops grown on Broadbalk. *Rothamsted Annual Report for 1968, Part* **2,** 50–62.

Jones, H.L. & Diamond, J.M. (1976). Short-time-base studies of turnover in breeding bird populations of the Californian Channel Islands. *Condor* **78,** 526–549.

Joos, F., Sarmiento, J.L. & Siegenthaler, U. (1991). Estimates of the effect of Southern Ocean iron fertilization on atmospheric CO_2 concentrations. *Nature* **349,** 772–775.

Jordan, A.M. (1986). *Trypanosomiasis Control and African Rural Development*. Longman, Harlow, England.

Jordan, C.F. (1989). *An Amazonian Rainforest*. UNESCO/Parthenon, Paris.

Jordan, W.R., Gilpin, M.E. & Aber, J.D., eds. (1987). *Restoration Ecology*. Cambridge University Press, Cambridge.

Jullien, M. & Thiollay, J.-M. (1996). [Effects of rain forest disturbance and fragmentation: changes in the raptor community in French Guiana]. *Journal of Biogeography* **23,** 7–25.

Kaiser, K.L.E. & Esterby, S.R. (1991). [The toxicity of 267 chemicals to six species and the octanol/water partition coefficient]. *Science of the Total Environment* **109/110,** 499–514.

Karr, J.R. & Roth, R.R. (1971). Vegetation structure and avian diversity in several New World areas. *American Naturalist* **105,** 423–435.

Kennedy, D.N. & Swaine, M.D. (1992). Germination and growth of colonizing species in artificial gaps of different sizes in dipterocarp rain forest. *Philosophical Transactions of the Royal Society B* **335,** 357–366.

Kime, D.E. (1999). A strategy for assessing the effects of xenobiotics on fish reproduction. *Science of the Total Environment* **225,** 3–11.

King, G.A. & Herstrom, A.A. (1997). Holocene tree migration rates objectively determined from fossil pollen data. In: *Past and Future Rapid Environmental Changes* (eds B. Huntley *et al.*), pp. 91–101. Springer, Berlin.

King, I.P. *et al.* (1997). Introgression of salt tolerance genes from *Thinopyrum bessarabicum* into wheat. *New Phytologist* **137,** 75–81.

King, M. (1995). *Fisheries Biology, Assessment and Management.* Fishing News Books, Oxford.

Kirschbaum, M.U.F. (1995). [The temperature dependence of soil organic matter decomposition]. *Soil Biology and Biochemistry* **27,** 753–760.

Klemmedson, J.O. & Smith, J.G. (1964). Cheatgrass (*Bromus tectorum* L.). *Botanical Review* **30,** 226–262.

Kogan, M. (1986). *Ecological Theory and Integrated Pest Management.* Wiley, New York.

Kohn, D.D. & Walsh, D.M. (1994). Plant species richness—the effect of island size and habitat diversity. *Journal of Ecology* **82,** 367–377.

Kooijman, S.A.L.M. (1987). A safety factor for LC_{50} values allowing for differences in sensitivity among species. *Water Research* **21,** 269–276.

Korning, J. & Balslev, H. (1994). Growth and mortality of trees in Amazonian tropical rain forest in Ecuador. *Journal of Vegetation Science* **5,** 77–86.

Krebs, C.J. (1994). *Ecology,* 4th edn. Harper Collins, New York.

Kundu, D.K. & Ladha, J.K. (1995). Efficient management of soil and biologically fixed N_2 in intensively cultivated rice fields. *Soil Biology and Biochemistry* **27,** 431–439.

Lachnicht, S.L., Parmelee, R.W., McCartney, D. & Allen, M. (1997). Characteristics of macroporosity in a reduced tillage agroecosystem with manipulated earthworm populations: implications for infiltration and nutrient transport. *Soil Biology and Biochemistry* **29,** 493–498.

Ladha, J.K., ed. (1995). Management of biological nitrogen fixation for the development of more productive and sustainable agricultural systems. *Plant and Soil,* **174,** 1–286.

Laevastu, T. & Favorite, F. (1988). *Fishing and Stock Fluctuations.* Fishing News Books, Oxford.

Lamar, R.T., Davis, M.W., Dietrich, D.M. & Glaser, J.A. (1994). Treatment of a pentachlorophenol- and creosote-contaminated soil using the lignin-degrading fungus *Phanerochaete sordida*: a field demonstration. *Soil Biology and Biochemistry* **26,** 1603–1611.

Lamb, D. (1990). *Exploiting the Tropical Rain Forest. An Account of Pulpwood Logging in Pupua New Guinea.* UNESCO/Parthenon, Paris.

Lamb, H.H. (1995). *Climate, History and the Modern World,* 2nd edn. Routledge, London.

Lamberson, R.H., McKelvey, R., Noon, B.R. & Voss, C. (1992). A dynamic analysis of the northern spotted owl viability in a fragmented forest landscape. *Conservation Biology* **6,** 505–512.

Lambert, D. (1989). *The Undying Past of Shenandoah National Park.* Rinehart, Boulder, CO.

Lampert, W. *et al.* (1989). Herbicide effects on planktonic systems of different complexity. *Hydrobiologia* **188/189,** 415–424.

Lamprey, H.F. (1963). Ecological separation of the large mammal species in the Tarangire Game Reserve, Tanganyika. *East African Wildlife Journal* **1,** 63–92.

Lanno, R.P., Stephenson, G.L. & Wren, C.D. (1997). Applications of toxicity curves in assessing the toxicity of diazinon and pentachlorophenol to *Lumbricus terrestris* in natural soils. *Soil Biology and Biochemistry* **29,** 689–692.

Larcher, W. (1995). *Physiological Plant Ecology.* 3rd edn. Springer. Berlin.

Larsen, C.P.S. (1997). Spatial and temporal variations in boreal forest fire frequency in northen Alberta. *Journal of Biogeography* **24,** 663–673.

Larsen, C.P.S. & MacDonald, G.M. (1998). An 840-year record of fire and vegetation in a boreal white spruce forest. *Ecology* **79,** 106–118.

Laurance, W.F. & Bierregaard, R.O., eds. (1997). *Tropical Forest Remnants: Ecology,*

Management, and Conservation of Fragmented Communities. University of Chicago Press, Chicago.

Laurance, W.F. *et al.* (1998). Effects of forest fragmentation on recruitment patterns in Amazonian tree communities. *Conservation Biology* **12,** 460–464.

Laurila, H.A. (1995). [Modelling the effects of elevated CO_2 and temperature on Swedish and German spring wheat varieties]. *Journal of Biogeography* **22,** 591–595.

Lawlor, D.W. (1995). Photosynthesis, productivity and environment. *Journal of Experimental Botany,* **46** (Special Issue), 1449–1461.

Lawson, G.J., Callaghan, T.V. & Scott, R. (1984). Renewable energy from plants: bypassing fossilization. *Advances in Ecological Research* **14,** 57–114.

Le Barbé, L. & Lebel, T. (1997). Rainfall climatology of the HAPEX-Sahel region during the years 1950–90. *Journal of Hydrology* **188/189,** 43–73.

Leach, G. & Mearns, R. (1988). *Beyond the Woodfuel Crisis: People, Land and Trees in Africa.* Earthscan, London.

Lee, M.D., Odom, J.M. & Buchanan, R.J. (1998). [New perspectives on microbial dehalogenation of chlorinated solvents]. *Annual Review of Microbiology* **52,** 423–452.

Lemon, R.W. (1994). Insecticide resistance. *Journal of Agricultural Science* **122,** 329–333.

Lenssen, N. & Flavin, C. (1996). Sustainable energy for tomorrow's world. The case for an optimistic view of the future. *Energy Policy* **24,** 769–781.

Lewis, D.C. (1968). Annual hydrologic response to watershed conversion from oak woodland to annual grassland. *Water Resources Research* **4,** 59–72.

Lieberman, D., Lieberman, M., Peralta, R. & Hartshorn, G.S. (1985). Mortality patterns and stand turnover rates in a wet tropical forest in Costa Rica. *Journal of Ecology* **73,** 915–924.

Likens, G.E. & Bormann, F.H. (1995). *Biogeochemistry of a Forested Ecosystem,* 2nd edn. Springer, New York.

Lincoln, R., Boxshall, G. & Clark, P. (1998). *A Dictionary of Ecology, Evolution and Systematics.* Cambridge University Press, Cambridge.

Lindenmayer, D.B. & Franklin, J.F. (1997). Managing stand structure as part of ecologically sustainable forest management in Australian mountain ash forests. *Conservation Biology* **11,** 1053–1068.

Lindenmayer, D.B. & Nix, H.A. (1993). Ecological principles for the design of wildlife corridors. *Conservation Biology* **7,** 627–630.

Lindow, S.E., Panopoulos, N.J. & McFarland, B.L. (1989). Genetic engineering of bacteria from managed and natural habitats. *Science* **244,** 1300–1307.

Lister, A.M. (1997). The evolutionary response of vertebrates to Quaternary environmental change. In: *Past and Future Rapid Environmental Changes.* (eds B. Huntley *et al.*), pp. 287–302. Springer, Berlin.

Long, S.P. & Mason, C.F. (1983). *Saltmarsh Ecology.* Blackie, Glasgow.

Longhurst, A.R. & Pauly, D. (1987). *Ecology of Tropical Oceans.* Academic Press, San Diego, CA.

Lovejoy, T.E. *et al.* (1986). Edge and other effects of isolation on Amazon forest fragments. In: *Conservation Biology* (ed. M.E. Soulé), pp. 257–285. Sinauer, Sunderland, MA.

Low, A.J. (1972). The effect of cultivation on the structure and other physical characteristics of grassland and arable soils (1945–70). *Journal of Soil Science* **23,** 363–380.

Lubow, B.C. (1996). Optimal translocation strategies for enhancing stochastic metapopulation viability. *Ecological Applications* **6,** 1268–1280.

Lugo, A.E. (1992). Comparison of tropical tree plantations with secondary forests of similar age. *Ecological Monographs* **62,** 1–41.

Lusk, C.H. (1995). Seed size, establishment sites and species coexistence in a Chilean rain forest. *Journal of Vegetation Science* **6,** 249–256.

Lynch, J.F. & Johnson, N.K. (1974). Turnover and equilibria in insular avifaunas, with special reference to the California Channel Islands. *Condor* **76,** 370–384.

Lynch, J.M. & Harper, S.H.T. (1985). The microbial upgrading of straw for agricultural use. *Philosophical Transactions of the Royal Society B* **310,** 221–226.

Mabberley, D.J. (1992). *Tropical Rain Forest Ecology*, 2nd edn. Blackie, London.

MacArthur, R.H. & MacArthur, J.W. (1961). On bird species diversity. *Ecology* **42,** 594–598.

MacArthur, R.H. & Wilson, E.O. (1967). *The Theory of Island Biogeography*. Princeton University Press, Princeton, NJ.

Mace, G.M. (1994). Classifying threatened species: means and ends. *Philosophical Transactions of the Royal Society B* **344,** 91–97.

Macilwain, C. (1994). Western inferno provokes a lot of finger-pointing but little action. *Nature* **370,** 585.

Mack, R.N. (1981). [Invasion of *Bromus tectorum* into western North America]. *Agro-Ecosystems* **7,** 145–165.

Mack, R.N. (1986). Alien plant invasion into the Intermountain West: a case study. In: *Ecology of Biological Invasions of North America and Hawaii* (eds H.A. Mooney & J.A. Drake), pp. 191–213. Springer, Berlin.

MacKenzie, D. (1997). Will rabies bite back? *New Scientist*, **156** (8 November), 24–25.

MacLean, D.A., Hunt, T.L., Eveleigh, E.S. & Morgan, M.G. (1996). The relation of balsam fir volume increment to cumulative spruce budworm defoliation. *Forest Chronicle* **72,** 533–540.

Macnair, M.R. (1993). The genetics of metal tolerance in vascular plants. *New Phytologist* **124,** 541–559.

MacNicol, R.D. & Beckett, P.H.T. (1985). Critical tissue concentrations of potentially toxic elements. *Plant and Soil* **85,** 107–129.

Majer, J.D. & Nichols, O.G. (1998). [Long-term recolonization patterns of ants in Western Australian rehabilitated bauxite mines]. *Journal of Applied Ecology* **35,** 161–182.

Malajczuk, N. (1979). Biological suppression of *Phytophthora cinnamomi* in eucalypts and avocados in Australia. In: *Soil-Borne Plant Pathogens* (eds B. Schippers & W. Gams), pp. 635–652. Academic Press, London.

Mann, K.H. & Lazier, J.R.N. (1996). *Dynamics of Marine Ecosystems*. Blackwell Science, Oxford.

Manokaran, N. & LaFrankie, V. (1990). Stand structure of Pasoh Forest Reserve, a lowland rain forest in Peninsular Malaysia. *Journal of Tropical Forest Science* **3,** 14–24.

Marini, M.A., Robinson, S.K. & Heske, E.J. (1995). Edge effects on nest predation in the Shawnee National Forest, southern Illinois. *Biological Conservation* **74,** 203–213.

Marrs, R.H. (1993). [Soil fertility and nature conservation in Europe]. *Advances in Ecological Research* **24,** 241–300.

Marrs, R.H. & Bradshaw, A.D. (1980). Ecosystem development on reclaimed china clay wastes. III. Leaching of nutrients. *Journal of Applied Ecology* **17,** 727–736.

Marrs, R.H., Bravington, M. & Rawes, M. (1988). Long-term vegetation change in the *Juncus squarrosus* grassland at Moor House, northern England. *Vegetatio* **76,** 179–187.

Marrs, R.H. & Hicks, M.J. (1986). [Study of vegetation change at Lakenheath Warren]. *Journal of Applied Ecology* **23,** 1029–1046.

Marrs, R.H., Hicks, M.J. & Fuller, R.M. (1986). Losses of lowland heath through succession at four sites in Breckland, East Anglia. *Biological Conservation* **36,** 19–38.

Marrs, R.H., Roberts, R.D. & Bradshaw, A.D. (1980). [Ecosystem development on reclaimed china clay wastes. I & II]. *Journal of Applied Ecology* **17,** 709–717 & 719–725.

Marschner, H. (1995). *Mineral Nutrition of Higher Plants*, 2nd edn. Academic Press, London.

Martin, J.H. *et al.* (1994). Testing the iron hypothesis in ecosystems of the equatorial Pacific. *Nature* **371,** 123–129.

Mason, C.F. (1996). *Biology of Freshwater Pollution*, 3rd edn. Longman, Harlow, England.

Mathys, W. (1975). Enzymes of some heavy-metal-resistant and non-resistant populations of *Silene cucubalis* and their interactions with some heavy metals *in vivo* and *in vitro*. *Physiologia Plantarum* **33,** 161–165.

Matlack, G.R. (1994). Plant species migration in a mixed-history forest landscape in eastern North America. *Ecology* **75,** 1491–1502.

Mattson, D.J. (1997). [Sustainable grizzly bear mortality calculated from counts of females]. *Biological Conservation* **81,** 103–111.

May, R.M. (1992). How many species inhabit the Earth? *Scientific American* **267** (4), 18–24.

McCarty, L.S. & Mackay, D. (1993). Enhancing ecotoxicological modeling and assessment. *Environmental Science and Technology* **27,** 1719–1728.

McCown, R.L. & Williams, J. (1990). The water environment and implications for productivity. *Journal of Biogeography* **17,** 513–520.

McEldowney, S., Hardman, D.J. & Waite, S. (1993). *Pollution: Ecology and Biotreatment*. Longman, Harlow, England.

McGuffie, K. *et al.* (1995). Global climate sensitivity to tropical deforestation. *Global and Planetary Change* **10,** 97–128.

McGuire, A.D., Melillo, J.M. & Joyce, L.A. (1995). The role of nitrogen in the response of forest net primary productivity to elevated atmospheric carbon dioxide. *Annual Review of Ecology and Systematics* **26,** 473–503.

McKenzie, J.A. & Batterham, P. (1994). The genetic, molecular and phenotypic consequences of selection for insecticide resistance. *TREE* **9,** 166–169.

McLaughlin, S.B. *et al.* (1987). An analysis of climate and competition as contributors to decline of red spruce in high elevation Appalachian forests of the eastern United States. *Oecologia* **72,** 487–501.

McLaughlin, T. (1971). *Coprophilia*. Cassell, London.

McLeod, A.R. & Long, S.P. (1999). Free-air carbon dioxide enrichment (FACE) in global change research: a review. *Advances in Ecological Research* **28,** 1–56.

McQuilkin, W.E. (1940). Natural establishment of pine in abandoned fields in the Piedmont Plateau region. *Ecology* **21,** 135–147.

Meffe, G.K. & Carroll, C.R. (1994). *Principles of Conservation Biology*. Sinauer, Sunderland, MA.

Meharg, A.A. & Macnair, M.R. (1990). An altered phosphate uptake system in arsenate-tolerant *Holcus lanatus*. *New Phytologist* **116,** 29–35.

Meharg, A.A. & Macnair, M.R. (1991). Uptake, accumulation and translocation of arsenate in arsenate-tolerant and non-tolerant *Holcus lanatus*. *New Phytologist* **117,** 225–231.

Memmott, J., Fowler, S.V. & Hill, R.L. (1998). The effect of release size on the probability of establishment of biological control agents: gorse thrips released against gorse in New Zealand. *Biocontrol Science and Technology* **8,** 103–115.

Merton, L.F.H. (1970). The history and status of the woodlands of the Derbyshire limestone. *Journal of Ecology* **58,** 723–744.

Metcalf, R.L. & Luckman, W.H. (1994). *Introduction to Insect Pest Management*, 3rd edn. Wiley, New York.

Metcalfe, J.L. (1989). [Biological water quality assessment of running waters based on macroinvertebrate communities]. *Environmental Pollution* **60,** 101–139.

Milchunas, D.G. & Lauenroth, W.K. (1993). Quantitative effects of grazing on vegetation and soils over a global range of environments. *Ecological Monographs* **63,** 327–366.

Milchunas, D.G., Lauenroth, W.K. & Burke, I.C. (1998). Livestock grazing: animal and plant biodiversity of shortgrass steppe and the relationship to ecosystem function. *Oikos* **83,** 65–74.

Miles, J. (1974). Effects of experimental interference with stand structure on establishment of seedlings in Callunetum. *Journal of Ecology* **62,** 675–687.

Milliman, J.D. & Meade, R.H. (1983). World-wide delivery of river sediment to the oceans. *Journal of Geology* **91,** 1–21.

de Miranda, J.C.C., Harris, P.J. & Wild, A. (1989). Effects of soil and plant phosphorus concentrations on vesicular–arbuscular mycorrhiza in sorghum plants. *New Phytologist* **112,** 405–410.

Mitchell, R.A.C. *et al.* (1995). [Effects of elevated CO_2 concentration and increased temperature on winter wheat]. *Plant, Cell and Environment* **18,** 736–748.

Møller, A.P. (1991). Clutch size, nest predation, and distribution of avian unequal competitors in a patchy environment. *Ecology* **72,** 1336–1349.

Monteith, J.L. & Unsworth, M.H. (1990). *Principles of Environmental Physics*, 2nd edn. Arnold, London.

Moore, N.W. & Hooper, M.D. (1975). On the number of bird species in British woods. *Biological Conservation* **8,** 239–250.

Moore, P.D., Chaloner, B. & Stott, P. (1996). *Global Environmental Change*. Blackwell Science, Oxford.

Morgan, J.M. (1983). Osmoregulation as a selection criterion for drought tolerance in wheat. *Australian Journal of Agricultural Research* **34,** 607–614.

Morgan, R.P.C., ed. (1986). *Soil Erosion and its Control*. Van Nostrand, New York.

Morgan, R.P.C. (1995). *Soil Erosion and Conservation*, 2nd edn. Longman. Harlow, England.

Morgan, R.P.C., Morgan, D.D.V. & Finney, H.J. (1984). A predictive model for the assessment of soil erosion risk. *Journal of Agricultural Engineering Research* **30,** 245–253.

Moriarty, F. (1988). *Ecotoxicology*, 2nd edn. Academic Press, London.

Morison, J.I.L. & Gifford, R.M. (1984). [Plant growth and water use with limited water supply in high CO_2 concentrations. I and II]. *Australian Journal of Plant Physiology* **11,** 361–374 & 375–384.

Morris, M.G. (1969). Differences between the invertebrate faunas of grazed and ungrazed chalk grassland. III. The heteropterous fauna. *Journal of Applied Ecology* **6,** 475–487.

Morris, M.G. (1971a). The management of grassland for the conservation of invertebrate animals. In: *The Scientific Management of Animal and Plant Commmunities for Conservation* (eds E. Duffey & A.S. Watt), pp. 527–552. Blackwell Science, Oxford.

Morris, M.G. (1971b). Differences between the invertebrate faunas of grazed and ungrazed chalk grassland. IV. Abundance and diversity of Homoptera-Auchenorhyncha. *Journal of Applied Ecology* **8,** 37–52.

Mundt, C.C. & Leonard, K.J. (1985). A modification of Gregory's model for describing plant disease gradients. *Phytopathology* **75,** 930–935.

Munro, J. (1967). The exploitation and conservation of resources by populations of insects. *Journal of Animal Ecology* **36,** 531–547.

Murdoch, W.W. & Briggs, C.J. (1996). Theory for biological control: recent developments. *Ecology* **77,** 2001–2013.

Murphy, D.D., Freas, K.E. & Weiss, S.B. (1990). An environment-metapopulation approach to population viability analysis for a threatened invertebrate. *Conservation Biology* **4,** 41–51.

Murray, F. & Wilson, S. (1990). Growth responses of barley exposed to SO_2. *New Phytologist* **114,** 537–541.

Myers, R.A. (1997). Comment and reanalysis: paradigms for recruitment studies. *Canadian Journal of Fisheries and Aquatic Sciences* **54,** 978–981.

Myers, R.A. (1998). When do environment–recruitment correlations work? *Reviews in Fish Biology and Fisheries* **8,** 285–305.

Myers, R.A., Hutchings, J.A. & Barrowman, N.J. (1997). Why do fish stocks collapse? The example of cod in Atlantic Canada. *Ecological Applications* **7,** 91–106.

Nagy, K.A. (1994). Seasonal water, energy and food use by free-living, arid-habitat mammals. *Australian Journal of Zoology* **42,** 55–63.

Nagy, K.A., Bradley, A.J. & Morris, K.D. (1990). Field metabolic rates, water fluxes, and feeding rates of quokkas and tammars in Western Australia. *Australian Journal of Zoology* **37,** 553–560.

Nagy, K.A. & Gruchacz, M.J. (1994). Seasonal water and energy metabolism of the desert-dwelling kangaroo rat. *Physiological Zoology* **67,** 1461–1478.

Nee, S. & May, R.M. (1992). Dynamics of metapopulations: habitat destruction and competitive coexistence. *Journal of Animal Ecology* **61,** 37–40.

Nelson, B.W. *et al.* (1994). Forest disturbance by large blowdowns in the Brazilian Amazon. *Ecology* **75,** 853–858.

Nepstad, D.C. *et al.* (1996). A comparative study of tree establishment in abandoned pasture and mature forest of eastern Amazonia. *Oikos* **76,** 25–39.

Newbery, D.M. *et al.* (1992). Primary lowland dipterocarp forest at Danum Valley, Sabah, Malaysia: structure, relative abundance and family composition. *Philosophical Transactions of the Royal Society B* **335,** 341–356.

Newhouse, J.R. (1990). Chestnut blight. *Scientific American* **263** (1), 74–79.

Newman, E.I. (1993). *Applied Ecology*. Blackwell Science, Oxford.

Newman, E.I. (1995). Phosphorus inputs to terrestrial ecosystems. *Journal of Ecology* **83,** 713–726.

Newman, E.I. (1997). [Phosphorus balance of contrasting farming systems, past and present]. *Journal of Applied Ecology* **34,** 1334–1347.

Newman, E.I. & Harvey, P.D.A. (1997). Did soil fertility decline in medieval English farms? Evidence from Cuxham, Oxfordshire 1320–1340. *Agricultural History Review* **45,** 119–136.

Newton, I., Bogan, J.A. & Haas, M.B. (1989). Organochlorines and mercury in eggs of British peregrines *Falco peregrinus*. *Ibis* **131,** 355–376.

Ng, F.S.P. (1978). Strategies of establishment in Malayan forest trees. In: *Tropical Trees as Living Systems* (eds P.B. Tomlinson & M.H. Zimmermann), pp. 129–162. Cambridge University Press, Cambridge.

Nicholson, S.E. (1989). Long-term changes in African rainfall. *Weather* **44,** 46–56.

Nobel, P.S. (1991a). *Physicochemical and Environmental Plant Physiology*. Academic Press, San Diego, CA.

Nobel, P.S. (1991b). Achievable productivities of certain CAM plants: basis for high values compared with C3 and C4 plants. *New Phytologist* **119,** 183–205.

Noble, C.L., Halloran, G.M. & West, D.W. (1984). Identification and selection for salt tolerance in lucerne. *Australian Journal of Agricultural Research* **35,** 239–252.

Noble, C.L. & Rogers, M.E. (1992). Arguments for the use of physiological criteria for improving the salt tolerance in crops. *Plant and Soil* **146,** 99–107.

Nolan, T. & Connolly, J. (1989). Mixed v. mono-grazing by steers and sheep. *Animal Production* **48,** 519–533.

Noon, B.R. & McKelvey, K.S. (1996). Management of the spotted owl: a case history in conservation biology. *Annual Review of Ecology and Systematics* **27,** 135–162.

Nowak, R.S. & Caldwell, M.M. (1984). A test of compensatory photosynthesis in the field: implications for herbivory tolerance. *Oecologia* **61,** 311–318.

Nudds, T.D. (1990). [The ecological effects of meningeal worms]. *Journal of Wildlife Management* **54,** 396–402.

Nye, P.H. & Greenland, D.J. (1960). *The Soil Under Shifting Cultivation*. Commonwealth Agricultural Bureau, Farnham Royal, England.

O'Brien, S.J. & Knight, J.A. (1987). The future of the giant panda. *Nature* **325,** 758–759.

O'Leary, J.W., Glenn, E.P. & Watson, M.C. (1985). Agricultural production of halophytes irrigated with seawater. *Plant and Soil* **89,** 311–321.

Oades, J.M. (1993). The role of biology in the formation, stabilization and degradation of soil structure. *Geoderma* **56,** 377–400.

Odell, R.T., Walker, W.M., Boone, L.V. & Oldham, M.G. (1982). *The Morrow Plots*. University of Illinois Agricultural Experiment Station Bulletin 775. Urbana-Champaign, IL.

Odell, R.T., Melsted, S.W. & Walker, W.M. (1984). Changes in organic carbon and nitrogen of Morrow plot soils under different treatments, 1904–73. *Soil Science* **137,** 160–171.

OECD (1997). *OECD Environmental Data 1997*. Organization for Economic Cooperation and Development, Paris.

Oesterheld, M., Sala, O.E. & McNaughton, S.J. (1992). Effect of animal husbandry on herbivore-carrying capacity at a regional scale. *Nature* **356,** 234–236.

Olsson, G. (1988). Nutrient use and productivity for different cropping systems in south Sweden during the 18th century. In: *The Cultural Landscape—Past, Present and Future* (eds H.H. Birks, H.J.B. Birks, P.E. Kapland & D. Moe), pp. 123–137. Cambridge University Press, Cambridge.

Oosting, H.J. (1942). An ecological analysis of the plant communities of Piedmont, North Carolina. *American Midland Naturalist* **28,** 1–126.

Oosting, H.J. & Humphreys, M.E. (1940). Buried viable seeds in a successional series of old field forest soils. *Bulletin of the Torrey Botanical Club* **67**, 253–273.

Opdam, P., Rijsdijk, G. & Hustings, F. (1985). Bird communities in small woods in an agricultural landscape: effects of area and isolation. *Biological Conservation* **34**, 333–352.

Otterman, J. (1974). Baring high-albedo soils by overgrazing: a hypothesized desertification mechanism. *Science* **186**, 531–533.

Otterman, J. *et al.* (1990). An increase in early rains in southern Israel following land-use change? *Boundary-Layer Meteorology* **53**, 333–351.

Owen-Smith, N. (1989). Megafauna extinctions: the conservation message from 11 000 years BP. *Conservation Biology* **3**, 405–412.

Oxley, D.J., Fenton, M.B. & Carmody, G.R. (1974). The effects of roads on populations of small mammals. *Journal of Applied Ecology* **11**, 51–59.

Pacey, A., ed. (1978). *Sanitation in Developing Countries*. Wiley, Chichester.

Packer *et al.* (1991). Case study of a population bottleneck: lions of the Ngorongoro Crater. *Conservation Biology* **5**, 219–230.

Palaniappan, V.M., Marrs, R.H. & Bradshaw, A.D. (1979). The effect of *Lupinus arboreus* on the nitrogen status of china clay wastes. *Journal of Applied Ecology* **16**, 825–831.

Paleg, L.G. & Aspinall, D., eds (1981). *Physiology and Biochemistry of Drought Resistance in Plants*. Academic Press, London.

Pardo, L.H., Driscoll, C.T. & Likens, G.E. (1995). Patterns of nitrate loss from a chronosequence of clear-cut watersheds. *Water, Air and Soil Pollution* **85**, 1659–1664.

Park, S.-Y., Shivaji, R., Krans, J.V. & Luthe, D.S. (1996). Heat-shock response in heat-tolerant and nontolerant variants of *Agrostis palustris*. *Plant Physiology* **111**, 515–524.

Parsons, A.J., Leafe, E.L., Collett, B. & Stiles, W. (1983a). [Leaf and canopy photosynthesis of continuously-grazed swards]. *Journal of Applied Ecology* **20**, 117–126.

Parsons, A.J., Leafe, E.L., Collett, B., Penning, P.D. & Lewis, J. (1983b). [Photosynthesis, crop growth and animal intake of continuously-grazed swards]. *Journal of Applied Ecology* **20**, 127–139.

Parsons, A.J. *et al.* (1994). Diet preference of sheep: effect of recent diet, physiological state and species abundance. *Journal of Animal Ecology* **63**, 465–478.

Parsons, A.J., Johnson, I.R. & Harvey, A. (1988). [Use of a model to optimize the interaction between frequency and severity of intermittent defoliation]. *Grass and Forage Science* **43**, 49–59.

Paul, E.A. *et al.* (1997). Radiocarbon dating for determination of soil organic matter pool sizes and dynamics. *Soil Science Society of America Journal* **61**, 1058–1067.

Pauly, D. & Christensen, V. (1995). Primary production required to sustain global fisheries. *Nature* **374**, 255–257.

Payne, C.C. (1988). Pathogens for the control of insects: where next? *Philosophical Transactions of the Royal Society B* **318**, 225–248.

Pearce, D. & Moran, D. (1994). *The Economic Value of Biodiversity*. Earthscan, London.

Peet, R.K. & Christensen, N.L. (1980). Succession: a population process. *Vegetatio* **43**, 131–140.

Penman, H.L. (1963). *Vegetation and Hydrology*. Commonwealth Agricultural Bureau, Farnham Royal, England.

Peoples, M.B., Herridge, D.F. & Ladha, J.K. (1995). Biological nitrogen fixation: an efficient source of nitrogen for sustainable agricultural production? *Plant and Soil* **174**, 3–28.

Perry, D.A. (1994). *Forest Ecosystems*. Johns Hopkins University Press, Baltimore.

Peterken, G.F. (1991). Ecological issues in the management of woodland nature reserves. In: *The Scientific Management of Temperate Communities for Conservation* (eds I.F. Spellerberg, F.B. Goldsmith & M.G. Morris), pp. 245–272. Blackwell Science, Oxford.

Peterken, G.F. (1993). *Woodland Conservation and Management*, 2nd edn. Chapman & Hall, London.

Peterken, G.F. (1996). *Natural Woodland: Ecology and Conservation in Northern Temperate Regions*. Cambridge University Press, Cambridge.

Peterken, G.F. & Game, M. (1981). Historical factors affecting the distribution of *Mercurialis perennis* in central Lincolnshire. *Journal of Ecology* **69,** 781–796.

Peterken, G.F. & Game, M. (1984). Historical factors affecting the number and distribution of vascular plant species in the woodlands of central Lincolnshire. *Journal of Ecology* **72,** 155–182.

Phillips, D.L. & Shure, D.J. (1990). Patch-size effects on early succession in southern Appalachian forests. *Ecology* **71,** 204–212.

Phillips, O.L. (1997). The changing ecology of tropical forests. *Biodiversity and Conservation* **6,** 291–311.

Phillips, O.L. *et al.* (1998). Changes in the carbon balance of tropical forests: evidence from long-term plots. *Science* **282,** 439–442.

Pickup, G. (1994). Modelling patterns of defoliation by grazing animals in rangelands. *Journal of Applied Ecology* **31,** 231–246.

Pickup, G. & Chewings, V.H. (1988). Estimating the distribution of grazing and patterns of cattle movement in a large arid zone paddock. *International Journal of Remote Sensing* **9,** 1469–1490.

Pigott, C.D. (1969). The status of *Tilia cordata* and *T. platyphyllos* on the Derbyshire limestone. *Journal of Ecology* **57,** 491–504.

Pimentel, D. & Pimentel, M. (1979). *Food, Energy and Society*. Arnold, London.

Pimm, S.L. (1980). Food web design and the effect of species deletion. *Oikos* **35,** 139–149.

Pimm, S.L., Jones, H.L. & Diamond, J. (1988). On the risk of extinction. *American Naturalist* **132,** 757–785.

Pitcher, T.J. & Hart, P.J.B. (1982). *Fisheries Ecology*. Croom-Helm, London.

Plumptre, A.J. & Reynolds, V. (1994). The effect of selective logging on the primate populations in the Budongo Forest Reserve, Uganda. *Journal of Applied Ecology* **31,** 631–641.

Polcher, J. & Laval, K. (1994). The impact of African and Amazonian deforestation on tropical climate. *Journal of Hydrology* **155,** 389–405.

Pollard, E. (1973). [Woodland relic hedges in Huntingdon and Peterborough]. *Journal of Ecology* **61,** 343–352.

Poole, J. (1997). *Elephants*. Colin Baxter, Grantown-on-Spey, Scotland.

Poorter, H. (1993). Interspecific variation in the growth response of plants to an elevated ambient CO_2 concentration. *Vegetatio* **104/105,** 77–97.

Porter, S.C. (1983). *Late-Quaternary Environments of the United States*, Vol. 1. *The Late Pleistocene.* Longman, London.

Potts, G.R. (1991). The environmental and ecological importance of cereal fields. In: *The Ecology of Temperate Cereal Fields* (eds L.G. Firbank, N. Carter, J.F. Darbyshire & G.R. Potts), pp. 3–21. Blackwell Science, Oxford.

Power, M.E. & Mills, L.S. (1995). The Keystone cops meet in Hilo. *TREE* **10,** 182–184.

Powlson, D.S., Pruden, G., Johnston, A.E. & Jenkinson, D.S. (1986). [The nitrogen cycle in the Broadbalk wheat experiment]. *Journal of Agricultural Science* **107,** 591–609.

Prestidge, R.A. & Ball, O.J.-P. (1997). [The utilization of endophytic fungi for pest management]. In: *Multitrophic Interactions in Terrestrial Systems* (eds A.C. Gange & V.K. Brown), pp. 171–192. Blackwell Science, Oxford.

Price, C. (1989). *The Theory and Application of Forest Economics*. Blackwell, Oxford.

Price, P.W., Westoby, M. & Price, B. (1988). Parasite-mediated competition: some predictions and tests. *American Naturalist* **131,** 544–555.

Primack, R.B. (1998). *Essentials of Conservation Biology*, 2nd edn. Sinauer, Sunderland, MA.

Primack, R.B. & Miao, S.L. (1992). Dispersal can limit local plant distribution. *Conservation Biology* **6,** 513–519.

Prince, R.C. (1992). Bioremediation of oil spills, with particular reference to the spill from the *Exxon Valdez*. In: *Microbial Control of Pollution* (eds J.C. Fry *et al.*), pp. 19–34. Cambridge University Press, Cambridge.

Putman, R. (1988). *The Natural History of Deer*. Christopher Helm, London.

Quay, P.D., Tilbrook, B. & Wong, C.S. (1992). Oceanic uptake of fossil fuel CO_2: carbon-13 evidence. *Science* **256,** 74–79.

Rackham, O. (1986). *The History of the Countryside*. Dent, London.

Rackham, O. (1990). *Trees and Woodland in the British Landscape*, 2nd edn. Dent, London.

Ralph, C.J. (1985). Habitat association patterns of forest and steppe birds of northern Patagonia, Argentina. *Condor* **87,** 471–483.

Rasmussen, P.E. & Parton, W.J. (1994). Long-term effects of residue management in wheat-fallow: I. Inputs, yield, and soil organic matter. *Soil Science Society of America Journal* **58,** 523–530.

Ratcliffe, D.A. (1970). Changes attributable to pesticides in egg breakage frequency and eggshell thickness in some British birds. *Journal of Applied Ecology* **7,** 67–115.

Ratcliffe, D.A. (1980). *The Peregrine Falcon*. Poyser, Calton, Staffordshire.

Rawes, M. (1981). Further results of excluding sheep from high-level grasslands in the north Pennines. *Journal of Ecology* **69,** 651–669.

Recher, H.F. (1969). Bird species diversity and habitat diversity in Australia and North America. *American Naturalist* **103,** 75–80.

Reichle, D.E. (1981). *Dynamic Properties of Forest Ecosystems*. Cambridge University Press, Cambridge.

Reiling, K. & Davison, A.H. (1993). Ozone sensitivity. In: *Methods in Comparative Plant Ecology: a Laboratory Manual* (eds G.A.F. Hendry & J.P. Grime), pp. 63–65. Chapman & Hall, London.

Reinert, R.E. (1972). [Accumulation of dieldrin in an alga, *Daphnia* and guppy]. *Journal of the Fisheries Research Board of Canada* **29,** 1413–1418.

Remmert, H., ed. (1994). *Minimum Animal Populations*. Springer, Berlin.

Risch, S.J., Andow, D. & Altieri, M.A. (1983). [Agroecosystem diversity and pest control]. *Environmental Entomology* **12,** 625–629.

Ritchie, J.C. & MacDonald, G.M. (1986). The patterns of post-glacial spread of white spruce. *Journal of Biogeography* **13,** 527–540.

Roberts, M.G. (1996). The dynamics of bovine tuberculosis in possum populations, and its eradicaton or control by culling or vaccination. *Journal of Animal Ecology* **65,** 451–464.

Roberts, R.D., Marrs, R.H., Skeffington, R.A. & Bradshaw, A.D. (1981). [Ecosystem development on naturally-colonized china clay wastes. I. Vegetation changes and accumulation of organic matter]. *Journal of Ecology* **69,** 153–161.

Robinson, P.J. & Henderson-Sellers, A. (1999). *Contemporary Climatology*, 2nd edn. Longman, Harlow, Essex.

Rogers, M.E., Noble, C.L., Halloran, G.M. & Nicolas, M.E. (1997). Selecting for salt tolerance in white clover: chloride ion exclusion and its heritability. *New Phytologist* **135,** 645–654.

Romme, W.H. & Despain, D.G. (1989). The Yellowstone fires. *Scientific American* **26** (5), 21–29.

Romme, W.H., Turner, M.G., Wallace, L.L. & Walker, J.S. (1995). Aspen, elk, and fire in northern Yellowstone National Park. *Ecology* **76,** 2097–2106.

Roper, M.M. & Ladha, J.K. (1995). Biological N_2 fixation by heterotrophic and phototrophic bacteria in association with straw. *Plant and Soil* **174,** 211–224.

Rose, F. (1976). Lichenological indicators of age and environmental continuity in woodland. In: *Lichenology: Progress and Problems* (eds D.H. Brown, D.L. Hawksworth & R.H. Bailey), pp. 279–307. Academic Press, London.

Rosenlund, G., Stoss, J. & Talbot, C. (1997). Co-feeding marine fish larvae with inert and live diets. *Aquaculture* **155,** 183–191.

Rothschild, B.J. (1986). *Dynamics of Marine Fish Populations*. Harvard University Press, Cambridge, MA.

Royama, T. (1984). Population dynamics of the spruce budworm. *Ecological Monographs* **54,** 429–462.

Rusch, G.M. & Oesterheld, M. (1997). Relationship between productivity, and species and functional group diversity in grazed and non-grazed pampas grassland. *Oikos* **78,** 519–526.

Russell, P.E. (1995). Fungicide resistance: occurrence and management. *Journal of Agricultural Science* **124,** 317–323.

Russell, W.M.S. (1967). *Man, Nature and History*. Aldus, London.

Salati, E. & Vose, P.B. (1984). The Amazon Basin: a system in equilibrium. *Science* **225,** 129–138.

Salt, D.E., Smith, R.D. & Raskin, I. (1998). Phytoremediation. *Annual Review of Plant Physiology* **49,** 643–668.

Sanford, R.L. *et al.* (1985). Amazon rain-forest fires. *Science* **227,** 53–55.

Sanginga, N., Vanlauwe, B. & Danso, S.K.A. (1995). Management of biological N_2 fixation in alley cropping systems: estimation and contribution to N balance. *Plant and Soil* **174,** 119–141.

Saulei, S.M. (1984). Natural regeneration following clear-fell logging operations in the Gogol Valley, Papua New Guinea. *Ambio* **13,** 351–354.

Saulei, S.M. (1985). The recovery of tropical lowland rainforest after clear-fell logging in the Gogol Valley, Papua New Guinea. PhD thesis, University of Aberdeen.

Saulei, S.M. & Swaine, M.D. (1988). Rain forest seed dynamics during succession at Gogol, Papua New Guinea. *Journal of Ecology* **76,** 1133–1152.

Savenije, H.H.G. (1995). New definitions for moisture recycling and the relationship with land-use changes in the Sahel. *Journal of Hydrology* **167,** 57–78.

Savill, P.S. & Evans, J. (1986). *Plantation Silviculture in Temperate Regions*. Clarendon Press, Oxford.

Scarre, C. *et al.* (1988). *Past Worlds. The Times Atlas of Archaeology*. Times Books, London.

Schaefer, M.B. (1954). Some aspects of the dynamics of populations important to the management of commercial marine fisheries. *Bulletin of the Inter-American Tropical Tuna Commission* **1,** 27–56.

Schaffers, A.P., Vesseur, M.C. & Sýkora, K.V. (1998). Effects of delayed hay removal on the nutrient balance of roadside plant communities. *Journal of Applied Ecology* **35,** 349–364.

Schemske, D.W. & Brokaw, N. (1981). Treefalls and the distribution of understory birds in a tropical forest. *Ecology* **62,** 938–945.

Schemske, D.W. *et al.* (1994). Evaluating approaches to the conservation of rare and endangered plants. *Ecology* **75,** 584–606.

Schmidt-Nielsen, K. (1997). *Animal Physiology*, 5th edn. Cambridge University Press, Cambridge.

Schrader, S. & Zhang, H. (1997). Earthworm casting: stabilization or destabilization of soil structure? *Soil Biology and Biochemistry* **29,** 469–475.

Schulze, R.E. & Kunz, R.P. (1995). Potential shifts in optimum growth areas of commercial tree species and subtropical crops in southern Africa due to global warming. *Journal of Biogeography* **22,** 679–688.

Schwartz, C.C. & Ellis, J.E. (1981). Feeding ecology and niche separation in some native and domestic ungulates on the shortgrass prairie. *Journal of Applied Ecology* **18,** 343–353.

Semmartin, M. & Oesterheld, M. (1996). Effect of grazing pattern on primary production. *Oikos* **75,** 431–436.

Semprini, L., Hopkins, G.D., McCarty, P.L. & Roberts, P.V. (1992). *In-situ* transformation of carbon tetrachloride and other halogenated compounds resulting from biostimulation under anoxic conditions. *Environmental Science and Technology* **26,** 2454–2461.

Shackleton, C.M. (1993). [Fuelwood harvesting and sustainable utilisation in a communal grazing land of eastern Transvaal]. *Biological Conservation* **63,** 247–254.

Shaner, D.L. (1995). Herbicide resistance: Where are we? How did we get here? Where are we going? *Weed Technology* **9,** 850–856.

Sharpley, A.N. *et al.* (1992). Transport of bioavailable phosphorus in agricultural runoff. *Journal of Environmental Quality* **21,** 30–35.

Shepherd, C.J. & Bromage, N.R., eds. (1988). *Intensive Fish Farming.* BSP Professional Books. Oxford.

Sheppard, L.J. (1994). [Causal mechanisms by which sulphate, nitrate & acidity influence frost hardiness in red spruce]. *New Phytologist* **127,** 69–82.

Sher, A.V. (1997). Late-Quaternary extinction of large mammals in northern Eurasia: a new look at the Siberian contribution. In: *Past and Future Rapid Environmental Changes* (eds B. Huntley *et al.*), pp. 319–339. Springer, Berlin.

Siddique, K.H.M., Tennant, D., Perry, M.W. & Belford, R.K. (1990). Water use and water use efficiency of old and modern wheat cultivars in a Mediterranean-type environment. *Australian Journal of Agricultural Research* **41,** 431–447.

Silvertown, J. & Smith, B. (1988). Mapping the microenvironment for seed germination in the field. *Annals of Botany* **63,** 163–167.

Simberloff, D. (1998). Flagships, umbrellas, and keystones: is single-species management passé in the landscape era? *Biological Conservation* **83,** 247–257.

Simberloff, D. & Stiling, P. (1996). Risks of species introduced for biological control. *Biological Conservation* **78,** 185–192.

Simberloff, D. & Stiling, P. (1998). How risky is biological control? Reply. *Ecology* **79,** 1834–1836.

Sims, P.L., Singh, J.S. & Lauenroth, W.K. (1978). The structure and function of ten western North American grasslands. I. Abiotic and vegetational characteristics. *Journal of Ecology* **66,** 251–285.

Sinclair, A.R.E. & Norton-Griffiths, M. (1979). *Serengeti: Dynamics of an Ecosystem.* University of Chicago Press, Chicago.

Skole, D. & Tucker, C. (1993). Tropical deforestation and habitat fragmentation in the Amazon: satellite data from 1978 to 1988. *Science* **260,** 1905–1910.

Slatyer, R.O. (1977). [Altitudinal variation in the photosynthetic characteristics of snow gum. III. Temperature response]. *Australian Journal of Plant Physiology* **4,** 301–312.

Sloof, W. (1983). Benthic macroinvertebrates and water quality assessment: some toxicological considerations. *Aquatic Toxicology* **4,** 73–82.

Smith, H., McCallum, K. & MacDonald, D.W. (1997). Experimental comparison of the nature conservation value, productivity and ease of management of a conventional and a more species-rich grass ley. *Journal of Applied Ecology* **34,** 53–64.

Smith, R.S. & Rushton, S.P. (1994). The effects of grazing management on the vegetation of mesotrophic (meadow) grassland in northern England. *Journal of Applied Ecology* **31,** 13–24.

Smith, S.V. (1984). Phosphorus versus nitrogen limitation in the marine environment. *Limnology and Oceanography* **29,** 1149–1160.

Snaydon, R.W. (1987). *Ecosystems of the World,* Vol. 17B: *Managed Grassland.* Elsevier. Amsterdam.

Snyder, N.F.R. *et al.* (1996). Limitations of captive breeding in endangered species recovery. *Conservation Biology* **10,** 338–348.

Somerville, L. & Greaves, M.P. (1987). *Pesticide Effects on Soil Microflora.* Taylor & Francis, London.

Sopher, D.E. (1980). Indian civilization and the tropical savanna environment. In: *Human Ecology in Savanna Environments* (ed D.R. Harris), pp. 185–207. Academic Press, London.

Soulé, M.E., ed. (1987). *Viable Populations for Conservation.* Cambridge University Press, Cambridge.

Soussana, J.F., Casella, E. & Loiseau, P. (1996). Long-term effects of CO_2 enrichment and temperature increase on a temperate grassland sward. II. Plant nitrogen budgets and root fraction. *Plant and Soil* **182,** 101–114.

Southwood, T.R.E., Brown, V.K. & Reader, P.M. (1979). The relationships of plant and insect diversities in succession. *Biological Journal of the Linnean Society* **12,** 327–348.

Speight, M.R. & Wainhouse, D. (1989). *Ecology and Management of Forest Insects.* Oxford University Press, Oxford.

Spencer, J.W. & Kirby, K.J. (1992). An inventory of ancient woodland for England and Wales. *Biological Conservation* **62,** 77–93.

Sprent, J.I. (1995). Legume trees and shrubs in the tropics: N_2 fixation in perspective. *Soil Biology and Biochemistry* **27,** 401–407.

Spurr, S.H. (1956). Forest associations in the Harvard Forest. *Ecological Monographs* **26,** 245–262.

Stage, A.R., Renner, D.L. & Chapman, R.C. (1988). *Selected Yield Tables for Plantations and Natural Stands in Inland Northwest Forests.* US Department of Agriculture Forest Service, Research Paper INT-394.

Stanhill, G. (1986). Water use efficiency. *Advances in Agronomy* **39,** 53–85.

Steffens, J.C. (1990). The heavy metal-binding peptides of plants. *Annual Review of Plant Physiology* **41,** 553–575.

Stevenson, F.J. (1986). *Cycles of Soil: Carbon, Nitrogen, Phosphorus, Sulfur, Micronutrients.* Wiley, New York.

Stewart, G. & Hull, A.C. (1949). Cheatgrass—an ecological intruder in southern Idaho. *Ecology* **30,** 58–74.

Stiling, P.D. (1996). *Ecology: Theories and Applications,* 2nd edn. Prentice-Hall, Upper Saddle River, NJ.

Stöhr, K. & Meslin, F.-M. (1996). Progress and setbacks in the oral immunisation of foxes against rabies. *Veterinary Record* **139,** 32–35.

Stouffer, P.C. & Bierregaard, R.O. (1995). Use of Amazonian forest fragments by understorey insectivorous birds. *Ecology* **76,** 2429–2445.

Stuart, A.J. (1991). Mammalian extinctions in the late Pleistocene of northern Eurasia and North America. *Biological Reviews* **66,** 453–562.

Sud, Y.C., Yang, R. & Walker, G.K. (1996). Impact of *in situ* deforestation in Amazonia on the regional climate: general circulation model simulation study. *Journal of Geophysical Research* **101,** 7095–7109.

Sumption, K.J. & Flowerdew, J.R. (1985). The ecological effects of the decline in rabbits due to myxomatosis. *Mammal Review* **15,** 151–186.

Sutherland, W.J. & Hill, D.A., eds. (1995). *Managing Habitats for Conservation.* Cambridge University Press, Cambridge.

Swain, A.M. (1973). A history of fire and vegetation in northeastern Minnesota as recorded in lake sediments. *Quaternary Research* **3,** 383–396.

Swank, W.T. & Douglass, J.E. (1974). Streamflow greatly reduced by converting deciduous hardwood stands to pine. *Science* **185,** 857–859.

Swetnam, T.W. & Lynch, A.M. (1993). Multicentury, regional-scale patterns of western spruce budworm outbreaks. *Ecological Monographs* **63,** 399–424.

Swinton, J. & Anderson, R.M. (1995). Model frameworks for plant-pathogen interactions. In: *Ecology of Infectious Diseases in Natural Populations* (eds B.T. Grenfell & A.P. Dobson), pp. 280–294. Cambridge University Press, Cambridge.

Sykes, M.T. (1997). The biogeographical consequences of forecast changes in the global environment: individual species' potential range changes. In: *Past and Future Rapid Environmental Changes* (ed B. Huntley *et al.*), pp. 427–440. Springer, Berlin.

Szeicz, J.M. & MacDonald, G.M. (1995). Recent white spruce dynamics at the subarctic alpine treeline of northwestern Canada. *Journal of Ecology* **83,** 873–885.

Szerbin, P. *et al.* (1999). Caesium-137 migration in Hungarian soils. *Science of the Total Environment* **227,** 215–227.

Taiz, L. & Zeiger, E., eds. (1998). *Plant Physiology,* 2nd edn. Sinauer, Sunderland, MA.

Te Beest, D.O., Yang, X.B. & Cisar, C.R. (1992). The status of biological control of weeds with fungal pathogens. *Annual Review of Phytopathology* **30,** 637–657.

Thiollay, J.-M. (1992). Influence of selective logging on bird species diversity in a Guianan rain forest. *Conservation Biology* **6,** 47–63.

Thomas, C.D. & Jones, T.M. (1993). Partial recovery of a skipper butterfly from population refuges: lessons for conservation in a fragmented landscape. *Journal of Animal Ecology* **62,** 472–481.

Thomas, D.S.G. & Middleton, N.J. (1994). *Desertification: Exploding the Myth.* Wiley, Chichester, England.

Thomas, J.A. (1991). Rare species conservation: case studies of European butterflies. In: *The Scientific Management of Temperate Communities for Conservation* (eds I.F. Spellerberg, F.B. Goldsmith & M.G. Morris), pp. 149–197. Blackwell Science, Oxford.

Thomas, R.S., Franson, R.L. & Bethlenfalvay, G.J. (1993). Separation of vesicular-arbuscular mycorrhizal fungus and root effects on soil aggregation. *Soil Science Society of America Journal* **57,** 77–81.

Thompson, H.V. & King, C.M. (1994). *The European Rabbit.* Oxford University Press, Oxford.

Thomson, B.D., Robson, A.D. & Abbott, L.K. (1992). The effect of long-term applications of phosphorus fertilizer on populations of vesicular-arbuscular mycorrhizal fungi in pastures. *Australian Journal of Agricultural Research* **43,** 1131–1142.

Thorarinsson, K. (1990). Biological control of the cottony-cushion scale: experimental tests of the spatial density-dependence hypothesis. *Ecology* **71,** 635–644.

Tilghman, N.G. (1989). Impacts of white-tailed deer on forest regeneration in northwestern Pennsylvania. *Journal of Wildlife Management* **53,** 524–532.

Tilman, D., May, R.M., Lehman, C.L. & Nowak, M.A. (1994). Habitat destruction and the extinction debt. *Nature* **371,** 65–66.

Tinner, W. *et al.* (1999). Long-term forest fire ecology and dynamics in southern Switzerland. *Journal of Ecology* **87,** 273–289.

Tisdall, J.M. & Oades, J.M. (1979). Stabilization of soil aggregates by the root systems of ryegrass. *Australian Journal of Soil Research* **17,** 429–441.

Tisdall, J.M. & Oades, J.M. (1980). The effect of crop rotation on aggregation in a red-brown earth. *Australian Journal of Soil Research* **18,** 423–433.

Tisdall, J.M. & Oades, J.M. (1982). Organic matter and water-stable aggregates in soils. *Journal of Soil Science* **33,** 141–163.

Tiver, F. & Andrew, M.H. (1997). Relative effects of herbivory by sheep, rabbits, goats and kangaroos on recruitment and regeneration of shrubs and trees in eastern South Australia. *Journal of Applied Ecology* **34,** 903–914.

Tomanek, G.W. & Albertson, F.W. (1957). Variations in cover, composition, production, and roots of vegetation on two prairies in western Kansas. *Ecological Monographs* **27,** 267–281.

Tomlinson, J.A. & Carter, A.L. (1970). [Studies in the seed transmission of cucumber mosaic virus in chickweed]. *Annals of Applied Biology* **66,** 381–386.

Tomlinson, J.A., Carter, A.L., Dale, W.T. & Simpson, C.J. (1970). Weed plants as sources of cucumber mosaic virus. *Annals of Applied Biology* **66,** 11–16.

Trainer, F.E. (1995). Can renewable energy sources sustain affluent society? *Energy Policy* **23,** 1009–1026.

Trout, R.C., Ross, J., Tittensor, A.M. & Fox, A.P. (1992). The effect on a British wild rabbit population of manipulating myxomatosis. *Journal of Applied Ecology* **29,** 679–686.

Trout, R.C., Tapper, S.C. & Harradine, J. (1986). Recent trends in the rabbit population in Britain. *Mammal Review* **16,** 117–123.

Tucker, C.J., Newcomb, W.W. & Dregne, H.E. (1994). AVHRR data sets for determination of desert spatial extent. *International Journal of Remote Sensing* **15,** 3547–3565.

Turner, M.G., Romme, W.H., Gardner, R.H. & Hargrove, W.W. (1997). Effects of fire size and pattern on early succession in Yellowstone National Park. *Ecological Monographs* **67,** 411–433.

Turner, N.C. (1997). Further progress in crop water relations. *Advances in Agronomy* **58,** 293–338.

Turner, R.K., ed. (1993). *Sustainable Environmental Economics and Management: Principles and Practice,* 2nd edn. Bellhaven, London.

Uhl, C., Buschbacher, R. & Serrao, E.A.S. (1988). Abandoned pastures in eastern Amazonia. I. Patterns of plant succession. *Journal of Ecology* **76,** 663–681.
UN Statistics Yearbooks. United Nations, New York.
UNEP (1991). *Environmental Data Report*, 3rd edn. *United Nations Environment Programme.* Blackwell, Oxford.
Usher, M.B. & Thompson, D.B.A. (1993). Variation in the upland heathlands of Great Britain: conservation importance. *Biological Conservation* **66,** 69–81.
Valiela, I. (1995). *Marine Ecological Processes*, 2nd edn. Springer, New York.
Van Alfen, N.K. (1982). Biology and potential for disease control of hypovirulence of *Endothia parasitica. Annual Review of Phytopathology* **20,** 349–362.
van Andel, T.H., Runnels, C.N. & Pope, K.O. (1986). Five thousand years of land use and abuse in the southern Argolid, Greece. *Hesperia* **55,** 103–128.
Van Andel, T.H., Zangger, E. & Demitrack, A. (1990). Land use and soil erosion in prehistoric and historical Greece. *Journal of Field Archaeology* **17,** 379–396.
van Campo, E. *et al.* (1990). [Comparison of terrestrial and marine temperature estimates for the past 135 kyr off southeast Africa]. *Nature* **348,** 209–212.
Vander Wall, S.B. & Balda, R.P. (1977). Coadaptations of Clark's nutcracker and the piñon pine for efficient seed harvest and dispersal. *Ecological Monographs* **47,** 89–111.
Verissimo, A., Barreto, P., Tarifa, R. & Uhl, C. (1995). Extraction of a high-value natural resource in Amazonia: the case of mahogany. *Forest Ecology and Management* **72,** 39–60.
Vickery, P.J. (1972). Grazing and net primary production of a temperate grassland. *Journal of Applied Ecology* **9,** 307–314.
Vitousek, P.M. (1991). Can planted forests counteract increasing atmospheric carbon dioxide? *Journal of Environmental Quality* **20,** 348–354.
Vitousek, P.M., Ehrlich, P.R., Ehrlich, A.H. & Matson, P.A. (1986). Human appropriation of the products of photosynthesis. *Bioscience* **36,** 368–373.
Vucetich, J.A., Peterson, R.O. & Waite, T.A. (1997). Effects of social structure and prey dynamics on extinction risk in gray wolves. *Conservation Biology* **11,** 957–965.
Waage, J. (1989). The population ecology of pest-pesticide–natural enemy interactions. In: *Pesticides and Non-Target Invertebrates* (ed P.C. Jepson), pp. 81–93. Intercept, Wimborne, England.
Waage, J.K. & Greathead, D.J. (1988). Biological control: challenges and opportunities. *Philosophical Transactions of the Royal Society B* **318,** 111–128.
Wadman, M. (1997). Dispute over insect resistance to crops. *Nature* **388,** 817.
Waites, G. (1998). *The Cassell Dictionary of Biology*. Cassell, London.
Walker, B.H., Ludwig, D., Holling, C.S. & Peterman, R.M. (1981). Stability of semi-arid savanna grazing systems. *Journal of Ecology* **69,** 473–498.
Walker, C.H., Hopkin, S.P., Sibly, R.M. & Peakall, D.B. (1996). *Principles of Ecotoxicology*. Taylor & Francis, London.
Walker, T.W. & Syers, J.K. (1976). The fate of phosphorus during pedogenesis. *Geoderma* **15,** 1–19.
Walsh, J.J. (1981). A carbon budget for overfishing off Peru. *Nature* **290,** 300–304.
Walters, C. & Maguire, J.-J. (1996). Lessons for stock assessment from the northern cod collapse. *Reviews in Fish Biology and Fisheries* **6,** 125–137.
Ward, D.M., Atlas, R.M., Boehm, P.D. & Calder, J.A. (1980). Microbial biodegradation and chemical evolution of oil from the *Amoco* spill. *Ambio* **9,** 277–283.
Wardle, P. (1991). *Vegetation of New Zealand.* Cambridge University Press, Cambridge.
Warner, A.C.I. (1981). Rate of passage of digesta through the gut of mammals and birds. *Nutrition Abstracts and Reviews Series B* **51,** 789–820.
Warren, M.S. & Key, R.S. (1991). Woodlands: past, present and potential for insects. In: *The Conservation of Insects and Their Habitats* (eds N.M. Collins & J.A. Thomas), pp. 155–203. Academic Press, London.
Warwick, S.I. (1991). Herbicide resistance in weedy plants: physiology and population biology. *Annual Review of Ecology and Systematics* **22,** 95–114.

Waters, E.R., Lee, G.J. & Vierling, E. (1996). Evolution, structure and function of the small heat-shock proteins in plants. *Journal of Experimental Botany* **47,** 325–338.

Watt, A.S. (1947). Pattern and process in the plant community. *Journal of Ecology* **35,** 1–22.

Watt, M., McCully, M.E. & Jeffree, C.E. (1993). [Plant and bacterial mucilages of the maize rhizosphere]. *Plant and Soil* **151,** 151–165.

Webb, N.R. (1998). The traditional management of European heathlands. *Journal of Applied Ecology* **35,** 987–990.

Webb, S.L. (1986). Potential role of passenger pigeons and other vertebrates in the rapid Holocene migrations of nut trees. *Quaternary Research* **26,** 367–375.

Weber, U.M. & Schweingruber, F.H. (1995). A dendroecological reconstruction of western spruce budworm outbreaks. *Trees—Structure and Function* **9,** 204–213.

Weischet, W. & Caviedes, C.N. (1993). *The Persisting Ecological Constraints of Tropical Agriculture*. Longman, Harlow.

Wellburn, A. (1994). *Air Pollution and Climate Change: the Biological Impact*. Longman, Harlow.

Wells, T.C.E., Sheail, J., Ball, D.F. & Ward, L.K. (1976). Ecological studies on the Porton Ranges: relationships between vegetation, soils and land-use history. *Journal of Ecology* **64,** 589–626.

West, L.T. *et al.* (1991). Cropping system effects on interrill soil loss in the Georgia Piedmont. *Soil Science Society of America Journal* **55,** 460–466.

Weste, G. (1986). Vegetation changes associated with invasion by *Phytophthora cinnamomi* of defined plots in the Brisbane Ranges, Victoria, 1975–85. *Australian Journal of Botany* **34,** 633–648.

Weste, G. & Marks, G.C. (1987). The biology of *Phytophthora cinnamomi* in Australian forests. *Annual Review of Phytopathology* **25,** 207–229.

Western, D. (1975). Water availability and its influence on the structure and dynamics of a savannah large mammal community. *East African Wildlife Journal* **13,** 265–286.

Whipps, J.M. (1997). Developments in the biological control of soil-borne plant pathogens. *Advances in Botanical Research* **26,** 1–134.

White, R.E. (1997). *Principles and Practice of Soil Science,* 3rd edn. Blackwell Science, Oxford.

Whitmore, T.C. (1984). *Tropical Rain Forests of the Far East,* 2nd edn. Clarendon, Oxford.

Whitmore, T.C. (1997). Tropical forest disturbance, disappearance, and species loss. In: *Tropical Forest Remnants: Ecology, Management, and Conservation of Fragmented Communities* (eds W.F. Laurance & R.O. Bierregaard), pp. 3–12. University of Chicago Press, Chicago.

Whitmore, T.C. (1998). *An Introduction to Tropical Rain Forests,* 2nd edn. Oxford University Press, Oxford.

Whitmore, T.C. & Brown, N.D. (1996). Dipterocarp seedling growth in rain forest canopy gaps during six and a half years. *Philosophical Transactions of the Royal Society B* **351,** 1195–1203.

Whitney, G.G. (1986). Relation of Michigan's presettlement pine forests to substrate and disturbance history. *Ecology* **67,** 1548–1559.

Whitney, G.G. (1990). The history and status of hemlock-hardwood forests of the Allegheny Plateau. *Journal of Ecology* **78,** 443–458.

Whitney, G.G. & Foster, D.R. (1988). Overstorey composition and age as determinants of the understorey flora of woods in central New England. *Journal of Ecology* **76,** 867–876.

Whittaker, R.H. (1956). Vegetation of the Great Smoky Mountains. *Ecological Monographs* **26,** 1–80.

Whittaker, R.H. (1966). Forest dimensions and production in the Great Smoky Mountains. *Ecology* **47,** 103–121.

Whittaker, R.H. (1975). *Communities and Ecosystems*. Collier-Macmillan, London.

Whittaker, R.H. *et al.* (1979). The Hubbard Brook ecosystem study: forest nutrient cycling and element behavior. *Ecology* **60,** 203–220.

Wierzchos, J., Ascaso Ciria, C. & Garzcia-Gonzalez, M.T. (1992). Changes in microstructure of soils following extraction of organically bonded metals and organic matter. *Journal of Soil Science* **43,** 505–515.

Wilcove, D.S. (1993). Turning conservation goals into tangible results: the case of the spotted owl and old-growth forests. In: *Large-Scale Ecology and Conservation Biology* (eds P.J. Edwards, R.M. May, & N.R. Webb), pp. 313–329. Blackwell Science, Oxford.

Wild, A., ed. (1988). *Russell's Soil Conditions and Plant Growth,* 11th edn. Longman, Harlow.

Wild, S.R., Berrow, M.L. & Jones, K.C. (1991). The persistence of polynuclear aromatic hydrocarbons in sewage sludge amended agricultural soils. *Environmental Pollution* **72,** 141–157.

Williams, M. (1989). *Americans and Their Forests.* Cambridge University Press, Cambridge.

Williams, M.A.J. & Balling, R.C. (1996). *Interactions of Desertification and Climate.* Arnold, London.

Williams, R.J. (1992). Gap dynamics in subalpine heathland and grassland vegetation in south-eastern Australia. *Journal of Ecology* **80,** 343–352.

Williams, R.J. & Ashton, D.H. (1987). Effects of disturbance and grazing by cattle on the dynamics of heathland and grassland communities in the Bogong High Plains, Victoria. *Australian Journal of Botany* **35,** 413–431.

Williams, R.J.P., ed. (1978). *Phosphorus in the Environment: its Chemistry and Biochemistry.* Excerpta Medica, Amsterdam.

Williamson, M. (1996). *Biological Invasions.* Chapman & Hall, London.

Wills, R.T. (1993). The ecological impact of *Phytophthora cinnamomi* in the Stirling Range National Park, Western Australia. *Australian Journal of Ecology* **18,** 145–159.

Wilson, E.O. (1992). *The Diversity of Life.* Bellkamp, Cambridge, MA.

Wilson, J.B. & Agnew, A.D.Q. (1992). Positive-feedback switches in plant communities. *Advances in Ecological Research* **23,** 263–336.

Wilson, J.B., Crawley, M.J., Dodd, M.E. & Silvertown, J. (1996). Evidence for constraint on species coexistence in vegetation of the Park Grass experiment. *Vegetatio* **124,** 183–190.

Winner, R.W., Boesel, M.W. & Farrell, M.P. (1980). Insect community structure as an index of heavy-metal pollution in lotic systems. *Canadian Journal of Fisheries and Aquatic Sciences* **37,** 647–655.

Wolf, C.M., Griffith, B., Reed, C. & Temple, S.A. (1996). Avian and mammalian translocations: update and reanalysis of 1987 survey data. *Conservation Biology* **10,** 1142–1154.

Wong, M., Wright, S.J., Hubbell, S.P. & Foster, R.B. (1990). The spatial pattern and reproductive consequences of outbreak defoliation in a tropical tree. *Journal of Ecology* **78,** 579–588.

Woodward, F.I., Lomas, M.R. & Betts, R.A. (1998). Vegetation-climate feedbacks in a greenhouse world. *Philosophical Transactions of the Royal Society B* **353,** 29–39.

World Resources Institute (1999). *World Resources 1998/9.* Oxford University Press, Oxford.

Wratten, S.D. & Powell, W. (1991). Cereal aphids and their natural enemies. In: *The Ecology of Temperate Cereal Fields* (eds L.G. Firbank, N. Carter, J.F. Darbyshire & G.R. Potts), pp. 233–257. Blackwell Science, Oxford.

Xu, D. *et al.* (1996). Expression of a late embryogenesis abundant protein gene from barley confers tolerance to water stress and salt stress in transgenic rice. *Plant Physiology* **110,** 249–257.

Yeaton, R.I., Travis, J. & Gilinsky, E. (1977). Competition and spacing in plant commmunities: the Arizona upland association. *Journal of Ecology* **65,** 587–595.

Yeo, A.R., Kramer, D., Läuchli, A. & Gullasch, J. (1977). Ion distribution in salt-stressed mature *Zea mays* roots in relation to ultrastructure and retention of sodium. *Journal of Experimental Botany* **28,** 17–29.

Yermiyahu, U. *et al.* (1997). Root elongation in saline solution related to calcium binding to root cell plasma membranes. *Plant and Soil* **191**, 67–76.
Zackrisson, O. (1977). Influence of forest fires on the north Swedish boreal forest. *Oikos* **29**, 22–32.
Zangger, E. (1992). Neolithic to present soil erosion in Greece. In: *Past and Present Soil Erosion: Archaeological and Geographical Perspectives* (eds M. Bell & J. Boardman), pp. 133–147. Oxbow Books, Oxford.
Zentmyer, G.A. (1985). Origin and distribution of *Phytophthora cinnamoni*. In: *Ecology and Management of Soilborne Plant Pathogens* (eds C.A. Parker, A.D. Rovira, K.J. Moore & P.T.W. Wong), pp. 71–72. American Phytopathological Society, St Paul, MN.
Zhong, Y. & Power, G. (1997). Fisheries in China: progress, problems and prospects. *Canadian Journal of Fisheries and Aquatic Sciences* **54**, 224–238.

Subject Index

This index includes references to some major groups of animal and plant, but individual species and genera are listed in the Species Index. The entries for continents, countries and regions cite only the more extensive passages.

Species Index

Either the English or Latin name is given here, or both, whatever is used in the text. The Glossary gives Latin equivalents of English names, and vice versa.